Hasan

URBAN THEORY

Urban Theory: New critical perspectives provides an introduction to innovative critical contributions to the field of urban studies. Chapters offer easily accessible and digestible reviews, and as a reference text *Urban Theory* is a comprehensive and integrated primer, which covers topics necessary for a full understanding of recent theoretical engagements with cities.

The introduction outlines the development of urban theory over the past two hundred years and discusses significant theoretical, methodological and empirical challenges facing the field of urban studies in the context of an increasing globally inter-connected world. The chapters explore twenty-four topics, which are new additions to the urban theoretical debate, highlighting their relationship to long-established concerns that continue to have intellectual purchase, and which also engage with rich new and emerging avenues for debate. Each chapter considers the genealogy of the topic at hand and also includes case studies that explain key terms or provide empirical examples to guide the reader to a better understanding of how theory adds to our understanding of the complexities of urban life.

This book offers a critical and assessable introduction to original and groundbreaking urban theory, and will be essential reading for undergraduate and postgraduate students in human geography, sociology, anthropology, cultural studies, economics, planning, political science and urban studies.

Mark Jayne is Professor of Human Geography at Cardiff University, UK.

Kevin Ward is Professor of Human Geography and Director of the Manchester Urban Institute at the University of Manchester, UK.

URBAN THEORY

New critical perspectives

Edited by Mark Jayne and Kevin Ward

Routledge
Taylor & Francis Group

LONDON AND NEW YORK

First published 2017
by Routledge
2 Park Square, Milton Park, Abingdon, Oxon OX14 4RN

and by Routledge
711 Third Avenue, New York, NY 10017

Routledge is an imprint of the Taylor & Francis Group, an informa business

British Library Cataloguing in Publication Data
A catalogue record for this book is available from the British Library

Library of Congress Cataloging in Publication Data
A catalog record for this book has been requested

ISBN: 978-1-138-79337-8 (hbk)
ISBN: 978-1-138-79338-5 (pbk)
ISBN: 978-1-315-76120-6 (ebk)

Typeset in Bembo
by Saxon Graphics Ltd, Derby

CONTENTS

FIGURES

CASE STUDIES

CONTRIBUTORS

Ben Anderson is a Reader in Human Geography at Durham University, UK. Recently, he has become fascinated by how emergencies are governed. He is currently conducting a genealogy of the government of and by emergency, supported by a 2013 Philip Leverhulme Prize. Other research has explored the implications of theories of affect for contemporary human geography. This work was published in a monograph in 2014: *Encountering Affect: Capacities, Apparatuses, Conditions*. He is also co-editor of *Taking-Place: Non-Representational Theories and Geography* (2010).

Richard Baxter is a Lecturer at the Department of Geography, Environment and Development Studies, Birkbeck, University of London. A social and cultural urban geographer, he has research interests in the modernist residential high-rise, geographies of architecture and geographies of home. His current research on the Aylesbury Estate in South London is exploring ideas of verticality, *un*making and the future. His PhD was completed at King's College London.

Jon Binnie is Reader in Human Geography at Manchester Metropolitan University, UK. His research focuses on the urban and transnational politics of sexualities. His most recent research focuses on geographies of transnational LGBTQ activism in Central and Eastern Europe and queer film festivals in Europe. He is the author of *The Globalization of Sexuality*; co-author of *The Sexual Citizen: Queer Politics and Beyond, Pleasure Zones: Bodies, Cities, Spaces*. He is also co-editor of *Cosmopolitan Urbanism* and special issues of *Environment and Planning A, Political Geography* and *Social and Cultural Geography*. Jon is currently working on a joint-authored monograph entitled *Sexual Politics Beyond Borders: Transnational LGBTQ Politics in Europe* for Manchester University Press.

Paul Chatterton is a writer, scholar and campaigner. He is Professor of Urban Futures in the School of Geography at the University of Leeds, UK, where he co-founded the 'Cities and Social Justice' Research Cluster and MA in 'Activism and Social Change'. He is currently Director of the University's Sustainable Cities Group. He has written extensively on urban change and renewal, civic experimentation and movements for social and ecological justice. He is co-founder of the Antipode Foundation, a public charity dedicated to research and scholarship in radical geography. Paul is also co-founder and resident of the pioneering low impact cohousing co-operative Lilac.

Mark Davidson is an Associate Professor in the Graduate School of Geography at Clark University, Worcester, Massachusetts, USA. He is an urban geographer whose research interests lie in three core areas: 'gentrification', 'urban policy, society and community' and 'metropolitan development, planning and architecture'. His research is international in scope, including work in Europe, North America and Australia. He has authored and co-authored papers in journals such as *Environment and Planning A, Ethics, Place and Environment, Transactions of the Institute of British Geographers* and *Urban Studies*. His current research includes a continued examination of new-build gentrification, a theoretical exploration of gentrification-related displacement and the empirically informed consideration of sustainability as a key policy concept. He has held fellowships at the Nelson A. Rockefeller Centre for Public Policy and Social Science, Dartmouth College, and the Urban Research Centre, University of Western Sydney. He holds a BA (Hons) and PhD in Geography from King's College London.

Tim Edensor teaches cultural geography at Manchester Metropolitan University, UK. He is the author of *Tourists at the Taj* (1988), *National Identity, Popular Culture and Everyday Life* (2002) and *Industrial Ruins: Space, Aesthetics and Materiality* (2005), as well as the editor of *Geographies of Rhythm* (2010). He is editor of peer-reviewed journal *Tourist Studies*. Tim has written extensively on national identity, tourism, industrial ruins, walking, driving, football cultures and urban materiality, and is currently investigating landscapes of illumination and darkness.

Ignacio Farías is a sociologist and an Assistant Professor of the Munich Center for Technology in Society (MCTS) and the Faculty of Architecture at the Technische Universität München, Germany. Ignacio works on science and technology studies, urban studies and cultural sociology with a focus on infrastructural transitions and participation. He is co-editor of *Urban Assemblages. How Actor-Network Theory Changes Urban Studies* (2009) and *Urban Cosmopolitics. Agencements, Assemblies, Atmospheres* (Routledge).

Mark Jayne is Professor of Human Geography at Cardiff University, UK. He is a social and cultural geographer whose research interests include consumption, the urban order, city cultures and cultural economy and he has published over 75 journal articles, book chapters and official reports. Mark is author of *Cities and Consumption* (2005), co-author of *Alcohol, Drinking, Drunkenness: (Dis)Orderly Spaces* (2011) and *Childhood, Family, Alcohol* (2015). Mark is also co-editor of *City of Quarters: Urban Villages in the Contemporary City* (2004), *Small Cities: Urban Experience Beyond the Metropolis* (2006) and *Urban Theory Beyond the West: A World of Cities* (2012).

Andrew E. G. Jonas is Professor of Human Geography at the University of Hull, UK. Andrew's research interests cover urban and regional development in USA and Europe. He is author or co-author of some 90 journal articles and book chapters. He has written a new textbook, *Urban Geography: A Critical Introduction*, co-authored with Eugene McCann and Mary Thomas. Andrew's edited books include *The Urban Growth Machine: Critical Perspectives Two Decades Later*, *Interrogating Alterity* and *Territory, the State and Urban Politics*.

Roger Keil is York Research Chair in Global Sub/Urban Studies at the Faculty of Environmental Studies at York University in Toronto, Canada. He researches global suburbanization, cities and infectious disease, and regional governance. As Principal Investigator of a Major Collaborative Research Initiative on *Global Suburbanisms: Governance, Land and Infrastructure in the 21st Century* (2010–2017) he works with 50 researchers and 18 partner organizations worldwide. He is the editor most recently of *Suburban Constellations* (2013) and co-editor of *Suburban Governance: A Global View* (2015).

Rob Krueger is the founding director of Worcester Polytechnic Institute's (USA) Environment and Sustainability Studies Program. He is a Guest Professor of Sustainable Spatial Development at the Institute of Geography and Spatial Planning at the University of Luxembourg. A broad, international audience recognizes Krueger's scholarship on urban sustainability. He has published dozens of papers in prestigious internationally peer-reviewed journals. He is co-editor of *The Sustainable Development Paradox*. His current co-authored book presents a novel approach to studying sustainable urbanism through cutting edge theories, field trips to 'sustainable cities' and workshops that allow students and faculty to directly engage in analysis or to develop their own field trips. He has been a Co-Principal Investigator on two U.S. Environmental Protection Agency (EPA) grants focused on environmental justice issues in Worcester. He serves as co-chair of the EPA's Environmental Justice Working Group.

Alan Latham is a cultural and urban geographer based at University College London. His research focuses on urban sociality, mobility and public-ness. He gained bachelor and master's degrees in New Zealand, before moving to the UK to study at the University of Bristol where he obtained his PhD. He has worked at the TU Berlin, and the Universities of Auckland and Southampton. He is the co-author of *Key Concepts in Urban Geography* and co-editor of *Key Thinkers on Cities*.

Loretta Lees is Professor of Human Geography and Director of Research in Geography at the University of Leicester, UK. Loretta is an urban geographer with international expertise in gentrification/urban regeneration, global urbanism, urban policy, urban public space, urban communities, architecture, urban social theory and the urban geographies of young people. Loretta is a Fellow of the Academy of Social Sciences (FAcSS) and a Fellow of the Royal Society of the Arts (FRSA). She has published 11 books and is working on 2 more, most of which have been about cities and urban theory. She was awarded the 2012 Antipode Scholar Activist Award and was invited to give a TEDx Brixton talk in 2014; she also co-organises *The Urban Salon: A London Forum for Architecture, Cities and International Urbanism.*

Melanie Lombard is Lecturer in Global Urbanism at the Global Urban Research Centre, University of Manchester, UK. Her research explores the everyday activities that construct neighbourhoods, through a focus on urban informality, place-making and land-tenure regularization, in Mexico, Colombia and the UK. She is Co-Director of the MSc Global Urban Development and Planning.

Robyn Longhurst is Pro Vice-Chancellor of Education and Professor of Geography at University of Waikato, New Zealand. Over the past five years she has been Editor of *Gender, Place and Culture* and Chair of IGU Commission on Gender and Geography. Robyn is author of *Bodies: Exploring Fluid Boundaries* (2001), *Maternities: Gender, Bodies and Spaces* (2008) and co-author of *Pleasure Zones: Bodies, Cities, Spaces* (2001) and *Space, Place, and Sex: Geographies of Sexualities* (2010).

Deborah G. Martin is a Professor at the Graduate School of Geography at Clark University in Worcester, Massachusetts, USA. Her research examines issues of place identity, urban politics, legal geographies, and urban activism. She has published in journals including *Annals of the Association of American Geographers, Antipode, Environment and Planning A, Human Ecology, International Journal of Urban and Regional Research, Mobilization, Professional Geographer, Progress in Human Geography, Transactions of the Institute of British Geographers, Urban Geography* and *Urban Studies.*

Colin McFarlane is Reader in Urban Geography at Durham University, UK. His research focusses on everyday life, urban infrastructure, and the politics of urban knowledge, especially in informal settlements. His recent work has examined the politics of urban sanitation in Mumbai through an ESRC project on 'everyday infrastructure'. He is author of *Learning the City: Knowledge and Translocal Assemblage* (2011) and has published widely in journals including *Public Culture, Society and Space, Transactions of the Institute of British Geographers, Antipode, Urban Studies,* and the *International Journal of Urban and Regional Studies.*

Paula Meth is a Senior Lecturer in Urban Studies and Planning at Sheffield University, UK. Her work focuses on gender, violence, and place with a particular emphasis on urban change in relation to informality and formalisation. She works primarily in urban South Africa but also in urban India and is the joint author of *Geographies of Developing Areas*, published by Routledge (2014, 2nd edition).

Steve Miles is Professor of Sociology at Manchester Metropolitan University, UK, and author of *Spaces for Consumption: Pleasure and Placelessness in the Post-Industrial City* (2010). His key publications include *Consumerism as a Way of Life* (1998) and *Youth Lifestyles in a Changing World* (2000) and he is co-author of *Consuming Cities* (2004) with Malcolm Miles. Miles has a particular interest in the role of consumption in the construction of identity, particularly in the context of regeneration and he has published on this topic in journals including *Urban Studies* and *Sociology*. Miles is currently editor of the *Journal of Consumer Culture* and his next book is entitled, *Retail and Social Change* and will be published by Routledge.

Andre Pusey completed his PhD in the School of Geography, University of Leeds, UK. His research utilises a form of 'militant ethnography' through the co-founding of the Really Open University, and his thesis evaluates the way in which the activism of the group created new radical values that attempt an exodus from the university-machine. He has published on radical education activism, autonomous social centres and the commons. He is currently a lecturer at Leeds Beckett University.

Jennifer Robinson is Professor of Human Geography at University College London, UK, and Visiting Professor at the African Centre for Cities, University of Cape Town. She has also worked at the University of KwaZulu-Natal, Durban, the LSE (London) and the Open University. Her book, *Ordinary Cities* (2006) developed a post-colonial critique of urban studies, arguing for urban theorizing that draws on the experiences of a wider range of cities around the globe. This project has been taken forward in her call to reinvent comparative urbanism for global urban studies in her recent, 'Cities in a World of Cities' article in *IJURR* and 'Thinking Cities Through Elsewhere', in *Progress in Human Geography*. Current projects include exploring transnational aspects of Johannesburg and London's policymaking processes and collaborative and community-based research comparing governance of large scale urban developments in London, Johannesburg and Shanghai (with Phil Harrison and Fulong Wu). She has also published extensively on the history and contemporary politics of South African cities, including *The Power of Apartheid* (1996).

Ugo Rossi received his doctorate degree in Development Geography from the University of Naples "L'Orientale", in Italy, in 2003. After struggling with academic precariousness for almost a decade, he got a faculty position as 'university researcher' in economic-political geography at the University of Turin, Italy. His main research interests variously relate to the field of urban politics, using both political economy approaches and ideas drawn from critical political theory. He is editor of *Dialogues in Human Geography* and *Archivio di Studi Urbani e Regionali*, an Italian-language journal of urban and regional studies. He is the co-author of *Urban Political Geographies. A Global Perspective* (2012). His work has appeared in a number of academic journals, including *Area, European Urban and Regional Studies, International Journal of Urban and Regional Research, Progress in Human Geography, Rivista Geografica Italiana, Social and Cultural Geography,* and *Urban Studies.*

Ola Söderström is Professor of Social and Cultural Geography at the Institute of Geography, University of Neuchâtel, Switzerland. He has published extensively on urban material culture, visual thinking in urban planning, and urban globalization. His current work focuses on the relational comparison of urban development and the urban geographies of mental health. His most recent book is *Cities in Relations: Trajectories of Urban Development in Hanoi and Ouagadougou* (2014).

Quentin Stevens is Associate Professor of Urban Design and Director of the Centre for Design and Society in the School of Architecture and Design at RMIT University, Australia. He is author of *The Ludic City,* co-author of *Memorials as Spaces of Engagement,* and co-editor of *Loose Space, The Uses of Art in Public Space* and *Transforming Urban Waterfronts.*

Joaquin Villanueva is Assistant Professor in the Department of Geography at Gustavus Adolphus College, Saint Peter, Minnesota, USA. Joaquin obtained his Ph.D. in Geography at Syracuse University. His research interests span the fields of political, urban, and penal geographies in order to examine contemporary crime-control policies implemented in disadvantaged neighbourhoods in Paris, France.

Kevin Ward is Professor of Human Geography and Director of the Manchester Urban Institute at the University of Manchester, UK. He is a geographical political economist whose current work involves rethinking what is meant by 'the urban' in urban politics, as elements of different places are assembled and reassembled to constitute particular 'urban' political realms. Kevin has published over a 100 journal articles and book chapters, and his books include the co-authored volumes *Urban Sociology, Modernity and Capitalism* (2002) and *Spaces of Work: Global Capitalism and the Geographies of Labour* (2004). Kevin is also editor of *Researching the City: a Guide for Students* (2013) and co-editor of *Neoliberalization: States, Networks, Peoples* (2007), *Mobile Urbanism: Cities and Policymaking in the Global Age* (2011).

David Wilson is Professor of Geography and African American Studies at the University of Illinois at Urbana-Champaign, USA. His research interests include the political economy of U.S. cities, urban political processes, cultural studies of U.S. cities, social theory and the built environment and qualitative methods. His publications include *Urban Inequalities Across the Globe* (2014) and *Finding the New Urban Frontier: America's New Redevelopment Machine* (2014).

Helen F. Wilson is a Senior Lecturer in Human Geography at the University of Manchester, UK. Her research interests include urban multiculturalism, intercultural dialogue, European cities and the geographies of encounter. She is working on a British Academy project on the mobilization of community intervention models and has published work on urban mobilities, multicultural schooling, the politics of tolerance and conflict management. She is currently working on a co-edited collection entitled *Encountering the City* (Routledge).

ACKNOWLEDGMENTS

The editors would like to thank Andrew Mould, Sarah Gilkes, Egle Zigaite, and Sophie Watson at Routledge for their patience and support and all of the contributors for ensuring that the compilation of this book was an enjoyable process.

Mark would like to thank his new colleagues at Cardiff University for making him feel at home so quickly, and the old ones at the University of Manchester who put up with him for a whole decade (2005–2015) and also those 'down the road' at Manchester Metropolitan University. Bethan Evans, David Bell, Gill Valentine, the Wollongong Boys, the Leyshons, Phil Hubbard, Sarah L. Holloway and Dr and Mrs Potts - Iechyd da!! Special thanks to First Class Daisy.

Kevin would like to thank the students on his second year course at the University of Manchester, *North American Cities*, who have been a joy to teach, the participants at the cities@manchester Summer Institute in Urban Studies in 2014 and 2015, whose energy, engagement and enthusiasm for all thinks critical and urban were wonderful pick-me-ups at the end of the academic year and to Tom Baker, Ian Cook, Mark Davidson, Iain Deas, Kim England, Graham Haughton, Stephen Hincks, Andy Jonas, Eugene McCann, Byron Miller, Cristina Temenos, Stephen Ward and David Wilson who have all co-authored with me in the last couple of years and with whom it has been great to work and write. Special thanks to Colette and Jack who continue to tolerate my interest in cities and my failings as a 'real geographer'.

The authors, editors and publishers would like to thank copyright holders for their permission to reproduce the following material:

Figure 5.2: 'Brother' and researcher Bronwyn McGovern at northern entrance to Cuba Mall in Wellington, New Zealand. Reproduced with permission of Belinda Brown Photography and Bronwyn McGovern.

Figure 7.1: Community identity in the landscape: Worcester, Massachusetts, USA. Reproduced with permission of Alexander Scherr.

Figure 7. 2: Traffic is one concern for community activists protesting land use changes: Worcester, Massachusetts, USA. Reproduced with permission of Alexander Scherr.

Figure 12.1: Anti-gentrification protests Taksim Square/Gezi Park, Istanbul, 2013. Reproduced with permission of Tolga Islam.

Figure 17.2: Foreign maid agency in central Singapore. Reproduced with permission of Alice Chen Huiyu.

1

A TWENTY-FIRST CENTURY INTRODUCTION TO URBAN THEORY

Mark Jayne and Kevin Ward

Introduction

Since the emergence of what are generally understood to be the first 'modern' cities in Europe and North America in the late-eighteenth and early nineteenth centuries, urban theory has been an important component of social science research. As we move through the second decade of what theorists, analysts and commentators term 'the urban age' or 'urban century', so it seems that it is becoming more important that urban theory generates new ideas and thinking in relation to the production of more inclusive and just urban futures. How we – as academics or students – go about doing this demands we name and pay attention to our theories. For theory is the lifeblood of the academic system, the tools of the trade through which complex ideas and understanding are generated, expressed and represented. Thought of in this way, theory in academia is equivalent to the language, lexicons and vocabulary used in many work places around the world: by car mechanics, computer programmers, doctors, heating engineers, lawyers, and so on. Professional worlds and their specific configurations can, of course, seem alien to those outside. Nevertheless, such things are the basis on which ideas, information and knowledge are communicated and exchanged. It is theory that enables the expression and performance of critical thought and intellectual advancement, important not only for the production of academic knowledge, but also in influencing how academics find out about cities through the methodologies they use, and, furthermore, in work which seeks to positively influence diverse audiences beyond universities (see Case study 1.1). For many undergraduates and graduate students, 'theory' is intimidating, one of the most challenging elements of acclimatizing to academic life. For those of you who can read, talk and write about theory, your studies are likely to be rewarding and, in all probability, successful. Theory should not be feared, and the challenge of getting to grips with the

CASE STUDY 1.1 WHAT IS URBAN THEORY AND WHY IS IT IMPORTANT?

Urban theory consists of ideas, languages and vocabularies that seek to explain cities and urban life. Urban theorists work to offer explanations of cultural, economic, political, and social practices and processes and their spatial manifestations. While it might seem obvious, theory always comes from somewhere and sometime. Urban theory can be grand with a high level of abstraction and/or draw on empirical, localized or contextual understanding of how cities are formed, how they function and how they change. Urban theory can be about 'big' urban questions as well as those focused on the 'ordinary' and everyday. However, with a concern to explain rather than describe, urban theorists seek to make the city understandable or legible. The challenges of explanation, none the less, ensure continual controversy and debate. Urban theory is thus an always-contested field, drawing on and/or critiquing, even eschewing, past theoretical work in order to offer new explanation of urban experience.

The critical perspectives offered by urban theory also play an important role in influencing how we find out about urban life through the methodologies we use to research the urban experience. Urban theory also resonates beyond academia. Important not just because of its intellectual value, urban theory is a resource utilized in many ways by a diverse range of actors, such as artists, musicians, planners, policymakers, politicians and writers.

Urban theory is the unique contribution that academics make to our understanding of cities both within and beyond universities. In these terms it is important to acknowledge how the emergence, evolution and impact of urban theory have informed how and what we think about cities. Highlighting weaknesses, omission and silences in the development of urban theory is, of course, the fertile ground that leads to the emergence of new critical perspectives. Contemporary urban theorists must rise to the challenge of how/ if their critical perspectives capture complexity and diversity, drawing on rich intellectual traditions of creative and innovative thinking in order to keep pace with the ongoing development of the urban experience.

complex, exciting and sometimes frustrating and annoying nature of urban thinking is likely to be the most intellectually rewarding element of academic study and research.

In this chapter we offer you a way to begin to better understand the meaning and value of urban theory; we first highlight some (but certainly not all) of the dominant currents of urban thinking, competing perspectives and long-standing fault-lines. In doing, so we also touch on how theory is related to the methods that have been used to categorize, label and study urban life. In drawing this introduction to a close we offer some insights into the twenty-six chapters of this book and how they might be read individually and together in order to gain an understanding of the theoretical terrain that underpins critical perspectives on the city.

A brief history of urban theory

Urban theory is constituted by work that takes place across a range of disciplines including: anthropology, architecture, cultural studies, economics, environmental studies, geography, history, planning, psychology, sociology and so on. Despite this inter-disciplinary nature, the contribution of these disciplines to urban theory has ebbed and flowed over time. While there might not be universal agreement, it is possible to argue that it is geographers and sociologists who have made amongst the most important contributions to advancing urban theoretical debates. Such comments notwithstanding, urban theorists, no matter what their disciplinary background, have invested considerable amounts of energy and time writing and debating the question of just what it is we should think about when we think about cities: a physical thing, a location in space, a political entity, a set of economic activities. Should we think of a city as being an administrative organization, a place of work, or a place to call home? Certainly we should consider how in a city we find ambivalence and boredom, love and hate, security and danger – and how cities can be places of hope and desire, pleasure and despair. As such, despite there being a set of different disciplinary contributions, it is often the case that theorists are (more often than not) speaking about the same things (see Case study 1.2).

Despite much common ground, a single and straightforward definition about what constitutes 'the city' is difficult to capture in simple terms, and furthermore any attempt to sketch out and depict the development and trajectories of diverse, complicated and contested notions of what constitutes urban theory over the past couple of centuries is similarly problematic. In this section we respond to such a challenge, but in doing so acknowledge the limits to adequately synthesizing and summarizing what is, when read together, a vast collection of contributions. In highlighting key ideas and people we acknowledge that any review of the history of urban theory is incomplete. It says as much about us as the authors, as about the field of work we purport to represent. As the reader it is important that you are not seduced by any 'neat' depictions or clearly laid out 'family tree' of thinkers or genealogy of ideas – it is, in sum, impossible to capture the complexity and depth of urban theory in this short introductory book chapter, or indeed in a series of books. For many academics, understanding, engaging with and keeping up to date with the development of urban theory is a lifelong challenge for their professional careers. Moreover, for those of you seeking an introductory understanding, do not think that as new sets of approaches, critiques or ideas emerge past work is left behind or lost forever. There are lots of examples of difference co-existing, or of traditions that were once out of favour being rescued.

However, it is also important to note, that for a significant proportion of academics who write about cities, advancing or contributing to the development

CASE STUDY 1.2 WHAT CONSTITUTES 'THE CITY'?

What do you think about when you think about 'the city'? Take a moment. What emotions, experiences, images, smells, sounds or words spring to mind? Lots, we are sure. The city is a complex mosaic, made up of probably everything you can think of – and more. However, despite popular and apparently shared understandings of what we understand to be a city, a consensus amongst urban theorists has been much more elusive.

Historically urban theorists differentiated the city, from towns or villages according to population size or density, built form, or economic, political or religious power. It was the concentration of people, living and working together, the agglomeration of buildings and infrastructure, institutions and organizations that defined urban spaces and places as cities, where a distinct 'way of life' emerged that was different from the 'countryside'. In other worlds the city was a physical space or territory, within which people and things in close proximity, with rhythms and tempos, led to the establishment of diverse and new urban cultures.

More recently, urban theorists have argued that distinctions between the rural and the urban have become irrelevant, or at least less relevant, and that the spread of urban life has led to the bleeding together of numerous cities, small towns, suburbs and so on to such a degree that it is more relevant to talk of city-regions, or globalized urban systems than discreet urban spaces. Such insights highlight how cities are not discrete bounded territories or individual settlements but, instead, that flows and juxtapositions, porosity and relational connectivity define urban life. As such, cities are now understood as existing in an era of increasingly geographic extended spatial flows – where relations stretch out across space. Cities now are considered as discontinuous, internally diverse, open and relational. In these terms all cities and expressions of territory are to varying degrees punctuated by and orchestrated thorough a myriad of non-territorial networks and relational webs of connectivity. Such insights suggest that if it ever was, it is now more difficult to offer a simple definition of 'the city'. Moreover, there is also a growing concern that urban theory has been overly dominated by a focus on a small number of large cities in Europe and North America, and that there is a need to develop a more comparative and cosmopolitan, inclusive and diverse urban theory if we are to better understand the proliferation of urban life around the world.

of urban theory is a minor or at best secondary concern. For example, the emphasis of many top-ranked urban studies journals that publish research and writing on cities is often weighed towards presenting empirical descriptions and explanations of practices and processes in cities. Similarly, highly popular textbooks aimed at students often tend to neglect the complexity and depth of interdisciplinary theoretical debate, instead identifying broad trends and key milestones in order perhaps to make content more accessible for readers. Many academic writers engaged in working beyond universities, and involved in policy and practice, may also not consider themselves to 'do' theory at all. Of course, there are many other journals and textbooks that do indeed lead with theoretically complex and/or groundbreaking empirical evidence. Moreover, for those seeking to make theoretical work legible for diverse audiences, including students, policymakers and practitioners, a key concern of their work is to apply critical theory to make positive changes in our cities and the lives of their citizens. While we will touch on such concerns, histories and debates about the relevance of urban theory throughout this chapter, we now turn to a review of developments and trajectories of urban theory – while reminding you to have at the forefront of your mind the many possibilities and pitfalls that are bound up with such a project.

While great cities have existed throughout human history it was in the late-eighteenth and nineteenth centuries that industrial revolution and rapid urbanization in the global north generated larger and more physically, politically, economically, socially, culturally, spatially and infrastructurally complicated cities than had previously existed. The foundations of the city that we seek to know today were laid; the kinds of grand (and not so grand) buildings, work and home lives of urban residents, models of citizenship, transport, infrastructure, leisure activities, social structures, fashion, identities, politics, consumption, and problems of health and disease, congestion, poverty, crime, homelessness and social divisions that are familiar to us today began to emerge at that time. For example, the birth of the modern city led theorists to develop the argument that dominant rural traditions were being replaced by new distinct forms of living in cities. Sociologists were keen to identify how a social continuum linked different types of community cohesion and social relations in urban and rural areas. Writing tended to be underpinned by a nostalgic view of a declining rural way of life (Tönnies 1887; Durkheim 1893 [1964]; Simmel 1903). Durkheim, for instance, suggested that beliefs, customs, dynamics, norms and routines that had previously created social cohesion based on likenesses and similarities between individuals were being eroded by the emergence of city-states and an urban market economy where social bonds were dominated by specialization, or in other words, a new kind of urban solidarity based on mutual interdependence. Georg Simmel (1903: 15) also argued that as urban modernity flourished and expanded at a rapid pace, city dwellers developed a 'blasé attitude' and 'aversion to strangers' that could lead to conflict and tension. Such a blasé attitude was described by Simmel as bystander indifference, founded on an abdication of responsibility and failure to care for our fellow citizens, leading to distanced people in cities. Many theorists at the time expressed the view that the newly developing city was cold, calculating and anonymous. For example, Engles (1844) highlighted how in the world's first industrial city, Manchester (UK), workers were exploited by capitalists and lived in inhuman conditions. It was also argued that, anonymity and a lack of social bonds relating to community, neighbourhood and kinship ensured life in these newly emerging cities was superficial and transitory, with urban citizens struggling to deal with alienation, anonymity and the isolation of the new modern city (Weber 1921; Wirth 1938; Durkheim 1893; Simmel 1903).

The urbanization of North America and the growing academic interest led to the establishment of the world's first Department of Sociology, in Chicago in 1892. Co-founders Robert E. Park and Ernest Burgess considered Chicago their 'urban laboratory' and sought to explore how humans adapted to new formations of urban life. The Chicago School's enduring contribution to urban theory has been the development of models of urban and social structures that sought to explain how the emergence of mechanisms of market economies were constituted by physical and social organization of the city. Indeed, the most famous legacy of the Chicago School has been representations of city in diagrammatic forms to explain land use, phases of invasion and occupation, zones of transition, social segregation and so on. While the urban-modelling advanced by the Chicago school was later seen as grossly oversimplified, such work was, none the less, an important springboard for measuring, modelling and representing the city. Moreover, many of the ideas that drove the work of the Chicago School were influenced by now-discredited concepts derived from eugenics and social Darwinism, which were applied in order to depict how pathologies of social life related to an urban ecology of competition and survival. Indeed, Hubbard (2006: 26) argues that 'seminal work of early sociologists on the psychosocial economies of urban space was later to inspire some relatively unsophisticated theories of the impact of the environment on behaviour specific forms of housing and criminal behaviour, health and moral virtue of dwellers'. Such important critique notwithstanding, representations such as Burgess's Concentric Zones Model and Hoyt's Sector Model are still the stuff of high-school classes. Moreover, despite some highly problematic theoretical foundations, the Chicago School importantly focused on what can be described as 'micro-sociology' and also pioneered methods, both of which have become mainstream academic activities today – participant observation and ethnographic techniques were utilized in order to study 'urban subcultures' such as gangs, homelessness and prostitution.

In Europe, social theorists were also theoretically engaged with the emergence of modern life and the cities that were beginning to form. For example, the Frankfurt School, set up in the early 1920s, founded by left-wing, middle-class German Jewish intellectuals took up the legacy of the writings of Karl Marx in order to advance understanding of social change. While not all thinkers associated with the Frankfurt School directly studied the city, their work has, none the less, had an enduring influence on urban theory. Writers such as Herbert Marcuse (1932) and Max Horkheimer and Theodor Adorno (1947) advanced Marx's concept of 'alienation' to emphasize how mass production of cultural artefacts such as music, fashion and literature produced a 'culture industry' that worked to integrate workers into capitalist society. They argued that the 'culture industry' led to a destruction of 'genuine' folk culture through the mass standardization of production and consumption, and further that a transformation of everyday culture had taken place, where workers were willing to accept boredom and exploitation at work because they could escape during their leisure hours to enjoy art, music, fashion, shopping and so on.

Other members of the Frankfurt School applied such thinking to the development of the modern city in a more focused way. Walter Benjamin (1999), who undertook research on the emergence of shopping arcades in Paris during the 1920s to 1940s described the ideological imposition of 'commodity fetishism', which was a 'trick' played by capitalism to disconnect the socially unjust conditions of the production of goods from the people consuming them, via their purchase in new spectacular urban spaces – shopping arcades – built using new technologies to construct grand 'artificial' covered shopping streets from glass and steel. Similarly, Siegfried Kracauer (1926) described the emergence of beautiful and grand new spaces of consumption such as theatres, libraries, town halls, cafes, department stores and so on as 'mass ornamentation', where spectacular architectural design, fashion cultures and 'window shopping' were examples of ludic pleasures of consumption that masked social inequalities (see Jayne 2005). While there has been significant criticism of the writing of the Frankfurt School for assuming that people were 'mindless dupes' buying into, rather than having a critical appreciation of, strategies of capital accumulation and the associated new urbanism, the ideas and topics studied by this influential group of scholars have had a long-standing influence on urban theory.

A further profoundly important thinker that began his writing career in parallel with the work of both the Frankfurt and Chicago Schools was Henri Lefebvre, a French sociologist and philosopher. Lefebvre's illustrious career began to flourish with his 'critique of everyday life' written in the 1930s. Also drawing on Marxist philosophy, Lefebvre argued that by colonizing everyday life, capitalism was able to reproduce itself. Working on similar terrain as members of the Frankfurt and Chicago Schools, Lefebvre discussed alienation, boredom, consumption, and the promises of freedom that emerged with increasing leisure time, which was increasingly shared across society. Lefebvre argued a consequence of the work and leisure rhythms of capitalism was reduced self-expression and a poorer quality of everyday life. Importantly, Lefebvre's 1984 critique advanced previous urban theoretical debate through his work on everyday life, by arguing that it was the progressive development of the conditions of human life, rather than control of capitalist production, through which humans could achieve a socially just world. Lefebvre's later work (1968, 1991), from the 1960s onwards, which had minimal impact at the time it was published has subsequently had profound influence on urban theory, focused on the 'social production of space' and the 'right to the city'.

While urban sociology was advancing new understandings of the emergence of the modern city and developing ideas of how to facilitate progressive urban change, geographers focused on applying and advancing the traditions of modelling and measurement initiated by the Chicago School. For example, post-war urban geography sought to establish itself as a spatial science, with calculators and computers being used to produce models through tested and verifiable hypotheses.

Geography, at the time, saw itself as a 'hard science' of the city, where mathematics and statistics underpinned a quantitative approach to studying urban life. In stark contrast to cutting-edge sociological urbanism, urban geographers were focused on 'accuracy and rigor'. This lent itself to applied writing and findings that provided 'objective' and 'value-free' perspectives that were particularly influential in urban planning and policy formation. Scientific concepts and methods allowed theoretical projects to model land use, migration and land values, residential behaviour and spatial arrangement, size, and number of settlements, hierarchies of cities, and so on. There was also a 'behavioural turn' in positivistic urban geography, which sought to strengthen urban models in order that planners and architects could take into account the irrationality of human behaviour – including a focus on how people thought about, and acted in cities, by introducing methods such as mental maps (Lynch 1960). This strand of positivist urban geography also utilized questionnaire surveys and represented data through computer simulation and modelling to highlight the ambiguous and contradictory nature of 'rational' decision-making by urban citizens.

However, from the 1970s, a fundamental shift in thinking occurred when geographers began to engage for the first time with the critical urban theory that had been generated out of the arts, the humanities and the social sciences. For example, initially geographers began to draw on long-standing sociological perspectives from the work of Karl Marx and Max Weber to consider how some social actors controlled urban assets and land markets. This 'urban managerialism' sought to consider the role of estate agents, mortgage lenders, planners and so on, in influencing urban change. However, in the context of broader civil and social unrest in cities and countries of the global north, many urban geographers considered that it was a mistake to continue to focus on urban managerialism and the use of positivist modelling and theorization. For example, David Harvey (1969), once a key protagonist of geography as a positivist spatial science, called for a more radical Marxist approach to understanding urban geography. Urban managerialism was dismissed as 'theoretically barren and politically quiescent … irrelevant to the problems of the modern city' (Harvey 1973: 34).

Such rallying calls took place at a time where the historic writing of Karl Marx was gaining favour through a critique of political economy inspired by a new generation of urbanists. David Harvey (1973), Manual Castells (1977), Peter Saunders (1986) – a mix of geographers and sociologist – to name but a few, were concerned to study how capitalism was produced and reproduced to occupy spaces and places in the city. Marx's legacy was to ensure a structural reading of society, and a concern to understand the influence of capital and class in cities gained prominence in radical and critical urban writing by geographers. While in fact Marx said little about cities themselves – viewing them simply as instruments of production – those who advanced such ideas, including Engels, the Frankfurt School, Lefebvre and so on, were now inspiring a new generation of urban geographers seeking to provide a clear statement on the connections between city life as a 'spatial fix' of capitalism. This new wave of thinkers promoted awareness of social agency through the expression of class consciousness in order to advance understanding of topics such as uneven economic development, inequality and injustice, property, land use and segregation, gentrification, state regulation and capital/labour conflicts, and in doing so focused on a variety of geographical scales from the household, to neighbourhoods, the city and regions. Underpinning the breadth and depth of Marxist-inspired urban theory has been the focus on the 'urbanization of injustice', with capitalism at the root of all urban problems. Since the late 1970s the emergence of Marxist thinking has continued to have a profound influence on urban theory, as new generations of what have been termed 'metromarxists' (Merrifield 2002) have sought to engage with complex practices and processes relating to capitalism and urban life.

For example, the economic crisis which began in Europe and North America in the late 1970s and 1980s was explained by Marxist urban theorists as a result of inherent weakness in capitalism and a tendency of over-accumulation leading to cycles of economic depression and 'creative destruction', through often dramatic and economically and socially catastrophic urban restructuring. Indeed, urban changes associated with the late 1970s and early 1980s era of de-industrialization led to high levels of unemployment, civic unrest, increased poverty and urban decay. However, by the end of the 1980s the emergence of new 'post-industrial' economic activity included increasing globalization, flows of capital, finance and the growth of multinational corporations. Such economic restructuring took place in parallel with political change through the 'hollowing out' of the nation state enabled by fiscal policy first championed by leaders such as Margaret Thatcher and Ronald Reagan, and the emergence of governmental agendas to pursue economic development and the success of markets above social justice. These changes had a profound effect on cities around the world, but particularly with regards to the 'creative destruction' of the industrialized urban global north. Indeed, following decay and crisis, a 'new' urbanism emerged through the regeneration and gentrification of business and residential areas, and the intensification of consumer culture and the politics of consumer sovereignty. Increased social and spatial segregation underpinned by increasing disparities between the rich and poor, as well as the physical boundaries of city and countryside, further gave way through intensified suburbanization.

It is no surprise of course that such dramatic cultural, economic, political, and social changes in the city led to intensive theoretical debate. Many Marxist theorists, including a number of writers who have been branded as the Los Angeles School, such as Jameson (1991), Soja (1989), and Davis (1990) argued that new cultural, economic and political forms and physical organizations of the city could be theorized as a new period of 'flexible' capitalism. On the other hand, many theorists argued that such conditions could not simply be reduced to changes in capitalism, but instead believed all that 'modernity' had promised the world; that science, rationality, planning, culture and government could improve the lives of all citizens of the city and had given way to a new epoch – a new period of history labelled as postmodern. Theorists such as Featherstone (1991), Berman (1992) and

Pred (1996) argued that postmodernity was a new era underpinned by the logic of consumer culture, an increasing 'aestheticization' of everyday life and a proliferation in the power of identity politics, where traditional social structures that had been constructed during modernity had dissolved away so that class, gender, age, sexuality, ethnicity and so on were less fixed than in previous generations – and now constituted by instability, malleability and fluidity. Perhaps the most influential theorist of the postmodern condition was Jean Baudrillard, who suggested that such urban change could be characterized by 'Disneyfication', and drew on the example of Las Vegas to theorize the emergence of new 'hyperreal' spaces, identities and social practices that were increasingly dominating urban life. Such writing importantly led to the popular use of urban 'semiotics' to read messages, signs and codes in order to connect political and economic systems to the 'things', social relations and activities that constitute postmodern urban spaces and places (Gottdiener and Lagopoulus 1986). While critiques of the debates that circulated around (post)modernity argued that a focus on cities such as Los Angeles and Las Vegas failed to account for the diversity of cities around the world, conversations about whether or not we had entered a new epoch of postmodernity, or were just experiencing a restructuring of capitalism were a key tenant of urban theory from the 1980s until the beginning of the new millennium (see, for example, Harvey 1991).

At the same time as debate about (post)modern urbanism was raging, a number of less high-profile but, none the less, profoundly important (and at times intersecting) threads of urban theory were emerging. For example, drawing on a longstanding tradition of measuring, modelling, ordering and ranking cities, different waves of theorists have sought to describe and explain 'an urban order' (see Bell and Jayne 2009). Inspired by the Chicago School and drawing inspiration from the geographer Walter Christaller's (1933) 'central pace theory', which focused on depicting 'urban systems', new work emerged which sought to theorize the impact of intensifying globalization through a focus on 'world cities' (see Friedmann 1986). The Globalization and World Cities Network (GaWC) portrayed a global hierarchy of cities based on concentrations of advanced producer services – financial services, law and accountancy, or on airport connectivity (see Beaverstock et al 1999). The counting and analysis of these measures of 'world-cityness' generated a schema that labelled cities in terms of their alpha, beta and gamma world city status or as 'relatively strong', 'some evidence' and 'minimal evidence' of world city formation. In a similar vein, urban planner Peter Hall (2004) identified global cities (with 5 million people or more), sub-global cities (with between 1 and 5 million people), regional cities (with a population of between 250,000 and 1 million) and provincial cities (with populations of between 100,000 and 250,000 people). Upon realizing that his work failed to account for cities throughout world, Hall then re-drafted his vision of the urban hierarchy with a specific focus only on Europe. He attempted to take account of functional and geographical approaches and presented an urban hierarchy in the following way – central high-level service cities (major cities and national capitals), gateway cities (sub-continental capitals), smaller capitals and provincial capitals, and finally county towns. This revised version was underpinned by measures of the geographical reach of cities, the agglomeration of economic and cultural activity, as well as air and rail connectivity.

Given the struggle to convincingly measure 'world-cityness' by pointing to hierarchies, ordering and schemas, a new branch of global-city research emerged which sought to overcome such weaknesses and to develop new ways of theorizing the rapid economic restructuring and associated speeding-up of processes of globalization during the 1980s. 'World-city' theorization gave way to research into the 'global-city' by measuring flows of capital, culture and people via the presence of: global multinational or national financial sector institutions, political command and control functions and culture industries and the tourist sector (see for example, Sassen 1994). Such writing was influential in highlighting the role of cities such as London, New York, Shanghai, and Tokyo as the new powerhouses of globalizing urbanism, and became influential in informing the work of urban planners and policymakers around the world who were seeking to achieve this status for their own cities. Later writing also saw the emergence of a theoretical successor of the world-city thesis that accounted for the emergence of 'global city-regions'. Such writing focused on *relationship between cities* within regions, and argued that increasing global urbanism could only be understood by focusing on 'mega' city-regions – as spatial nodes, made up of specific and unique political, economic and cultural milieus that allowed 'city-regions' such as London, New York, Tokyo, Shanghai to succeed in a globalizing work (Scott *et al* 2001).

Around the same time a diverse raft of theoretically complex and critically focused writing emerged from frustration with what was argued to be an increasingly ideological dogma that some thought had come to characterize Marxist urbanism and its privileging of structure over agency, a rigid view of the city that failed to account for the diversity and difference of urban life. For example, critical dialogue between Marxist urbanists and feminists (Walby 1997; Valentine 1997) highlighted the importance of understanding gender inequalities underpinning patriarchy as a vital theoretical project in its own right and not just reducible to a function of capitalism. In a similar vein, a profound critique of Marxist urbanism emanated from postcolonial theorists. Postcolonial thinking highlighted a multidimensional critique of the hegemony of western geographical imaginations and the idea that the study of the city could not be reduced to ideas and case studies generated in 'the western' urban capitalist world (Fanon 1961; Said 1978; Spivak 1993).

Critical Marxist, feminist, postcolonial and post-structural thinking (which will be introduced shortly) has often also been constituted by, and intersected with a diverse raft of theory that has influenced our understanding of cities. For example, a broader 'cultural turn' in the social sciences, partially emerged out of urban theory. Many thinkers were determined to escape long-standing intellectual traditions where 'culture' was seen as less important than the realms of political and economic life, or indeed that we need to develop better understanding of the 'fuzzyness' of categories such as 'politics, 'the economy' and 'culture' as being easily separate and distinct categories (see Amin and Thrift 2002). Key contributions include the work of cultural studies theorists such as those at the Centre for Contemporary Cultural Studies (CCCS) at the University of Birmingham, UK, that opened up dialogue at the intersections of anthropology, history, literary criticism and sociology (Stuart Hall 1973; Hebdige 1979). Rather than focus on 'high' culture, theorists considered instead popular culture such as chart music, television programmes and advertising. For example, Stuart Hall emphasized the critical opportunities involved in the encoding/decoding of cultural texts (which later influenced postmodern semiotics) to critique one-way power relations between producers and consumers that were evident in the work of the Frankfurt School. Work from the CCCS also importantly advanced, but also critiqued, the work of the Chicago School by using ethnography to develop progressive understanding of urban subcultures based on music and fashion, which allowed marginalized urban groups to resist powerful structures (Hebdige 1979).

The influential work of Pierre Bourdieu has similarly had an important impact on the development of urban theory. Bourdieu was concerned to unpack the dynamics of power in society, and in particular the diverse and subtle (and not so subtle) ways in which social order was maintained within and across generations (see for example his writing on habitus). In *Distinction: A Social Critique of the Judgment of Taste* (1984) Bourdieu highlighted how fashion, manners, tastes and so on are related to social position, or more precisely, are themselves key moments in acts of social positioning. He outlined how class distinction was maintained by a combination of economic, social and cultural capital in order to understand the experience of subjects within objective social structures. Bourdieu combined qualitative and quantitative research methods to evidence his ideas: combination social theory with data from surveys, photographs and interviews. Bourdieu's work also emphasized the corporeal nature of social life and highlighted the role of practice and embodiment in social dynamics. Bourdieu's thinking has been central to a diverse range of writing about cities with regards to theoretical and empirical engagement with topics such as gentrification, gender, class, sexuality, consumption, fashion and so on.

Other long-standing topics have also been reimagined and reinvigorated in order to develop understanding of the cultural (and political and economic) lives of cities. For example, the work of Ervin Goffman (1959) – whose writing on the 'presentation of self' highlighted the 'theatrical' managing of bodies in urban spaces not only as a way of 'seeing and being seen' but also as a key mechanism to manage and overcome previous depictions of an alienating urban experience – has had a long-lasting effect on urban studies. Similarly, the work of Judith Butler (2006) which draws on political philosophy, psychoanalysis, ethics and gender theory has highlighted the discursive construction of gender, not as a 'natural' effect of biology, but as a social and cultural construction and most evident as a 'performance' of identity. Such writing, when read hand in hand with other critical Marxist, feminist, postcolonial and post-structural writing, has influenced theoretical focus on intersections of class, ethnicity, age, gender, sexuality (particularly queer theory), disability and so on, as constituting inequality and enabling challenges to powerful structural and experiential life in our cities (see for example, Bell and Binnie 2004; Imrie 1996).

In a similar vein, work which seeks to challenge classed, gendered and racist assumptions that underpinned urban planning theory and practice often draws on the work of Jane Jacobs, and in particular her book *The Death and Life of Great American Cities* (1961) as a key moment where theoretical agendas have impacted on planning and policy beyond universities. With no professional background in planning, Jacobs argued that 1960s urban renewal in the USA was failing to respect the needs of most city-dwellers and successfully contributed to grassroots protests in New York city to protest against 'slum' clearance and the planning of freeways through deprived areas of the city. There is now a long tradition of writing that draws on Jacobs' view of neighbourhood life, and the role of 'street life' in enabling vitality, social mixing and diversity to thrive in our cities. For example, the work of Sharon Zukin (1988, 2010) which has been profoundly important in generating interest in 'the symbolic economy' of cities in order to explore how people give neighbourhoods their 'sense of place', but also how an emphasis on 'distinctiveness' is increasingly used as a tool of elites to enable the privatization and commercialization of public and residential space (see Jayne 2016).

The final body of writing that we focus on to draw this section of the chapter to a close is the diverse body of writing that has been described as post-structural urban theory through an emphasis on language, representation and power underpinning the production of space, which can be interrogated through a focus on immaterial and material forces. For example, Michel Foucault, a French philosopher, highlighted how relationships between power and knowledge underpin social control in complex visible and hidden ways. Foucault's (1979, 2008) writing is voluminous and diverse but his work on power and discipline, governmentality of institutions and of 'the self' has inspired urban writers focused on a broad range of issues such as health, crime and 'deviance', sexuality, networks of power, public and private institutions, policing and policy. Foucault's thinking has also influenced the development of post-structural methods such as discourse and textural analysis that work to uncover hidden power relations. Other important post-structuralist writing, of theorists such as Deleuze and Guattari (1987) and Latour (2005), has highlighted how human and non-human actors come together in urban political and social struggles and has also been particularly important in the development of urban theory (McFarlane 2011a; Farías 2011; Dovey 2011). For example, work on urban networks, practices and spaces has drawn on poststructuralist thinking in order to interrogate 'not to the static arrangement or a set of parts, whether organized under some logic or collected randomly, but to processes of arranging, organizing, fitting together... where assemblage is a whole of some sort that expresses some identity and claims to a territory' (McCann and Ward 2011: 12). Similarly, Amin and Thrift's (2002) book *Cities: Reimagining the Urban* argues that too much contemporary urban theory is underpinned by nostalgia for a humane, face-to-face and bounded city. Amin and Thrift suggest that the traditionally understood divide between the city and the rest of the world has been perforated through urban sprawl, and urbanization 'as a way of life'. In doing so Amin and Thrift outline an innovative urban geography focused on 'the multiple spatial networks that any city is embroiled in, and to… [allowing consideration of] the full force of those networks and their juxtaposition in a given city upon local dynamics' (2002: 112). It has been argued that engagement with poststructuralist thought has dramatically 'changed urban research' (Farías and Bender 2009), with the emergence of a focus on the interaction of human and non-human actors in order to better understand the complexity and openness of cities and the ways in which they are assembled and disassembled.

Urban theory: new critical perspectives

Given the amount of material that has just been covered, let us take a moment to pause and to reflect on our whistle-stop tour that has highlighted some of the key contributions to the internally heterogeneous and multi-disciplinary field that is urban theory. Now, of course, is a good time to ask yourself some important questions such as: what is the current state of urban theory? Is contemporary urban theory capable of capturing the complexity of contemporary cities around the world? How is the contemporary city being studied? Using what methods? Are urban theorists doing enough to influence politics, policy-making and popular debate? We ask you to consider these questions as you read through the chapters of this book, both individually and collectively as a whole volume. As you will undoubtedly have noticed, either implicitly or explicitly, authors build on strengths, and respond to weaknesses in historic theoretical traditions, often drawing on diverse bodies of writing to develop overlapping and intertwining narratives. Chapters highlight work that responds to the challenge to offer critical interventions into long-standing debates, or seeks to develop points of departure and new ways of thinking about cities.

As we have seen throughout this introductory chapter, critical Marxist, feminist, postmodern, postcolonial and post-structural theory, as well as a diverse, overlapping intersection of other bodies of writing are now a resource for urban theorists to draw upon in order to advance an understanding of the urban world. For example, recent innovative theoretical work relates to theorists' attempts to develop a balanced understanding of territorial urban spatial fixes of capitalism while attending to increasing understanding of relational urbanism (see McCann and Ward 2010, 2012; Jayne et al 2011, 2013). This seeks to understand how 'all contemporary expressions of territory … are to varying degrees, punctuated by and orchestrated through a myriad of trans-territorial networks and relational webs of connectivity' (MacLeod and Jones 2007: 1186). Emerging from 'the cultural turn' some urban theorists have shown how study of the city can contribute to broader theoretical interest across the social sciences on topics such as emotions, embodiment and affect, or more-than representational theory to consider mundane, everyday practices that shape the conduct of human beings towards others and themselves in particular sites (Thrift 2004b). A raft of new thinking has also emerged to respond to the challenges and opportunities offered by postcolonial theory, and weaknesses that ensue from a dominance of theory formulated by focusing too much on large cities in Europe and North America. Such writing suggests that we have only scratched the surface of the diversity of urban experiences around the world and there is a need now, more than ever, for a more diverse, inclusive and comparative urban theory (see, for example, Bell and Jayne 2006, 2009; Robinson 2006; Parnell and Robinson 2012; Isoke 2013; Roy and Ong 2011; Edensor and Jayne 2012; Ward 2010; Jayne 2013). Other theorists are also offering new innovative ways of doing theory that do not implicitly draw on 'western' traditions of knowledge production and genealogies of ideas, key thinkers and writing conventions (see for example, Simone 2004, 2001). This plurality – the absence of a taken-for-granted single theorization of the city – is both empowering for the field but also troubling, at least according to some (Scott and Storper 2015).

It will be clear from reading this introductory chapter that contemporary urban theory is constituted by a diverse range of approaches, methods and theories. It thus follows that there are multiple views on the efficacy or otherwise of the current state of urban theory. It is worth remembering that many of the theoretical innovations in urban theory that we have discussed came about because of dissatisfaction with the development of ideas. For example, over the past few decades a number of different critiques have emerged such as Manual Castells' (2000) view that there has been a relative lack of radical impulses for change that once characterized urban sociology in the 1960s and 1970s. Geographer Nigel Thrift (1993) has pointed to the ways in which 'urban' has often become hidden amongst other parts of the discipline such as cultural, economic, feminist, population, transport geography and so on, so that studies of topics such as consumption, music, leisure, identity technology and travel are eschewing, or not taking seriously the backdrop of urban life, and hence urban thinking is not always at the cutting edge of geographical writing. Other criticism has highlighted that research methods have not always kept pace with advancement of theoretical developments, ensuring a disjuncture between what we know about the world and how we can research the urban world (see Ward 2014). The extent to which urban theory is being translated and communicated to diverse audiences in order to ensure that urban theory contributes to political debate, policy and popular debate has been an ongoing concern (Harvey 1974; Ward 2005, 2007). Disciplinary conflict also emerges where sociologists argue that other theorists don't take social relations seriously, geographers argue that other urban disciplines fail to understand the role of space and place in a sophisticated way, or that academic planners fail to account for sociological or geographical urban theory in their work.

Discussion has also circulated around the proliferation of diverse urban theories over the past thirty years, including the new critical perspectives covered in this book. For some, that there is now an 'urban theory of everything' highlights diversity that is not a strength, but rather a hindrance to the ability of contemporary urban theory to talk about 'the city' in a way that 'matters'. In a recent debate high-profile 'MetroMarxists' have made this very case. For example, Neil Brenner (Brenner *et al* 2011: 227), when discussing urban assemblage thinking and research on materialities, argues that such theory must take more seriously underlying logics and inequalities of capitalist accumulation. While Brenner welcomes the 'innovative, intellectually adventurous impulse behind recent assemblage-theoretical interventions', and is cautiously optimistic about the empirical foci, he suggests that assemblage/materialities writing downplays the 'context of contexts' and thus fails to adequately grasp how capitalism shapes contemporary urbanization. In a similar vein, Erik Swyngedouw (2015) reflecting on 'urban theory for the twenty-first century' welcomed the theoretical rigour and empirical richness of new urban critical perspectives but described such innovations as 'staccato theory', having a 'machine gun' effect on the coherence of urban thinking. Such insights highlight a view that urban theory has moved too far from 'meta-theories' that can offer universal explanations of urbanism, to a proliferation of theories that offer perspective on the diversity and complexity of urban life (see Curry 2005).

Perhaps the disagreements that underpin such perspectives are best highlighted through the detail of a 'round-table' debate amongst urban theorists working from a variety of theoretical traditions. Discussion has emerged around the question of what should constitute the centre ground or 'cutting edge' of urban theory? In essence a number of long-standing fault lines came to the fore in arguments regarding the best routes for theorists to contribute to spatial transformations and to work towards socially just cities (Marcuse *et al* 2014). In a round-table debate there was an acknowledgement of clearly distinct traditions of urban theory that draw predominately, on the one hand, on the readings of Marx and Engels, Lefebvre and Harvey, and their notion of capitalist urbanization of injustice, and on the other hand those who are aligned with theorists such as Bourdieu, Baudrillard, Deleuze, Foucault, Latour, Thrift and numerous others. In sum, a long-standing tension between Marxist political-economy readings of urban change and development and other, post-structural critical perspectives that suggest that not everything in the city can be understood in terms of the power of capital, continues to characterize contemporary urban theory.

Before you feel the need to throw your academic mortarboard into one corner of the ring or another in this enduring battle, readers should be aware, and look for examples throughout the chapters of this book, of what was acknowledged to be the case in the round-table discussions that many contemporary theorists do not necessarily see themselves as inhabiting completely separate camps (see Marcuse *et al* 2014). For example, take the two of us! In broad terms our work could be defined by the approaches of these two differentiated approaches. However, in the detail of our work one of us is often found applying social and cultural theory to understanding urban political and economic practices and processes – and the other vice versa. Perhaps this 'mix and match' approach is one that now is characteristic of urban theory in a manner that does not compromise historic traditions of urban theory, but instead adds value to theorists' work by interrogating the critical tension between different theoretical traditions and drawing on ideas and thinking that offer the best theoretical terrain to develop critical understanding of our cities. For those of you reading this book who have already, or in future will embark on a professional life of academic work, it is you of course who will also get drawn into these debates and have to make your own mind up about the most progressive way to generate 'new critical perspectives' on urban theory.

This is also an important time to remind us all that academics are not working in a 'vacuum' or the overused term of an intellectual 'ivory tower'. Urban theory, as it has always done, takes as its focus – as its stimulus – the world around us. Over the past two hundred years, academic life, just like other professional worlds has, and will continue to be, constituted by fashions and trends, new innovations and of course some dead ends. Indeed, as Hubbard (2006: 248) states, theoretical trajectories can quickly become 'tired, worn-out and well, *boring*' or underpinned by 'recycled critiques and endless circulating same messages' (2006: 57). However, Hubbard goes on to warn against the simple 'out with the old, in with the new approach' where established lexicons and orthodoxies are replaced with neologisms. Hubbard further urges us to be reflexive regarding how, what and why we study, acknowledging that an understanding of cities can only ever be partial, and, though some might try, there can never be a unified theory of cities.

With such warnings ringing in our ears we encourage readers to judge the *new critical perspectives* that constitute this book with reference to the long-standing theoretical debates, points of departure from established thinking and controversies, conflicts and tensions identified in this introductory chapter. To that end, we want to finish with a few handy tips on how to read the content of this book. The boxed case studies in each chapter offer insights into key components of the argument. The key readings at the end of the chapters are the 'go to' sources that provide the reader with a means of engaging with each topic in more detail. Above all, we encourage you to take up the challenge of going to the original sources, no matter how far back in time they go. The chapters in this book are designed to open up a world of ideas to you, rather than simply being a sampler of topics where each chapter will tell you all you need to know. And finally, please read and think across the content of the chapters. Look to understand and engage with the rich diversity of ideas and thinking; look for difference and similarities across the chapters. In doing so you will have made important strides in getting to grips with intellectual traditions that have emerged over two centuries, and as the chapters of this book will now go on to show, there are so many new fruitful avenues for theory to continue to guide critical contemporary and future understanding of cities and urban life.

Key reading

The following books offer a student friendly broad introduction to interdisciplinary urban theory by summarizing trajectories and genealogies of key ideas.

Harding, A. and Blockland, T. (2014) *Urban Theory: A Critical Introduction to Power, Cities and Urbanism in the 21st Century*, London: Sage.

Hubbard, P. (2006) *City*, London: Routledge.

Jonas, A.E.G., McCann, E. and Thomas, M. (2015) *Urban Geography: A Critical Introduction*, Oxford: Wiley Blackwell.

Parker, S. (2015) *Urban Theory and the Urban Experience: Encountering the City* (2nd Edition), London: Routledge.

Short, J.-R. (2014) *Urban Theory: A Critical Assessment*, London: Palgrave Macmillan.

2

AFFECT

Ben Anderson

Introduction

Claims about urban life have long involved claims about affective life. Myriad transformations in and to cities have been cast in affective terms, whether implicitly or explicitly. Think, for example of Simmel's (1903) description of the European metropolis in terms of the oscillation between experiences of shock and a blasé attitude, in contrast with Crary's (2013) attention to today's hyperactive 24-hour capitalist city, animated by a background of restlessness and felt in the intensification of stress and in the creation of spaces of respite and relaxation. Whilst such diagnoses of the affective life of cities have long been part of urban theory, in the last ten years or so affect has become an explicit topic of interest and focus. There are numerous reasons for this, many extending beyond urban theory and implicating changing ideas of the human and social and cultural life that are part of what has been dubbed the 'affective turn' or 'emotional turn' (see Gregg and Seigworth 2011). Beyond the observation that cities are 'roiling maelstroms of affect' (Thrift 2004b: 57), has been a recognition that attempts are made to manipulate urban affective life in all kinds of ways, from the design of the seductive façade of consumption spaces (see Chapter 9, this volume) to enactments of legislation that prescribe how bodies can inhabit public space in order to create and maintain specific atmospheres (see Chapter 5). At the same time, and as a counter to any sense of affective life as fully determined by attempts to manipulate it (Barnett 2008), there has been a renewed attentiveness to the conditions of formation of urban experience and the particularities of different experiences. Where urban experience runs from the pre-personal to the transpersonal; that is, from the sometime non-cognitive background feelings of stress that surface for some during regular commutes (Bissell 2014), to give one example, to the atmospheres of security that envelop particular sites, such as train stations, where bodies move and congregate, to give another example (Adey *et al* 2013).

How, then, are affects part of urban life? And what implications does learning to sense and disclose affective life have for urban theory? There is no one agreed-upon definition of 'affect' given the complexity of work across the social sciences and humanities. In the first section of the chapter, I offer a working definition of 'affect'. As the chapter proceeds, I draw out some differences between how affect has been understood and researched in urban theory. The next section then argues for and offers a conceptualisation of the affective life of cities that centres the question of how urban affects are patterned. To this end, subsequent sections explore how specific urban affects emerge and how urban affects are unevenly distributed. The conclusion suggests some implications of learning to attune to the affective life of cities.

What is affect?

The term 'affect' has been used by geographers and others in the study of cities to attune to a range of intensities and atmospheres felt through bodies: hostility, as racism intensifies in an encounter between racially marked bodies (Swanton 2010); the surprising force of a sexually charged glance (Lim 2007); the civility of tolerance (Wilson 2013); the seductions of urban public space (Allen 2006); 'cruel optimism' (Berlant 2011), as people are almost sustained by promises that bring them harm, to give some examples. These examples remind us that affects are not the reserve of one special part of urban life. They are as much part of trans-local geo-economic processes as they are part of the particularities of sites; as much elements in the politics and governance of cities as they are folded with the hesitant formation of informal practices; as much part of the infrastructural backgrounds that sustain urban circulations as they are intensified in punctual encounters across social differences as bodies meet and affect one another. Always imbricated in urban processes that are more than affective, the affective life of cities is sensed when we register the atmospheric qualities of urban life: vibrant, lively, scary, edgy, deadened, and so on.

Whilst theories of affect vary, use of the term affect signals an attention to lived experience. Work with an interest in affect shares, then, a promise and an imperative (Anderson 2014: 9–11). The promise is of an urban theory that senses and discerns the subtle, elusive, dynamics of everyday living and touches how abstract socio-spatial processes are felt and lived. The imperative is to understand how forms of power work through affect (Thrift 2004b). For some, this involves diagnosing the contemporary urban condition as one in which affect has become a new 'object-target' as forms of power emerge and change, for others it involves (re)thinking cities as affectively imbued scenes of multiple attachments, events and forces.

There is no single definition of affect, or associated terms such as emotion, feeling or mood. A genealogy of affect as word, term and concept remains to be written, although Ngai (2005) identifies one origin in the third-person observation by psychologists of emotional states. Interest in affect is typically associated with the emergence of non-representational theories (McCormack 2013; Thrift 2004b), a

---

CASE STUDY 2.1 AFFECT AND URBAN GOVERNANCE

Urban life is affective – in the sense that 'transience', or 'insecurity', or 'hope', or 'excitement', or other affects – are not only part of cities, but elements in diverse processes of urbanization. For one example of the imbrication of affects in urban processes consider the importance of the creation of distinct 'atmospheres' to particular modes of urban governance. David Harvey (1989a) reflects on a shift in urban governance: from managerialism to 'initiatory' and 'entrepreneurial' modes of action across Western cities. In a mood of political and economic uncertainty in capitalist economies post the recession of 1973, and as part of a complex shift to forms of flexible accumulation with ambivalent political effects and opportunities, the new modes of action involve zero-sum inter-urban competition for various forms of resources. Harvey stresses that the new modes of action are 'speculative' (ibid: 7) in orientation: bound up with future-oriented hopes for a financial return, for economic development, and so on. One generic 'entrepreneurial' mode of action is based on 'attracting' or 'enticing' consumers by improving the 'quality of life' and delivering 'spectacular' events. In short, what is acted upon is the affective life of cities. The hope being that something like the atmosphere of a place will draw in consumers and perhaps further investment. Harvey (ibid: 9) summarizes what a city must become in strikingly affective terms: 'Above all, the city has to appear as an innovative, exciting, creative, and safe place to live or to visit, to play and consume in.'

---

body of thought that is itself plural. But work on affect in urban theory owes much to a number of precursors in geography and elsewhere. In addition to humanist phenomenology, research on affect is indebted to Feminist scholarship that showed how emotions matter, disrupted the marginalizing or silencing of emotional experience, traced how emotions 'get into' urban life and vice versa, and experimented new forms for representing lived experience (e.g. Valentine 1989). Exemplary here is a long tradition of research on the gendering of urban 'fear of crime' (see Pain 1991). What this work did so well was to shuttle between the lived experience of urban life and the manner by which fears attach to, cluster around, and intensify in relation to certain places and gendered bodies. In doing so, the affective life of cities was, first, revealed as always-already fractured and, second, as differentially related to and lived in ways that both enacted and reproduced social differences. In short, urban life cannot be reduced to a single, dominant mood. Non-representational theories of affect share with work on emotional geographies the insights that affects express relations, and that the rational and emotional/affective are complexly enfolded. From these starting points, work on affect asks a series of questions: how are affects emergent from and part of the billions of encounters that compose cities?; and how, given the contingency of urban life, is affective life mediated and organized in relation to the trans-local processes that cross between cities and draw them into relation?

These questions are refracted through different theories of affect. Thrift (2004b) identifies four theories that offer different versions of what affect is: phenomenological approaches that describe the bodily processes and states through which affects occur in everyday life; psychoanalytic theories where affects are vehicles or manifestations for the drives; naturalist accounts, often deploying the brain sciences, that treat affects as biologically rooted universals; a Spinozan-Deleuzian notion of affect as always emergent capacities to affect and be affected. Despite a nascent resurgence of interest in psychoanalysis in geography (Pile 2011), and the continued importance of phenomenological approaches, it is Spinozan-Deleuzian theories that have been central to the interest in affect. There, affect refers principally to a body's or bodies' 'capacity to affect and be affected'. A body's 'charge of affect' is two-sided; that is, what a body can do and does do is a matter of intensive 'capacities to affect' *and* 'capacities to be affected'. On this understanding, cities shape and fold into affects in all manners of dramatic and mundane ways. Miller (2014), for example, shows how the shopping malls that are now such commonplace features of urban landscapes are engineered with the aim and hope of folding specific 'capacities to affect and be affected', such as comfort and excitement, into the practice of consumption and the activity of purchase. Malls might be thought, then, as an urban technology for producing a direct link between bodily capacities and capital accumulation. Normally this use of the term affect involves invoking an analytic distinction between affect and emotion: emotion being used to refer to the ways in which affects are named, interpreted and reflected. By contrast, affect refers to the intensity of experience: a quality that provides something close to the background sense of an event or practice or space. We might contrast, say, the background atmosphere of a particular urban space in which people dwell with the way in which that atmosphere is translated into particular feelings about the space. This is not to say that affect is cleaved from emotion. In what has become a well-cited passage, Massumi (2002) describes their enfolding:

> An emotion is a subjective content, the sociolinguistic fixing of the quality of an experience, which is from that point onward defined as personal. Emotion is qualified intensity, the conventional, consensual point of insertion of intensity into semantically and semiotically formed progressions, into narrativizable action-reaction circuits, into function and meaning.
>
> *(Massumi 2002: 28)*

For now, we should note that instead of affect and emotion existing in separate unconnected levels (contra Pile 2011), Massumi and others stress multiple relations between emotion as 'subjective content' and affect as bodily capacities. Emotion as qualified personal content feeds back into the emergence and organization of affect, whilst at the same time being the most intense expression of the capture of affect *and* of affect's ongoing escape (Massumi 2002: 35). Working with this distinction, which should be stressed is but one way of understanding the relation and has been subject to critique for introducing an analytic separation between modalities (Pile 2011), opens up questions about how affect/emotion are patterned and how the ongoing process of ordering affective life is connected to the (trans) local processes that ensure complex commonalities and differences across and between cities. It is to this question that the following section turns.

Patterns of affect

All cities can be thought of as made up of multiple, diverse, affects. Affects that resonate together to give cities, or parts therein, something like a background atmosphere; distinct affective qualities that envelope places. Such an attention to the ongoing-ness and diversity of affective life avoids assuming that cities have, or could have, a single or dominant identifiable, affective register based, for example, on the production of homogenized spaces of deadened affect. In short, when it comes to affective life no one city can function as a 'paradigmatic city' that would allow us to generalize about the urban affective condition in or across the global north or south. For evidence of both the need for an urbanism attentive to diverse affects and its difficulty, consider the case of the urbanization of nature in New York city (Gandy 2003). Gandy's magisterial account demonstrated that specific affects and emotions constantly play a part in the production of urbanized nature. For example, New York's then current water crisis is linked to declining public confidence in the safety of drinking water (p. 60), cultural anxieties fold into the mid-18th-century growth of new parks (p. 84), and Puerto Rican communities inhabit landscapes of despair (p. 161). Whilst it is not Gandy's project, despite the obvious importance of 'confidence', 'anxieties' and 'despair', it is not clear what these affects do – how they function, as part of the production of an urban nature. Nor is it clear how this affect or that emotion emerges and changes as it becomes part of the processes that produce New York's nature. What is clear, however, is that the urbanization of nature is, in part, an affective process, as are all urban processes. Recall, for example, how the entrepreneurial forms of action that Harvey (1989a) identified as part of the shift to flexible accumulation implicate affects such as hopes in numerous ways (see Case study 2.1).

The example of New York and the urbanization of nature reminds us that there is no single 'Big Story' or 'Big Picture' about affect and urban life today. Despite, that is, attempts to offer such a generalizing account that reads for interurban homogeneity and identifies something like a nameable and knowable urban mood. Consider, for example, claims that link what Hubbard (2003) describes as an 'ambient fear' or 'low grade fear' that supposedly saturates the spaces of everyday urban life to epochal diagnoses that, today, 'capitalist cities' are marked by a 'culture of fear'. Generalizing beyond any one city, a generic class of city – 'capitalist cities' in the global north – are given a characteristic collective affect: fear or anxiety. The affective life of cities becomes a second-order expression of the mood of a historical epoch that extends beyond and conditions the everyday life of cities. To avoid the

---

CASE STUDY 2.2 HOPE AND A SENSE OF TRANSIENCE

For an example of how urban life is made and remade through innumerable ordinary affects consider the example of Rio de Janeiro's main garbage dump. Informal work at the dump is enveloped by what Kathleen Millar (2014) calls a 'sense of transience'. The roughly two-thousand workers who earn a living by collecting and selling recyclables at the dump come and go as they move in and out of the formal economy. Ordinary hopes for participation in Rio's formal economy pervade the dump; these hopes are frequently expressed by workers, and fold into a sense of the transience of work on the dump. Nevertheless, and despite the repetition of a hope and desire for paid, formal work, and the hard struggle to make a living from the dump, workers who 'get a job' in the formal economy frequently reappear at the dump and resume informal labour. Millar's argument is that work at the dump, even though it is precarious, allows workers to cope with the myriad predictably unpredictable 'everyday emergencies' that make up the lived condition of poverty. Millar (2014) summarizes this sense of multiple reinforcing precarities: 'unstable daily living destabilizes work' (referring to formal labour). The dump, then, becomes a contradictory affective space: simultaneously a refuge linked to the hope of getting by in the midst of insecurity, and a site and source of suffering.

---

homogenization of this kind of approach, work on urban affects can learn from the recent critique of universalistic modes of urban life and the embrace of the material and lived diversity of 'ordinary cities' where 'diverse ranges of relational webs coalesce, interconnect and fragment' (Amin and Graham 1997: 417). Affects, on this understanding, would not only overlap with those 'relational webs', but also be elements in the production of the particularity of urban life and a way in which particularity is registered. Nevertheless, whilst remaining cautious about any grand diagnosis, we should recognize that trans-urban processes are themselves affective and, in part, create affects. In short, it is not only that the study of affect has to be particularistic and result in numerous case studies of affective differences. Urban theory concerned with affect might become attuned to how the diversity of urban life is felt affectively and how affects are conditioned, without being determined, by intra-urban processes that are themselves always-already affective. How, then, to understand the patterning of affect whilst starting from the lived and material diversity of cities that are always in relation to multiple outsides? And how to focus on the irreducibility of urban affects to various urban processes whilst staying attuned to the surprise of affective life amid the coexisting differences that make up cities? (See Case study 2.2). The following two sections sketch an approach to urban affective life that asks how affects emerge and how affects are distributed.

Emergence

Consider Katz's (1999) evocative description of being 'pissed off' while driving in the city of Los Angeles. Being 'pissed off' is both a routine and exceptional part of urban life in Los Angeles. Routine, as it happens as part of the innumerable encounters between people and cars that the city is organized around. But exceptional, as 'being pissed off' is a momentary occurrence that disrupts normalized ways of relating and doing something. Katz shows how 'being pissed off' emerges from the fracturing of a series of layered entanglements between the driving body and the technology of the car. The dramatic moment that occurs in the everyday practice of driving, a rise and decline of anger, is conditioned by a host of heterogeneous phenomena that, although not all specific to Los Angeles, come together to make Los Angeles a particular city: the position of the automobile and oil in America's macroeconomic policies, the freedoms and feelings afforded by car ownership, a limited system of public transportation, the historical geographies of suburbia, the organization of work, and the economic geography of Los Angeles, among much more. Although there is more that could be said about this anger – and Katz describes how its embodiment in 'giving the finger' is vital to its difference in degree and kind from other angers – Katz's salient point is to describe how whilst anger is always associative and collective it also remakes relations, if only momentarily. Bodies are realigned, for a while, in an angry moment. And this realignment, whilst momentary, happens thousands of times every day. We might say that flashes of anger, flashes that recede or dissipate, are a part of Los Angeles' infrastructure alongside the other affects of driving that were briefly interrupted in the angry moment – not least memories of the feeling of the promise of freedom of movement.

Focusing on the emergence of specific affects in relation to a complex set of specific proximate and distant relations moves us far away from 'big picture' urbanism. Instead, it affords a view of cities as ongoing, only ever partially stabilized, constellations riven by multiple layered affects that come and go, emerge and stay for a while. 'Being pissed off' is but one affect; you will be able to think of many others. As well as understanding the efficacy of affects as urban life is remade and made, emphasis is placed on understanding their 'conditions of emergence', that is their connection to other urban processes. The example of 'being pissed off' gives us a good sense of how affects are 'irreducible' (Anderson 2014) to a set of proximate and distant relations. By which I mean that affects are only ever forged in relation to much that is not affective – in the example this includes the material infrastructure that grids Los Angeles – but nevertheless cannot be reduced to a simple effect of those relations. Affects have a force. 'Being pissed off' is a minor element in the infrastructure of mobility that makes Los Angeles one ordinary city amongst others.

'Conditions of emergence' may also include affective conditions: the moods that constitute something like a taken-for-granted background to urban life. Take Bissell's (2014) analysis of the formation of stressed bodies in urban life. Recalling a long tradition of work on the affective adjustments to the density of cities, Bissell shows how the emergence of stress is connected to the intensification of a set of urban processes that result in an extension of commuting distances and journey times (including the concentration of economic activity in cities and the distance between affordable homes and location of work). Typically seen as an affective effect of disruptions to the infrastructures that constitute the background of urban life, Bissell instead shows how stress cannot simply be 'read off' from the form of urban disruption, even though it is not unconnected. Instead, stress is a matter of what Bissell (2014: 191) carefully calls 'subtle, slow creep transformations that necessarily build to tipping points over time'. And one of the elements that becomes part of the dynamic of slow creep/tipping point are a range of affective conditions refracted through the lives and bodies of individuals – the 'quiet' atmosphere of a train carriage, the sense of 'insecurity' afforded by casual work, amongst others in the case of Bissell's interviewees. Stress surfaces and intensifies, then, in the midst of a city's other affective conditions. We might think of the emergence of affects as happening in the midst of constellations of individuated and collective affects, some of which come and go in moments, others of which form atmospheres or moods that sink into urban spaces.

Distributions

Affects are unevenly distributed and imbricated in processes of distribution. For example, debates about the privatization of public spaces in Western cities implicitly and explicitly assume a distribution of affects and emotions. The emergence of a splintered urban form is assumed to be bound up with collectives who feel fear (white, middle classes) for their personal safety and express that fear through an urban architecture bound up with a logic of containment, prohibition, and control (e.g., guards, gates, and other human and non-human forms of surveillance: Graham and Marvin 2001; Macleod 2002). Fear becomes attached to certain bodies (the poor, homeless, etc.) and is expressed in a form of spatial organization. Part of a particular process of urban transformation, privatization, is a distribution in which certain affects and emotions are entangled in relations of conflict, and co-operation, within and between human (often raced and classed bodies made present as looming threats) and non-human bodies (specific forms of 'gated' architecture).

The key broader problem is to understand how the uneven distributions of affects fold into provisional, and spatially nuanced, topologies of power. Strangely, work on affect has so far been rather one-dimensional when talking about power – note, for example, Thrift's (2004) vocabulary of 'the powerful' and their 'manipulation' or 'engineering' of affects and emotions. Following Allen (2003: 2), instead of seeing power as centred or decentred, we need to identify the 'diverse and specific modalities of power' and thus attend to how affects/emotions function differently in relation to the varied kinds and mixtures of powers, including, for example, domination, authority, manipulation, coercion, and seduction. For one example of the intimacy between forms of power and processes of distribution consider the 'stigmatization' of particular urban peoples and places. Slums, ghettos and other marginalized sites have long held a spatial 'taint' or 'blemish'. Note how Wacquant *et al* (2014) give a causal role to something like affect (alongside other processes) in the guise of 'fears' and 'fantasies' in the process whereby 'disgrace' attaches to particular places:

> Indeed, it is well established that the 'accursed share' of urban society has had its special wards, the *bas-fonds*, slums and rookeries, precincts of the *submundo* and *Unterwelt*, since the mid-19th century, as a result of the confluence of urbanization, industrialization, and upper-class fears as well as fantasies about the 'teeming masses' rallying the city.
>
> *(Wacquant et al 2014: 1273; emphasis added)*

Wacquant *et al* (2014) go on to show how contemporary stigmatization of urban boroughs operates, at least in part, as an affective process that violently distributes bodies and places into hierarchies of value. And it is an affective process in a number of ways. Stigmatised neighbourhoods are viewed as 'threats' to 'the fabric of the nation' (p. 1273). A small number of urban districts are 'reviled' and 'renowned'. Labels with an affective charge, such as 'ghetto' or 'sink estate', are attached to the districts. Incidents of deviance and violence are 'sensationalized' (p. 1274) and residents branded as 'outcasts', with fears and anxieties attaching to racialized bodies. Finally:

> the stigmatized districts of dispossession in the postindustrial city elicit overwhelmingly negative emotions and stern corrective reactions driven by fright, revulsion, and condemnation, which in turn foster the growth and glorification of the penal wing of the state in order to penalize urban marginality ... Long gone are the ambivalent fascination and lurid attraction that political and cultural elites felt for the sordid bas-fonds of the emerging industrial city.
>
> *(Wacquant et al 2014: 1274)*

Wacquant *et al*'s claim is of a background transformation in the dominant affective relation with now-stigmatized places and of a connection between a form of intervention – the penal wing of the state – and 'fright, revulsion, and condemnation'. There is much more to be said about the effects for residents of stigmatization. And, of course, the affective geographies of such places extend beyond stigmatization. Consider the police activity described by Fassin's (2010) powerful ethnography of a rapid response 'anti-crime squad' in a precinct in a Parisian suburb. Animated by racialized stigmatization, the police bring multiple violences to residents who, consequently, come not only to fear the police's arrival, but also to conduct themselves with restraint in an often-doomed attempt to foreclose police violence. So, the affective geographies of 'advanced marginality' are multiple. But what this example and the example of privatization show are the twofold ways in which affects are distributed/distribute as they emerge. On the one hand, specific collective affects have a quasi-causal role in processes of privatization and stigmatization. They 'get into' those processes, folding into their other non-affective components. On the other hand, affects attach to places and bodies and so condition without determining how those places and bodies are and can be related to. Witness, for example, the mixture of fear, revulsion and renown that envelops some stigmatized urban districts and perpetuates and intensifies processes of marginalization.

Conclusion

What implications does this emphasis on the emergence and distribution of affects have for the conduct and practice of urban theory? A starting point of the chapter has been that affect cannot be safely separated off from other more apparently material geo-economic or geo-political processes that shape cities. Affects are parts of, but irreducible to, the various processes through which cities are assembled (social-spatial polarization, industrial restructuring and so on). Hopes, for example, are part of the cycles of investment and disinvestment that remake and unmake urban landscapes (Miyazaki 2006). Nor, importantly, is a focus on affect a synonym for particularistic studies of the specificity of this or that urban place that would emphasize subjective experience at the expense of extensive, trans-local, processes. It should be clear that affects are never purely personal and never only a matter of the idiosyncratic differences between this or that city (even though affects can have a particularizing effect when, for example, the 'atmosphere' or 'mood' of a city is felt and named). Rather, to attune to emergences and distributions of affect is to sense how pre-personal and trans-personal processes come together as urban spaces come to be felt as having a particular mood. It is to understand how the features that are taken to constitute and to some extent define the specificity of cities – density, changeability, coexistence of differences, for example – are as immediately affective as they are more conventionally material: changeability, for example, perhaps being felt in the 'aliveness' of a city, or a space or place therein. And it is to understand how urban affects are at once singular and general as trans-local processes are articulated and assembled in specific ways.

This means that urban theory concerned with affect might do some of the following. First, trace how affects are elements in, but irreducible to, the trans-local processes, economic or otherwise, which constitute cities and draw cities into relations of commonality and difference. Second, understand how those trans-local processes are articulated through urban contexts that are already enveloped and animated by myriad affects that compose something like an affective background to urban life. Third, understand the uneven distribution of affects and how uneven affective geographies – say the absence of hope in particular places or the intensification of stigmatization – enact, express and reproduce processes and patterns of urban inequalities. Fourth, stay attuned to those innumerable affects that come and go, flash forth for a moment, before dissipating; whether that be through the uneventful or eventful happenings in which alternative ways of being and living are felt, or through the temporary creation of other spaces that disrupt, if only in minute ways, normal, normalized, urban life. The task for urban theory, in short, is to learn to attune to how affects are part of urban processes and forms and, by doing so, avoid any reductive claim that a single affect could characterize today's urban condition.

Key reading

Gregg, M. and Seigworth, G. (2011) *The Affect Theory Reader*, Durham, NC: Duke University Press.

An edited collection summarizing differences and tensions within the emerging field of 'affect theory'.

Pain, R. (1991) 'Space, sexual violence and social control: Integrating geographical and feminist analyses of women's fear of crime', *Progress in Human Geography*, 15 (4): 415–431.

An important work on the gendered distribution of fear of crime, showing how fear is both an expression of, and constitutive of, social-spatial relations.

Thrift, N. (2004) 'Intensities of feeling: towards a spatial politics of affect', *Geogr. Ann.*, 86 (1): 57–78.

Offers a theoretically sophisticated understanding of ordinary cities as always-already affective, and diagnoses some of the ways in which urban affects are manipulated.

3

ARCHITECTURE

Richard Baxter

Introduction

Architecture, like people, is a central element of cities. Buildings always surround us and human bodies are constantly engaging with them. Despite architecture's importance to the city, it has historically not occupied a principal position in urban theory. Perhaps this reflects the wider tendency to omit the 'ordinary' dimensions of cities (Robinson 2006) or because many social scientists have felt that buildings are best left to architects. However, this is beginning to change as a new wave of urban scholars turn their attention to architecture and the built environment. This chapter introduces the three main approaches that have developed to study architecture in urban theory. The first is work that has explored buildings as symbol and the second is scholarship that has investigated architecture in the everyday. The third is research that has turned towards the architects themselves and calls for more urban researchers to consider and study architecture. It also argues that now is the time to examine more fully the relationship between architectural design and human response because of the wider advancements made in urban theory.

Architecture as symbol

Like other fields, scholarship on architecture has been shaped, and helped to shape, broader methodological debates in urban theory. The first main way that architecture was studied was informed by research on landscape, such as Raymond Williams' (1973) influential *The Country and the City*. This approach understood landscape as something that symbolized wider social, economic and political contexts. Transferring this to the built environment, researchers were, therefore, interested in the symbolic meanings of architecture. This meant that the actual building was less important than the social context in which it was situated.

Utilizing an understanding of wider social and economic conditions, the researcher's role was one of interpretation. This did not, at least at first, require any knowledge of architectural design. As Lees (2001: 55) argued, buildings were seen 'as signs or symptoms of something else – be it class, culture, capitalism, or resistance to them'.

Given its importance in geography in the 1970s–80s, it is not surprising that early work on architecture using this approach was informed by Marxist theory. Some of the research in Cosgrove and Daniels' (1988) *Iconography of Landscape*, sought to reveal the class politics embedded in architecture. For example, Woolf (1988) critiqued the vast sums of public money spent on Paris's opulent opera house, the Palais Garnier, a building constructed for the Parisian elite as opposed to the poor (see Woolf 1988). Perhaps the most well-known Marxist account is geographer David Harvey's (1979) historical analysis of the Sacré-Coeur at Montmartre in Paris. Ignoring the lure of the basilica's aesthetic beauty, Harvey preferred to examine the wider political context of its construction that informed its meaning. He argued that it was built to symbolize and naturalize the power of church and state following the brutal suppression of the communist uprising, the Paris Commune, at Montmartre in 1871. Far from an innocent religious artefact, the Sacré-Coeur is, therefore, implicated in bloodshed and the history of Parisian class warfare.

Providing an excellent class-based historical analysis of Paris at the time, one of the problems with Harvey's approach is that it ignored the Sacré-Coeur's design. Responding to this critique, Hopkins (1990) later used a Marxist approach to study the West Edmonton shopping mall in Canada, the largest in North America. However, he firmly implicated the architectural form of the mall within the capitalist logic of consumption and profit. Postmodern architecture had been used at the mall to create simulated 'elsewhere' worlds, such as European boulevards and Caribbean pirate seascapes. Hopkins argued that this aimed to attract consumers to the mall, encourage them to remain and spend, and divert attention away from its main purpose as a site of consumption. The beginning of exploring the relationship between architecture and selling, this has informed the study of signature architecture in global cities (McNeill 2007) and hyperreal environments in Las Vegas (Fox 2007).

Reflecting the growth of postmodernism in urban theory in the 1980s-90s, some authors became more interested in the study of cultural dynamics and cultural theory. Originating in the work of anthropologist Clifford Geertz (1988), semiotics, which understands the world as constituted by webs of meaning, became influential. Drawing on this work, Barnes and Duncan (1992) argued that buildings were artefacts that communicated something about the wider cultures that had produced them. The researcher's role was, therefore, to 'read' hidden cultural codes. Some argued that the rejection of the political economy was naïve and damaging (Harvey 1989c), but the study of culture has often revealed a cultural politics. Researching Vancouver, Canada, Ley (1993) illustrated how the architectural forms of the north and south banks of the river at False Creek were the result of different planning ideologies at different periods; modernism versus postmodernism. These different cultures in the architectural and planning departments had implications for the residents who lived there, for example he argued that the postmodern architecture at South False Creek created more humane, diverse and localized neighbourhoods.

Interested in joining a concern for the workings of capital and culture, Lees and Demeritt (1998) argued that cultural images, or narratives, are fundamental in the regeneration of working class areas. Studying the same city, Vancouver, they argued that dystopian images of the 'sin city', involving drug use and crime, and the 'sim city', which depicted ordered, clean and prosperous futures, enabled and justified spatial transformation and displacement. As more recent research illustrates, such simplified images remain central in the gentrification of historic low-income inner city locations (see Campkin 2013; Lees 2014b). However, it should be mentioned that, informed by poststructuralist ideas, some authors argued against the reduction to singular meanings, or narratives, in this broad body of work. Critiquing the politics of including certain interpretations at the expense of others, it was argued that accounts need to explore the multiple meanings associated with architecture (see Barnes and Duncan 1992). Arguably the landmark study for this polyvocal approach is Domosh's (1989) examination of Pulitzer's 'New York World Building', which attended to the economic, technological and cultural rationales behind its design.

CASE STUDY 3.1 *HOLLOW LAND*

Emerging scholarship on architecture is revealing dimensions and processes that have not previously been considered. Putting forward the 'politics of verticality', Weizman's (2007) book *Hollow Land: Israel's Architecture of Occupation* argues that a horizontal focus has subsumed urban theory. Omitting the vertical dimensions of space, he argues that cities should be conceived as a volume since this reveals the three-dimensional geometries by which power is exercised over impoverished groups. Weizman's case study is the West Bank and Gaza, where verticality is central in Israeli domination and Palestinian oppression. For example, he illustrates how archaeological excavations under Jerusalem selectively explore ancient Jewish settlements, but ignore Palestinian ones, thus naturalizing and justifying the Israeli occupation. The modern vertical architecture of Israel is also designed to segregate Palestinians and exclude them from modern infrastructure. In this way, walled concrete flyovers pass over Palestinian urban space ensuring the smooth transport of Israelis from one Israeli space to another. However, as Harker (2014) argues, this debate about the vertical meaning of the built environment omits ordinary experiences and relations in the city. Weizman's work, Harker posits, 'has a tendency to expunge Palestinians from the spaces within its purview by rhetorically hollowing out Palestinian spaces and landscapes of more intensive relations' (ibid: 322). This argument about everyday experience points to the importance of the next section.

Research seeking to explore architecture as symbol has unsurprisingly been diverse and varied. (See Case study 3.3). One of the first to incorporate time into the analysis, King's (1986) well-known examination of the bungalow illustrated how wider economic, political and social transformation can alter the physical form and meanings associated with architecture. Beginning as an indigenous structure called a 'banggolo' in Bengal, the bungalow developed into an opulent dwelling, and became implicated in racist narratives, as it was appropriated under British colonial rule. King argued that not only does architecture change with society, but it can also facilitate its transition. Similarly recognizing the role of architecture in shaping society, feminist urban historian and architect Dolores Hayden (1997) later critiqued the design of Western suburbia for reflecting and supporting gender inequality. Moving beyond analysis, under Homemakers Organization for a More Egalitarian Society (HOMES) she proposed alternative suburban neighbourhoods that were organized through small-scale co-operatives. Such work was ahead of its time since it illustrates the benefits of blurring the disciplinary boundaries between the social sciences and humanities, and architectural practice.

Architecture in the everyday

Informed by postmodernism, the turn of this century was characterized by growing interest in everyday life in urban theory. This was not the same as the ethnographic research of inner city neighbourhoods by the Chicago School, urban sociologists who examined the social structures that informed community experience and behaviour in the 1920s-30s. Although sharing a similar concern for the everyday, this new wave of work drew on an array of post-structuralist ideas that encompassed the broad agenda of the 'practice turn' (see Cetina *et al* 2000) and moved towards greater theoretical complexity. In human geography Nigel Thrift (2008) developed 'non-representational theory', which sought to attend to an everyday world beneath human consciousness (see Chapter 2, this volume). Focusing on human practice, it was less interested in the 'why' than the 'how'. The first author to adopt this approach in architecture was Lees (2001), through an investigation of the Vancouver public library. Moving beyond past studies that situated architecture in its wider social, economic and political context, she argued that architecture was a lived and dynamic entity involving everyday inhabitation. As she stated: 'I have moved from an abstract contemplation of architecture as representation to a more active and embodied engagement with the lived building' (ibid: 75). Using participant observation, she therefore described the multiple activities of building users, such as children playing on the escalators and a homeless woman in the toilets. However, differing from latter accounts, it should be noted that Lees explored both the symbolic and non-representational to provide a multi-layered understanding of the library. Based on the coliseums of ancient Rome, she also argued that the library's design could be construed as racist since it was insensitive to the legacy of colonialism and the increasingly high Asian immigrant population.

This turn from representation to the everyday complexities of architecture, such as human inhabitation, has used a variety of theoretical approaches. Building on Lees' work, Kraftl and Adey (2008) introduced the idea of affect to study architecture (see Chapter 2 this volume). A contested term, this is the non-cognitive 'push' between things with the potential to induce an effect, such as an emotional response. Increasingly, powerful actors, such as the state or private corporations, are engineering affect to influence the experiences and behaviours of human bodies in cities (Thrift 2004b). As a result, Kraftl and Adey (2008a) explored how architecture is designed to produce affects. Exploring the Nant-y-Cwm Steiner School in Wales, they argued that the school's architecture aids in the creation of certain subjects. Winding woodland paths, natural materials and a protective interior design create a 'magical childlike world' (ibid: 944) that enables children to progress through desirable psychological developmental stages. Thinking in more sinister terms, Adey (2008) illustrated how architectural affect is used by architects and airport governance to induce feelings of pressure, encourage passengers to move in certain directions and aid in surveillance. This positions architecture as another technology implicated in disciplining and control.

Unsurprisingly, materiality has always been present in these accounts since this is a primary component of architecture. However, emerging urban theory began to critique an over-focus on the human at the expense of the non-human. Drawing on sociologists such as Bruno Latour and theories like actor network theory (ANT) (Latour 1993), and later assemblage theories (see Edensor 2011a; and Chapter 4, this volume), some authors therefore began to foreground the non-human in buildings, like architectural technology, and the 'actor-networks' that hold them stable. One of the main advocates of this approach has been Jane M. Jacobs through research on the modernist high-rise Red Road estate in Glasgow, Scotland (UK). She argued that happenings on the estate, or 'building events', were co-produced through alliances formed between social and technological actors (Jacobs *et al.* 2007). In this way, the practice of viewing through high-rise windows 'is complex and contingent. It depends upon an array of other factors, both material and immaterial: existing window discourses, design specifications, regulations, residents' (Jacobs *et al.* 2010: 256). Some critics argued that ANT approaches are overly complex, do not address inequality and lack policy relevance (see Bloor 1999), but it has been used to provide policy recommendations about architecture (Baxter and Lees 2008).

The emphasis of materiality resulted in a backlash, with some authors re-asserting human qualities absent in ANT research. Exploring a shopping mall in Milton Keynes, UK, Rose *et al.* (2010) argued that investigations of architecture need to pay more sustained attention to human subjectivity. This includes feelings, specifically 'the feel *of* buildings, feelings *in* buildings and feelings *about* buildings' (ibid: 334), plus people's reflective thoughts and their pre-cognitive affects. Providing a fuller account of human subjectivity in buildings poses challenges for research methods and, moving towards innovation, Rose *et al.* (2010) deployed 'walk-along' interviews, whereby the researcher observes and interviews participants as they are walking. Similarly observing the omission of emotion, Lees and Baxter (2011) examined the emotional intensity of a fear event in a high-rise lift involving a knife. Attempting to mediate these debates, they argued that the study of the immaterial, in this case emotion, is not incompatible with attention to the material, or even an actor network theory approach. Although a difficult task in the context of empirical work, as Lees and Baxter (2011: 117) stated:

> keeping the material and the immaterial in balance, to get to the sensual world of materiality and emotion, to combine the functional and the emotional, however, calls for a conceptual juggle that informs our understanding of the relationship between 'the self and the places of our (en)actions'.

Given growing interest in the sensory dimensions of the body in the social sciences (see Pink 2009), it is not surprising that research on architecture and the senses has emerged. Paterson (2011) argued that 'encounters with architectural space have unsurprisingly betrayed the visual bias present within the discipline at large'. He aimed to develop a more fully embodied, more-than visual appreciation of the built environment that incorporates touch, or the haptic. This does not just involve the touching of materials by the body, but also the felt manifestations, such as muscular sensations, within the body as it moves through architectural space. Paterson argued that this provides a more non-representational understanding of buildings than found in the work of Lees (2001), but it does omit sound. Responding to this with the 'Kilmahew Audio-Drift', Gallagher (2014) sought to capture some of the sounds at an abandoned modernist seminary called St Peters in Glasgow. However, as he reminded us, such recordings are only ever representations, themselves partial and always situated in wider social contexts. They can communicate something about architecture that is absent from text, but they are not the sounds as they happened in the world. An emerging and under researched aspect of architecture, this is a worthwhile area for future research (see Case study 3.2). Other senses also need attention, such as smell and taste.

CASE STUDY 3.2 ARCHITECTURE AND DANCE

Contrary to an argument made by Thrift (2004b), the engineering of affect can involve a positive and emancipatory politics. Interested in the relationship between architecture and dance, Merriman (2010) explored the style 'contact improvisation' which was developed by Anna Halprin in the 1970s. This postmodern form critiqued rule-bound and elitist 'formal' dance styles, such as ballet. Rather than being choreographed, contact improvisation involves improvisation, spontaneity, the sensations and a fluidity of movement. For Merriman, contact improvisation is strongly related to architecture and landscape: Halprin encouraged 'dancers' to establish an embodied dialogue with architecture. Called 'becoming aware of space', dancers allow themselves to feel and be affected by the environment, with this informing their movement. Anna Halprin's husband, landscape architect Lawrence Halprin designed a dance deck in their back garden so that dancers could move 'with the ever-changing natural surroundings – light conditions, temperatures, air currents, seasonal foliage – lending a dynamic quality to the architecture of this performance space' (ibid: 433). In this way, opening the body to architecture becomes an emancipatory act, which, by moving in harmony with the dynamic affectivities of place, frees the self from society's disciplining movements. Critics responded that dance does little to challenge powerful structural forces, such as capitalism (Harvey 1989c), but it can aid in pragmatically showing alternative ways of dwelling, and can help to mitigate the trauma inflicted on the disciplined body.

FIGURE 3.1 Photograph of undergraduates on the module 'Geography, Architecture and the City' learning contact improvisation under direction of Dr Libby Worth (Source: Richard Baxter).

Architecture and human response

In recent times, a small amount of work has attended to the relationship between architecture and human response, such as interaction, emotion and movement. Comparing Milton Keynes and Bradford town centres in the UK, Degen and

Rose (2012) examined the connection between the built environment, sensory experience and practices of walking. They found that not only did the built environment inform sensory experience, for example participants stated that Milton Keynes smelt of donuts and soap, but the different architectures resulted in alternative ways of walking. Based on linear modernist planning, walking in Milton Keynes tended to be fast and automatic. By comparison, walking in Bradford involved ambling at different speeds as the body reacted to a more diverse architectural environment. Working in the field of creative economies, Rantisi and Leslie (2010) further argued that architecture can influence creative working practices. Exploring the bohemian quarter of Mile End in Montreal, Canada, they illustrated how the materiality of buildings encourages creativity through functional and symbolical means. For example, old warehouses provide the large space required for fashion designers and their shops, while big windows let in the natural light needed when working with colour. The grungy aesthetic of old buildings also felt authentic and raw, thus inspiring creativity. As they stated:

> material factors in Mile End provide the conditions for creative practices (within a primary field and across fields) by mediating the risks of experimentation, providing aesthetic stimuli, and balancing the aesthetic and commercial dimensions of creative work.
>
> *(ibid: 2863)*

An important and interesting area of architectural study, more research is needed on the relationship between architecture and human response. Otherwise urban theory risks providing 'thin' accounts that do not fully understand the dynamic, interactive and co-affecting relationships between humans and architecture. This being said, building on previous work (e.g. Degen and Rose 2012; Rantisi and Leslie 2010), the nature of the relationship between design and human response needs more theoretical attention. This has not advanced since the heated debates of the 1970s–90s when some authors argued for a strong correlation between design and human action, such as crime and anti-social behaviour (Newman 1972; Coleman 1985) or community (Mickelson 1970; Duany and Plater-Zyberk 1992). By comparison, others claimed that these design-determinist approaches were naïve arguing that human response was the result of social factors, such as class or race (Talen 1999; Spicker 1987). This needs to be studied now not so much because architecture is increasingly being engineered to manipulate the human body (Kraftl and Adey 2008), but because there is limited understanding about whether this matters. Currently, social scientists are unable to provide convincing answers about the extent to which, how and under what limitations architecture influences experience, emotion, movement and interaction etc.

Compared to research in the 1970–80s, an advantage of studying this relationship at the current time is the theoretical advancements that have been made in urban theory. A variety of post-structuralist informed theories, such as actor network theory (Latour 1993) etc., can now help to inform the debate. Drawing on such work, future investigations must move beyond a dichotomy between design and society to provide more sophisticated and accurate theoretical accounts of how design and the social interrelate to affect human response. For example, one possibility is to use Ingold's (2008) work to develop an 'architecture as meshwork' which illustrates how a phenomenon like social interaction is the result of varying intensities of elements, including design and social ones, that intermix and intensify at specific sites. Taking both design, and social and cultural theory seriously, it is these kinds of arguments that will help to open up new ways of thinking about architecture and the city.

Making architecture

All the literature discussed so far explores architecture after its construction. Seeking to build on this, emerging scholarship has begun to study architecture before it is built, specifically focusing on architectural firms. This has been informed by both the symbolic and practice-based approaches. Adopting the latter, one of the most influential is Albena Yaneva's (2009) study of architect Rem Koolhaas's Office for Metropolitan Architecture (OMA) in Rotterdam, the Netherlands. Arguing that 'one particular subject still seems to be left aside – the actual dynamics of architectural design process and its material, cognitive and cultural dimensions' (ibid: 3) – she provided an ethnography of architects' practices at the OMA. Informed by ANT, she argued that buildings are made through small active moments in everyday life that involve both human and non-human actors. These include talking, thinking, writing, drawing, plus sketches, paper, digital screens and scaled models of the design. Involving gradual translation or change, buildings become known by architects through a series of such micro-interactions. Yaneva called this the 'making of a building', although the process is one of making, *un*making and remaking as older ideas, designs and models are dismissed, dismantled, rethought and reshaped. In this way, she foregrounds the architectural practices that are important in shaping the urban landscape.

Many architectural firms are not just involved in the practicalities of design, but produce large numbers of digital visualizations for the purpose of selling future projects to prospective clients, policymakers or the public. This was first explored by Houdart (2008) through an investigation of perspective drawings for a World Fair, the 2005 Expo in Nagoya, Japan. The architect used multi-layered hybrid images, involving photographs, computer graphics and elements from databases, to convince people that the Expo could exist. However, this meant that the aesthetic 'look' of the images was more important than their 'fit' with reality, with adjustments made to improve their appeal. As Rose *et al.* (2014) discussed, such Computer Generated Images (CGIs) are increasingly important in the marketing and shaping of large-scale urban redevelopments. Moving beyond research that has examined their content or affective power, they also used an ANT approach to explore how the 'CGIs travel extensively through a network of different offices, servers, and screens' (ibid: 391). Beginning as detailed drawings by architects in

Computer Aided Design software, they are stripped and made to look more aesthetically pleasing by 'visualizers'. Then they are sent to 'render farms' to make them appear more like photographs and, finally, worked on in photo-manipulation software such as Photoshop etc. As such, they understand CGIs as a network, which are acted on and transformed at a number of 'interfaces'. A process that has received less attention in urban theory, this illustrates the amount of work, interactions and technologies involved in the regeneration of cities and implicitly in the production of inequality (e.g. Lees 2014).

Less interested in the practice-turn, some authors have continued to use the symbolic approach to investigate architectural firms. For example, McNeill (2007, 2009) has produced a variety of work on the development of tall office architecture in global cities designed by signature architects. He argued that these offices have tended to be narrated as part of a global architectural aesthetic, which has emerged from a global flow of specialized design knowledge transferred through 'elite' architectural firms (see Olds 2001). However, this ignores how other powerful agents restrict architectural autonomy. Researching Sydney's skyline, he illustrated how architects Foster and Partners and Renzo Piano's Building Workshop had to respond, and adjust their designs, to the planning approval process, developers' needs and the demands of future clients. For developers, who are paying for projects, it is precisely the ability of signature architects to negotiate these diverse interests that make their services desirable. Differing from existing work on architects' practices (Yaneva 2009; Rose et al. 2014), this research highlights the complex power relations in which architectural firms are situated. Overall, the 'making of architecture' is an innovative emerging sub-field, but, mainly focused on commercial and important public buildings; more attention can also be paid to less spectacular urban architectures, such as residential homes and neighbourhoods in the city.

Conclusion

To some extent, urban scholarship is at the beginning of fully engaging with architecture. This makes it a very exciting and creative time to be researching architecture and the built environment. This chapter has illustrated some of the main ways that urban theory has studied architecture so far. Calling for more scholars to become interested in this form, which is such a central part of the city, it has also argued that now is the time to investigate the relationship between architecture and human response because of wider theoretical developments in urban theory. However, there are numerous avenues ripe for further exploration, such as the connection between architecture and temporality (Till 2000), the myriad of ways that architecture is unmade (DeSilvey and Edensor 2013; Dylan 2006) and further examination of architecture and urban inequality (Kraftl 2010). Related to this, some UK universities, such as the University of Manchester, are beginning to integrate social science disciplines and architectural schools. It would be interesting to witness this replicated across the globe since it will result in new knowledge and new ways of designing and inhabiting buildings. Perhaps more urban scholars will then take up the challenge of not just researching architecture, but also designing and producing a more liveable and fairer city.

Key reading

Hayden, D. (2002) *Redesigning the American Dream: the Future of Housing, Work and Family Life*, New York: W. W. Norton.
   Hayden illustrated how US suburbia has supported gender inequality and examined alternative designs that promote a more just city.
Lees, L. (2001) 'Towards a critical geography of architecture: the case of an ersatz colosseum', *Ecumene: A Journal of Cultural Geographies*, 8: 51–86.
   This paper moves beyond just the representation of architecture to explore also human experience and practice within buildings.
Rantisi, N. M. and Leslie, D. (2010) 'Materiality and creative production: the case of the Mile End neighborhood in Montreal', *Environment and Planning A,* 42: 2824–2841.
   Studying a neighbourhood in Montreal, Canada, Rantisi and Leslie illustrated the important role that architecture plays in the creative industries.
Weizman, E. (2007) *Hollow Land: Israel's Architecture of Occupation,* London: Verso.
   Understanding the city as a three dimensional space, *Hollow Land* documents how the Israeli state uses vertical architectures to control and dominate the Palestinian population.
Yaneva, A. (2009) *The Making of a Building*, Oxford: Peter Lang.
   Rather than explore buildings, Yaneva used ethnography to investigate the human and non-human architectural practices that produce them.

4

ASSEMBLAGES

Ignacio Farías

Introduction

The concept of assemblage comes from philosophy. It was introduced by Deleuze and Guattari (1987), for whom assemblage designates one of the two main ways in which the real is formed. Its conceptual companion is the stratum. Strata define layers or levels of reality constituted by homogeneous types of elements, structures and codes. They distinguish three fundamental strata: the inorganic stratum, the organic stratum and the alloplastic stratum, which includes anthropomorphic forms, both technological and linguistic. Strata thus have a clear unity of composition. They enter into relations of mutual support and collision, but each stratum retains its own forms and substances. Strata are empirically observable, but certainly insufficient to understand how things take shape, how cities and urban phenomena are structured.

Assemblages involve fundamentally different form-giving processes, not based on a unity of composition, but on the co-functioning of ontologically heterogeneous terms, such as materials, technological artefacts, bodies, texts, concepts and symbols. 'The difficult part', explains Deleuze, 'is making all the elements of a non-homogeneous set converge, making them function together [...] The assemblage is co-functioning, it is a 'sympathy', symbiosis' (Deleuze and Parnet 1987: 51). The French term *agencement*, which is what we translate here as assemblage, literally means 'putting together', 'arrangement', 'laying out' or 'fitting' of such heterogeneous terms (Wise 2005). One could think of assemblages as inter-stratic phenomena or as metastrata. But it goes beyond that, as the elements that constitute an assemblage are not just stratified entities, but also qualities, intensities or speeds. Assemblages thus involve and bring together elements that are commonly conceptualized as belonging to incommensurable levels of reality. Assemblages also involve *topologies* of relationships that do not depend on physical proximity among their components. Instead of a discrete or bounded whole, assemblages involve formations that challenge topographic understandings of near and far, as well as the idea of geographical scale: 'geometric scale and the idea that actors move up and down them, or "jump" them even, is somewhat misplaced' (Allen 2011: 157). Thus, the assemblage concept enables us to explain the holding together of heterogeneous things without actually ceasing to be heterogeneous (ibid).

Perhaps their most important feature is that assemblages are made of emergent relationships that are external to their components (DeLanda 2006). Assemblages rely upon the *capacities* of entities, from rocks to concepts, to affect and be affected; capacities that come to the fore in interactions among components. Assemblages do not necessarily determine or transform the *properties* of their various components, which maintain a relative independence that allows them to simultaneously participate in various assemblages in different capacities. Hence, DeLanda (2006) opposes assemblages to totalities, in which each component is fully determined by their relational position in the whole. In assemblages, components are rather contingent, in the sense that they are neither necessary nor fully random. This certainly doesn't mean that anything goes, but only that assemblages could be made differently.

Assemblages 'are never fully stable and well-bounded entities, they don't have an essence, but exist in a state of continual transformation and emergence' (Ureta 2014: 232). They are always moving between organized and stabilized matter and the unfixed, the indeterminate, the shifting, between actual arrangement and virtual possibilities, and thus challenging simplistic binaries opposing practices and structures, change and obduracy, or mess and order. Considering that the term is often used one-sidedly to emphasize the fluidity of the existing, it becomes crucial to stress that assemblages create and claim territories and places of identity. Indeed, when Deleuze and Guattari define the notion of assemblage, the first feature they underline are the processes of territorialization.

But, more crucially, the identity of assemblages is shaped by how they function and what they do. Indeed, as Callon has noted: '*agencement* has the same root as agency: *agencements* are arrangements endowed with the capacity of acting in different ways depending on their configuration' (Callon 2007: 320). By referring to Foucault's work, Deleuze and Guattari (1987: 551) suggest that there are constellations of assemblages that share principles of functioning and shape what could be called a 'culture' or an 'age'. Discipline or governmentality could be understood as abstract diagrams shaping the ways in which assemblages act.

In urban studies, the notion of assemblage has been generally welcomed, as it has allowed conceptualizing the city and the urban in more complex ways. By emphasizing the investigation of heterogeneous arrangements of natural, technical and social entities, assemblage approaches have challenged definitions of the city that emphasize its spatial form, political economy or cultural meaning and practice. The assemblage concept has been deployed to think of urban natures, infrastructures, policies and socialities as co-constitutive, challenge structuralist notions of urban space and scales and unveil multiple time regimes. However, and despite such common emphases, it seems necessary to account for profound differences in the way assemblage is deployed.

In this chapter, I revise three major ways of using the concept of assemblage in urban studies. The first section revises three research traditions focusing on infrastructural, metabolic and policy processes in the city that have incorporated the notion of assemblage as a descriptive device. Next, I present current developments in urban theory that expand the Deleuzian perspective with actor-network theory, reimagining the city as a multiplicity enacted in evolving urban assemblages. Finally, I address the critical capacities and ethical commitments enabled by the notion of urban assemblages when conceived as matters of care. A short conclusion emphasizes the need of using assemblage as a departure point for empirical explorations and theoretical work.

Assemblage as a descriptive device

The notion of assemblage has entered the field of urban theory as a descriptive term to emphasize the heterogeneous, multi-local and processual nature of urban infrastructures, metabolisms and, most recently, policies.

Since the early 2000s, notions of assemblage have been mobilized in two interrelated literatures focusing on the political economy and ecology of urban infrastructures. First, in the major work *Splintering Urbanism*, Graham and Marvin (2001) propose describing cities as sociotechnical assemblies or machinic complexes to emphasize the centrality of infrastructures in urban condition and urban dynamics. Taking inspiration from the history of large technical systems (Hughes 1983), as well as from relational research of information infrastructures (Star and Ruhleder 1996), Graham and Marvin featured urban infrastructures as encompassing – besides technical artefacts, processes or expertise – organizational arrangements, sociocultural practices and meanings, and political struggles. The huge body of urban research converging in, but also triggered by this book, underscores the intermingling of the techno-scientific and the socio-political in urban assemblages. Second, the notion of a socio-natural assemblage has been used in the burgeoning field of urban political ecology (subsequently referred to as UPE) to describe the urban as shaped by metabolic processes (Heynen *et al* 2006). The central proposition to this line of thought is that the city can no longer be thought of as in opposition to nature, but as a process through which both the social and the biophysical are co-produced (Gandy 2004). UPE has pioneered in showing that nature does not designate a realm outside the urban, relevant for the acquisition of resources and waste disposal, and in arguing that social histories, power structures and capital accumulation dynamics shape urban socio-natural environments (Wachsmuth 2012).

In these interconnected branches of literature, infrastructures and metabolisms are often described as socionatural and sociotechnical assemblages. As a descriptive term, however, the notion of assemblage appears as exchangeable with other terms capturing the same hybridity, such as cyborg urbanization (Gandy 2005; Swyngedouw 2006a). Most importantly, the notion of assemblage is neither used to generate new research questions nor to provide an analysis of urban processes. This is particularly evident in the case of UPE, where the Marxian understanding of metabolism entails studying the urbanization of nature in terms of labour/capital contradictions, politico-economic stratification and class struggles. UPE's attention to socio-natural assemblages only aims at a methodological enhancement of Marxian-Lefebvrian urbanization theory rather than at a decentring of the city, which is still imagined as 'a metabolic circulatory process' (Swyngedouw 2006a: 33; see critique of 'methodological cityism' by Angelo and Wachsmuth 2014). As Holifield has demonstrated, 'the significance of nonhuman agents here [in UPE] lies in their "social mobilization"' (2009: 646), not in their capacities to shape urban assemblages. Even in studies that more strongly acknowledge their capacity to, for example, resist commoditization and thus change capitalist accumulation patterns, there is a tendency to imagine the economy 'as an already constituted structural unity *that only consequently comes into contact with a recalcitrant non-human nature*' (Braun 2008: 669). Thus, whereas the assemblage concept is used to emphasize the heterogeneous components of urban processes, it is not used as a concept to reimagine the urban at large. The notion of assemblage is limited to a descriptive function and subordinated to other theoretical understandings of the dynamics of urban processes.

Descriptive uses of assemblage can also be found in recent contributions focusing on the translocal and mobile character of urban policies and political formations. A key work is Saskia Sassen's (2006) book *Territory, Authority, Rights*, which looks at the transformation of modern political formations from national to global assemblages. Historically reconstructing how power assemblages are formed and transformed, Sassen underscores the different intersections between territoriality, authority and right in national sovereignty regimes and in global financial urban centres. A second deployment of assemblage, more decidedly attuned to urban political formations, can be found in the work of McCann and Ward (2011, 2012, 2013) on the multi-local mobilization of urban policy. Against traditional policy transfer literature that focuses on prescriptive classifications of transfer actors and institutions and the idea that policies 'emerge in full-form from a specific place or that they circulate unchanged' (2013: 8), their research stresses the global interconnections and flows through which policies are reshaped. Their approach also aims at destabilizing linear understanding of a policy cycle, by showing that it cannot simply be understood as programs of governmental action operating as external forces upon the urban, but as relational processes of assembling materials and resources, legal bodies and social practices, as well as expert and lay knowledge. Even forms of policy resistance begin here to be seen as constitutive parts of a policy assemblage, thus stressing the co-production of urban policies and cities: 'We tend to use "assemblage" in a descriptive sense to encourage both an attention to the composite and relational character of policies and cities and also to the various social practices that gather, or draw together diverse elements of the world into relatively stable and coherent "things"' (McCann & Ward, 2012: 24).

This descriptive use also aims at combining an assemblage perspective with research questions and concerns of classical political economy: 'We tended toward

Neo-Marxian political economy, but through our use of "assemblage" we also take seriously many poststructuralist insights' (McCann 2011: 145). Such an approach involved above all understanding that policies 'are assembled in particular ways and for particular interests and purposes' (McCann and Ward 2013: 8). From this perspective, the politics of assemblage that needs to be unveiled in critical research involves the production of exemplars, a process that involves the hierarchization and blackboxing of policy assemblages, leading to the emergence of 'clearly identifiable 'models' of practice often tagged with city or country names' (McCann 2011: 144), and their global circulation.

Assemblages and the multiple ontology of cities

A second way of using assemblage in urban studies has aimed at renovating urban theory by posing questions concerning the plural ontologies of the urban. The origins of this approach can be linked back to early problematizations of urban theory's focus on political economy and Thrift's (1993) early concern with what he called an urban impasse. It is indeed fair to say that most urban studies literature is still living from the radical theoretical innovations introduced by Lefebvre, which made possible it to think about the urban from a political economy perspective. The ontological use of the notion of assemblage in current urban theory is an invitation to think not against, but beyond the Lefebvrian matrix in order to radically 'reimagine the urban'. This invitation is in the subtitle of Amin and Thrift's (2002) path-breaking book *Cities*, arguably the first to introduce the Deleuzian concept of 'machinic assemblages' in the contemporary theorizing of cities and urban life. Following in their steps, cities have begun to be reimagined as 'mecanospheres', that is, as force-fields composed of constantly evolving machinic assemblages.

This approach has been further developed in current contributions that propose a more direct dialogue with actor-network theory (ANT) to approach 'urban assemblages' (Farías and Bender 2009). Relying on ANT has allowed a number of important features implicit in the notion of assemblage to be made explicit and researchable. The first one is that ANT offers an inspiring conceptual repertoire based on the notions of actants, translation processes and modes of ordering that enables us to distinguish among the various ways in which humans and non-humans shape urbanization processes. Here it is important to clarify that the principle of generalized symmetry proposed by ANT does not aim to equalize the actual capacities of different components of an assemblage, but to force a detailed empirical account of how asymmetry, difference and exclusion are produced in urban assemblages; how specific human or non-human actants become obligatory passage points for a whole network; how are the contributions of many heterogeneous actants black-boxed; etc. ANT challenges urban students to answer these questions without assuming pre-existing power asymmetries among actors deriving from underlying socio-economic structures (Callon 1986), to pay attention to how asymmetries and inequalities are moulded in assemblages.

The second insight that ANT makes apparent is that urban assemblages cannot be thought of as constituted or contained within a city, as they delineate a multi-local topology connecting elements and actors from specific sites within the city and across the globe. Importantly, ANT allows stressing the fact that urban assemblages are not constituted at a particular scale, but as interconnections of sites. At the same time, ANT emphasizes that the shaping capacities of elements and actors from far away do not rely on some form of immaterial power, but are always mediated by material flows and artefacts. Thus, through ANT, urban assemblages can be seen as material circulations producing a spatial topology and thus producing the city. Similarly, ANT also emphasizes how time is produced, multiplied and folded in specific assemblages. Interestingly, it is not just that timing, speed, memory, anticipation and rhythm are shaped differently in different urban assemblages, leading to a plurification of urban temporalities, but also that the past and future, history and prognosis are reimagined as actual forces shaping the present. 'Assemblage thinking is concerned with how different spatio-temporal processes are historically drawn together at a particular conjuncture' (McFarlane 2011a: 209). In this sense, ANT allows emphasizing that cities are not spatio-temporal coordinates containing assemblages, but rather that cities are enacted by assemblages in different ways.

This latter point illustrates perhaps the most important contribution that follows from ANT for the study of the city, namely, the idea that the city is a multiplicity. The work of Mol (2002) has been crucial in thinking of objects as multiplicities enacted in non-coherent ways through different sociotechnical arrangements. Hence the notion of urban assemblages in the plural is crucial in grasping the city as a multiple object, in conveying a sense of its multiple enactments. From this perspective, the city is not an object 'out-there' existing independent of and even prior to specific networks of actants and practices, but constituted, made and unmade, in such networks. As multiplicities, cities are more than one singular reality, but also less than many. Multiplicity is woven in the inter-actions and intra-actions of the sheer variety of entities and forces of urban assemblages, so that multiple enactments are always in contact, requiring coordination, articulation and care.

This approach partly contrasts with DeLanda's (2006) use of assemblage theory to consider the reality of cities. In a paradoxically stratified description, DeLanda imagines the city as a specific level of emergence of social assemblages composed of neighbourhood assemblages, which in turn would be made up of family assemblages, formed by person assemblages etc. City assemblages would also be key components of nation-state assemblages, which again would be components of world region assemblages. Instead of using assemblage theory to problematize the reality of the city, such uses risk limiting themselves to translating traded understanding of cities into an assemblage vocabulary.

It could thus be argued that the most interesting uses of assemblage as an ontological concept are those less oriented towards pointing to existing assemblages than towards opening up new questions, concepts and insights into the reality of

CASE STUDY 4.1 POLITICS OF ASSEMBLAGE COORDINATION

In *The shelter that wasn't there*, Ureta (2014) provides one clear example of the multiplicity of the urban that results from the relations of exteriority constituting urban assemblages. In his work, he focuses on an empirical controversy around a bus shelter that was planned to be built in front of the National Museum of Art in Santiago de Chile. By following this controversy in a non-reductive way – that is, focusing on the objects at stake – Ureta shows how the bus shelter is a multiple object enacted differently in each of the urban assemblages at stake. For the engineers and experts at Coordinación de Transantiago, the office engaged in reassembling the public transport system, shelters would play a crucial new role, as the public bus network was to be reconfigured into a feeder-trunk system needing smooth and easy bus transfer. The position, size and design of the shelter as a transfer station had been determined in accordance with systemic considerations involving the whole network. When the shelter plans arrived at the National Monuments Council (NMC) for their approval, a techno-legal controversy arose. For the NMC, the bus shelter interfered with one of the most important patrimonial buildings in the city, the Biblioteca Nacional. Interestingly, the city enacted by the NMC is a city of buildings, monuments and other 'mutable immobiles' (Guggenheim, 2009), not a city of traffic and flows. It is precisely because buildings are mutable that they require protection, and this protection 'did not [just] refer to the material structure of the building, but [also] to its *view*' (Ureta 2014: 9). Unpacking this controversy and its resolution, Ureta moves from simply attesting the existence of multiple urban assemblages to studying practices of coordination among competing or incompatible assemblages. More generally, Ureta makes apparent that it is in such coordination practices between multiple urban assemblages that issues of power and inequality in urbanization processes appear in all their force.

cities. DeLanda's (2000) own work on mineralization and geological infiltration of urban assemblages, in which he shows how quantum leaps in urban form have been enabled by new minerals entering the urban exoskeleton, is indeed an early fascinating example of what an assemblage perspective on the city can do. It also resonates with current work by, for example, Edensor (2013) on the vitality of brick-stones and their capacities to enable sensual and imaginative experiences. One could argue that if one takes assemblage as an ontological departure point, urban research should then go beyond the mere identification of an urban assemblage. The discussion on the multiplicity of cities and the ensuing coordination work is a good example of this. Other examples would include McFarlane's work on urban dwelling (2011b), as well as on urban learning (2011c) which use assemblage thinking to elucidate the various ways in which the reality of cities is woven together in everyday practices (see Case study 4.1). Corsín and Estalella's (2014) work on assembling neighbours is also a case in point, as they apply an assemblage perspective to describe neighbour assemblies as a method for urban conviviality and dialogue. Färber's (2014) work on low-budget urbanism also shows how an assemblage perspective destabilizes the monolithic descriptions of the so-called austerity urbanism. There are indeed many other brilliant examples of thinking urban processes with assemblages that cannot be rapidly summarized here (Angell *et al* 2014; Blok 2013; Denis and Pontille 2014; Höhne 2011; Jacobs 2012a; Manuel Tironi 2009; Martin Tironi 2014).

Assemblage as a critical practice

McFarlane's (2011a) seminal paper on 'Assemblage and critical urbanism' has been crucial in initiating a debate about the critical capacities resulting from thinking about the city from an assemblage perspective. This is indeed a timely and much-needed conversation, as students of urban assemblages have been criticized for de-politicizing urban analysis, without quite understanding how assemblage thinking produces a redistribution of the political. Assemblage thinking certainly undoes a humanistic version of urban politics, but not in order to accept uncritically or at worst to justify what exists, but to reimagine the matters of urban politics (see Farías 2011). Seen from this perspective, it is not surprising that critical scholars argue 'for a narrower, primarily methodological application' (Brenner *et al* 2011: 230), as the ethico-political consequences of thinking with assemblage debunk the premises of the classical critique of ideology.

The first of such premises is that urban politics results from struggles among humans over the appropriation of urban space (Brenner *et al* 2011: 236). In this context, the city is conceived as a '*point of collision* between the mobilizations of the deprived, the discontented and the dispossessed on the one side and, on the other, ruling class strategies to instrumentalize, control and colonize social and natural resources' (Brenner *et al* 2010: 182). Assemblage thinking complicates such analytical prognosis by, first, reconceiving agency as a capacity distributed in sociotechnical networks, so that if humans appear to 'have' agency and enter into struggles, this is because they are 'equipped' with the necessary sociotechnical devices. Politics is thus not primarily a matter of human agency per se, but shaped by innumerable objects and technologies which, as Latour nicely put it, involve a form of politics by other means. Including objects in the realm of politics also implies moving from a conflict-based model of politics based on the existence of structural contradiction towards a controversy-based model of politics based on the irruption of ontological uncertainties. Such a shift is expressed in the use of the notion of concern to understand the affective engagement with things which are considered to be of trouble, worry or care, instead of the more abstract notion of interest defined by a particular position in a power field. Accordingly, assemblage thinking allows politics to be conceived less as a struggle among structurally defined groups or classes than as an ontological politics, in which what is at stake is the construction of a common world (Farías 2011, 2014).

The second challenge involves the task of the critical scholar, which in critical urban studies is described as deciphering the hidden structural contradictions and injustices, unveiling the ideologies of the ruling class and enlightening people about the structural forces behind their matters of concern. Against such romantic deployment of the critical intellectual, assemblage thinking makes a plea for empirical, careful and respectful engagements with the various actors involved in urban politics, including financial capitalists and neoliberal technocrats, in order to 'not impose 'ready-made explanations' upon the cartographies of actors and networks' (Puig de la Bellacasa 2011: 88).

The key feature of assemblage thinking that sustains its critical capacity is its double emphasis on the material, actual and assembled, on the one hand, and the emergent, the virtual and the multiple, on the other (Farías 2009). McFarlane points precisely to the 'disjunctures between the actual and the possible, between [for example] how urban inequality is produced and lived and how relations might be produced otherwise' (2011a: 210) to ground the critical potential of assemblage thinking. Thus, going back to Deleuze's notion of 'left assemblages' (Tampio 2009), the critical potential of assemblage would involve constituting a 'political subjectivity oriented towards the actualisation of ideals and the realisation of potential' (McFarlane 2011a: 205). Potentiality is indeed crucial in McFarlane's critical rendering of assemblages, as it points to the excessive forces that overflow actual arrangements, opening up a space for imagining and practicing urban life differently. The critical potential of assemblages thus requires a commitment to making, to assembling alternatives that might reshape the *commons* (see Chapter 6, this volume).

It seems crucial here to follow Maria Puig de la Bellacasa (2011) when she proposes treating sociotechnical assemblages as matters of care. Instead of a radical politics based on the sudden realization of the virtual, a politics of care involves first posing critical questions about who actually does the devalued doings necessary to sustain urban assemblages, visibilizing the human labour sustaining 'smart' infrastructures. But the point is not just visibilizing care, but generating care by contributing to articulating matters of concern, as well as maintaining a speculative commitment to the possible becomings of things. Indeed, the fundamental ethico-political question resulting from understanding urban assemblages as matters of care is not *whose* assemblage or *for whom* to care, but *how* to care, *how* to carefully assemble urban life.

Conclusion

The recent career of the urban assemblages concept in urban theory has been quite impressive. Whereas a few years ago it was difficult to fill an edited volume with authors who would work along these lines, today the challenge seems to be quite the opposite: to know where to start reading the bulk of books and articles dealing with assemblage urbanism and to learn to distinguish the fundamental differences between the different uses of it. This chapter has attempted to provide urban students with some tools for starting to engage with assemblage thinking in urban theory. It is, I believe, an important task, for the set of ontological problematizations, empirical sensibilities and ethico-political commitments that this recent literature is bringing into urban studies is here to stay, independent of the future fortune of the concept of assemblage itself. In doing so, I have attempted to make clear, assemblage thinking is not simply about redescribing urban phenomena and processes as urban assemblages, but about developing new conceptual repertoires to address assemblage processes and formations in cities. Taking assemblage thinking not as an arrival, but as a departure point for urban theory does not just offer protection against the fluctuations of intellectual fashions, but also involves extending an invitation to continue reimagining the urban.

Key reading

Amin, A. and Thrift, N. (2002) *Cities: Reimagining the Urban*, Cambridge Press: Polity.
Probably one of the most important theoretical interventions in the field of urban studies since Lefebvre, opening the way for a radical redescription of the urban.
Farías, I. and Bender, T. (eds.) (2009) *Urban Assemblages: How Actor-Network Theory Changes Urban Studies*, London: Routledge.
A collection of articles and interviews exploring the challenges that actor-network theory poses to common imaginations of urban natures, technologies and spatial structuration.
McFarlane, C. (2011) 'Assemblage and critical urbanism', *City*, 15(2): 204–224.
A key intervention that triggered debate about the fundamental ontological, epistemological and political differences across assemblage and critical urbanism.

5

BODIES

Robyn Longhurst

Introduction

'The body', over the past twenty years, has become a central tenet of social science research, including studies of the urban. In my own discipline of human geography this trend of focusing on the body began largely in feminist geography, on account of women in many western contexts being historically associated with the body. Urban geographers and others in cognate disciplines have also increasingly drawn on perspectives such as feminism, psychoanalysis, postcolonialism and poststructuralism, to examine 'cities and bodies'. For example, Steve Pile (1996) considered the mutually constituted relationship between the body and the city. Iris Young (1990) focused on urban social justice and ways in which this is constructed through certain kinds of embodiment. Phil Hubbard (2012) paid attention to 'sex and the city'.

There is no one 'neat and tidy' definition of the body. On the one hand, it seems counter-intuitive to even attempt this given that we all *have* bodies, or rather we are *all* bodies. We cannot exist, materially at least, in cities (or anywhere else for that matter) without them. On the other hand, defining the body has been a pursuit of philosophers across the centuries from the Ancient Greeks to the postmodernists, although there has been little agreement as to what constitutes the body. In actuality, there is no *the* body. Bod*ies* come in a variety of forms that exist in a variety of temporal and spatial contexts. *The* body is an illusion. Bodies are gendered, sexed, raced, aged, imbued with culture, a particular size and shape and so on. They are inscribed and inhabited by intersecting sets of power relations. Their materiality, cultural inscriptions, and psychological dimensions make them complex entities which have been theorized differently at different times. Steve Pile and Nigel Thrift (1995: 6) argue the meaning of the body is 'equivocal, often ambiguous, sometimes evasive and always contested by those who attempt to understand it'.

This chapter focuses on the body, or more accurately, on bod*ies* and the ways in which they have been researched, understood and represented in urban theory over the past few decades. First, I discuss the concept of geographical scale – from bodies to the globe. Second, drawing on a range of examples, I explain some of the ways in which bodies over successive decades, from the 1990s until now, have been examined in urban studies, especially urban geography. Following this is a discussion of human–animal embodied subjectivities since it is not only human bodies but also animal bodies which have been examined in recent years, as the binary division between culture and nature has been deconstructed. The chapter concludes that many urban theorists who have chosen to work at the scale of the body have done so in order to highlight the power relations that operate in different urban spaces.

Geographical scale

Bodies are the most micro, or 'intimate', scale of analysis for studying the urban. Different scales – bodies, homes, communities, cities, regions, nations, and globe – are appropriate for analysing different kinds of urban spaces and activities with each representing an intersection between a range of different people and places. Sally Marston (2000: 219; italics in original) argues that scales are socially constructed: 'Over the last ten years, scholars in human geography have been paying increasing theoretical and empirical attention to understanding the ways in which the production of *scale* is implicated in the projection of *space*.' Scale is not meant to be read as absolute, as in cartographic scales that are used to draw maps, but rather as overlapping and 'nested' (Smith 1992b).

This means that urban theorists while focusing on the scale of the city also pay attention to other scales including broader ones such as the globe and narrower scales such as bodies. Examining international mobility and interconnectedness can tell us a great deal about the form, function and politics of the urban, but so too can examining the experiences of individuals and groups in local spaces (see Chapter 17, this volume). Scales overlap, they are fluid and contingent. Focusing on bodies does not necessarily mean ignoring other scales of analyses although there may be instances where the body is the most useful platform for thinking about specific kinds of actions and interactions in cities. These instances are often ones where it is useful to highlight to power, politics, embodied difference and subjectivity. Reading bodies through the lens of gender, sex, race, class and other social categories can help show up socially produced inequalities in urban space. Cities are not, however, just spaces of inequality and oppression, they are also spaces of resistance and liberation. There is nothing inherently 'good' or 'bad' about using particular scales of analysis, but about the strategic use that different scales, and intersecting scales, can be put to.

Co-producing bodies and spaces

Urban theorists and geographers in the 1990s came to call in their analyses terms such as the body, bodies, embodiment, corporeality, performativity and lived space. Their research has been informed by a variety of theoretical approaches including psychoanalysis, phenomenology (the 'lived body') and social constructionism (the body as a surface of inscription), the latter being the most popular. The social constructionist, or cultural approach as it is sometimes referred to, draws upon poststructuralist theory to make the case that bodies are constructed by cultural and social systems or discourses. Feminist theorist Elizabeth Grosz, in 1992, argued that there is a two-way linkage between bodies and cities that can be defined as an *interface*, or maybe even a co-building. In this important early work Grosz (1992: 248) explains: 'The city in its particular geographical, architectural, spatializing, municipal arrangements is one particular ingredient in the social constitution of the body.' In other words, bodies and urban spaces construct each other in complex ways meaning that it is impossible to think about bodies without thinking about space and vice versa. Heidi Nast and Steve Pile (1998: 1) note: 'we live our lives – through places, through the body'. Bodies are 'performed' not simply in but *through* space. An example of early research on the 'interface between bodies and cities' can be found in my examination of pregnant women's experiences of a shopping mall in Hamilton, New Zealand (see Case Study 5.1).

CASE STUDY 5.1 PREGNANT BODIES IN CENTREPLACE, HAMILTON, NEW ZEALAND

Inspired by Grosz's argument, in the mid-1990s, I sought to examine how the spaces of a particular shopping mall, Centreplace, in Hamilton, New Zealand constructed pregnant women's bodies in particular ways. Centreplace covered 5 acres (or 2.2 hectares) and comprised 3 levels, included facilities such as a multi-storey car park, crèche, toilets, food court, tavern, cinema and approximately 90 shops (namely boutiques), fashion accessory stores, a perfumery, a craft store, jewellers, an appliance centre, and a sewing machine centre to name just a few. It did not include a large 'anchor' or department store. The main participants and targets of the consumer activity in Centreplace were middle-class women, and yet, more than half of the 31 pregnant women interviewed for this research indicated that they did *not* feel like they belonged in the mall. All of the participants in the study were visibly pregnant with their first baby. They were therefore for the most part quite conscious of their changed corporeality, a corporeality which meant they did not feel welcome in the space of the mall despite having consumer dollars to spend. The participants reported that the toilet cubicles were too small for their growing bodies and that there were not enough of them; they frequently had to queue. The seats in the cinema were cramped and uncomfortable for their pregnant bodies; there was little lumbar support. The mall's floors became slippery when water was traipsed in on wet days. Heavily pregnant women reported that the escalators and stairs were difficult to negotiate. The spaces of the mall had not been designed and built with pregnant bodies in mind. The notion that visibly pregnant women belong at home rather than in public space, an idea that still resonated with some in Hamilton in the mid-1990s was reinforced in the concrete construction of this urban space. Approximately a decade later shopping malls and centres in Hamilton began to realize that pregnant women and parents with young children are potentially important customers. The discourse of pregnant women belonging at home also began to shift. While still only a few businesses have designated parking spaces for pregnant women, many have now assigned spaces close to mall entrances for parents with young children, for example, Westfield Shopping Centre (see Figure 5.1) (but still not Centreplace).

In the mid-1990s the cultural environment of Centreplace shopping mall continued to reinforce dominant understandings of pregnant women with some participants reporting being welcomed into spaces previously largely unknown to them, such as Pumpkin Patch (a shop selling soft toys), Wool Warp and Weft, Extra Elegance and Kooky Garments (the last two shops mentioned sell clothes for 'larger' and pregnant women), on account of their new-found status as 'becoming mothers'. In other stores, such as designer boutiques that do not stock clothing above size 14, the women felt less welcome. One participant claimed that she received a 'frosty' response from a worker in a lingerie shop when she wanted to purchase, not a feeding bra, but 'sexy' lingerie. In this case study it can be seen that dominant understandings of pregnancy, maternal roles, gender, and sexuality were embedded not just in the bricks and mortar of the space, but also in the culture of the space.

FIGURE 5.1 Making space for parents with young children, Westfield Shopping Centre, Hamilton, New Zealand (Source: Robyn Longhurst).

Also in the 1990s philosopher Iris Young (1990) addressed the relationship between bodies and cities. She drew on anthropologist Mary Douglas's notion of 'abjection' (described by Young (1990: 124) as 'unconscious fears and aversions that continue to define some groups as despised and ugly bodies') in order to argue that cities are structured by homophobia, ageism, ableism, racism, and sexism. Unconscious actions, judgments, manners, feelings, ways of speaking, bodily reactions to 'others', jokes and images all matter in the production of bodies and cities. Young expands on this idea claiming that 'cultural imperialism' structures urban life. She describes 'cultural imperialism' as one group projecting their own values and perspective as the norm or universal. In the process victims of 'cultural imperialism' are rendered invisible at the same time as they are marked out as different and Other. Young's focus on the body (e.g. what it means to have black skin), while at the same time addressing broader spatial scales (e.g. cities and nations), adds weight to her argument about theories of justice and 'the emancipatory city'.

(Gendered) embodied subjectivities

In the early 2000s many researchers, especially feminist researchers, began to draw on the work of theorists such as the aforementioned Iris Young and Elizabeth Grosz to explore body-space relations in more depth. For example, in 2001 I argued that 'boundaries', whether they be bodily boundaries, disciplinary bodies, or conceptual boundaries are a useful motif for thinking about (gendered) body-space relations (Longhurst 2001). I drew on an example of business managers who work in the Central Business Districts (CBDs) of Auckland, New Zealand and Edinburgh, Scotland arguing:

> highly tailored, dark coloured (often black, dark grey or navy) business suits function to seal the bodies of (men and women) managers. Firm, straight lines and starched creases give the appearance of a body that is impervious to outside penetration. They also give the appearance of a body that is impervious to the dangers and threats of matter that is inside the body making its way to the outside. It is considered inappropriate for matter to make its way from the inside to the outside of bodies (for example, farting, burping, urinating, spitting, dribbling, sneezing, coughing, having a 'runny nose', crying, and sweating) in most inner city work places. This suited, professional, respectable body, however, can never be guaranteed. Like all bodies it is continually monitored and disciplined but inevitably proves itself to be insecure.
>
> *(Longhurst 2001: 7)*

Not only do managers discipline their own bodies but also the bodies of their employees. Although the prescriptions for the ideal presentation of employees vary somewhat (depending on the work), all managers play a role in disciplining or shaping the bodies of 'their' workers, their 'look', dress, level of fitness and behaviour being just a couple of examples.

Michael Brown (2000) who also focused on body-space relations discussed the invisibility of gay sex acts ('bodies in the closet') in the city of Christchurch, New Zealand. Brown conducted his research in 1996 Christchurch at the same time that Christchurch was named in a national news programme as 'the sex capital of New Zealand' and yet gay venues remained largely invisible. Unlike straight venues they carried no or minimal signage indicating their whereabouts. This is a useful example of how focusing at the scale of the body, as well as the club, street, or park, can be useful in furthering understanding of the sexual production of urban space.

Carolyn Prorok (2000) in her analysis of urban space addressed not so much the materiality or representation of bodies but their spirituality. She conducted nearly three years of fieldwork in a Puerto Rican Espiritismo-worshipping community in New York City concentrating on the central figure in the community, Maria, the priestess, who negotiates the exigencies of a world of 'poor health, lost love, and economic deprivation' (p. 57) through a world of spirits and trance:

> Her [Maria's] articulation of Espiritismo uses the social boundaries between self and community, male and female, Catholic and non-Catholic, the human world and the spirit world, and racial identities as points of departure to solve problems of well-being … for herself and her community of devotees. Maria's body is the primary site where these issues are resolved.
>
> *(pp. 57–58)*

These examples of research I have cited from the early 2000s illustrate that bodies and urban spaces are socially constructed and inextricably linked. 'On the ground' case studies gained in popularity at this time although some urban theorists and geographers remained reticent about discussing the materiality of bodies. Words that highlight the 'real' fleshiness and fluidity of bodies such as menstruate, ejaculate, bleed, vomit, and urinate still remained absent from many (but not all) of these discussions of people's experiences of the urban. There may be many reasons for this (see Longhurst 2001) such as that the aforementioned words can prompt discomfort in 'respectable' places such as university staff offices and in 'respectable' disciplines such as urban theory and geography.

Intersecting embodied subjectivities

In the mid-2000s research on body-space relations began to increase. Whereas in the 1990s and early-2000s the reporting of research findings on this topic was often confined to one or two sessions at conferences, a couple of articles in an occasional journal and a few edited and authored books, by the mid-2000s focusing on a variety of spatial scales including 'the body' had become much more commonplace. Social constructionist approaches continued to be popular with some researchers focusing on materiality while others continued to avoid discussing 'real' bodies. The emphasis on gendered bodies tended to remain in the mid-2000s and beyond but there was also a trend towards considering others aspects of subjectivity such as body size, life stage and migrant status. Researchers also became increasingly interested in thinking through how various subjectivities intersect.

An example of work that focused on subjectivity beyond the usual trinity of 'gender, race and class' is written by Rachel Colls (2006). Colls examined the lives of women considered to be 'fat' or obese. She was particularly interested in the ways in which these women's bodies are produced through the consumption of clothing. Colls accompanied three research participants as they shopped, writing up the experience in a research diary and conducting an interview immediately or a day after. Colls paid attention to the women's experiences of what it feels like to shop for 'big clothes', and how they evaluate the (un)suitability of clothing for their (un)suitable bodies. I also conducted work on 'fat' bodies, including my own. Using autobiography as a method of inquiry I interrogated my experience of becoming smaller through dieting, that is of losing more than a third of my body weight. For me the experience was paradoxical on several accounts including that I am a feminist scholar who critiques discourse around women and slimness and yet at the same time desired to be slim (Longhurst 2012).

Other research in the 2010s that highlights dimensions of embodied subjectivity and the ways in which they intersect is Lesley Kern and Beverley Mullings's (2013) study of violence and fear in Kingston, Jamaica and Toronto, Canada. They sketch the gendered but also racialized and classed dimensions of violence and fear in these neoliberal cities (see Chapter 18, this volume). 'Discourses of fear and insecurity ... encourage the embodiment of danger in specific 'others' (racialized and gendered inner-city poor in the case of Toronto, and the police and government officials in [the] case of Jamaica)' (Kern and Mullings 2013: 39). Kern and Mullings (2013: 29) make a case that much research on 'the right to the city' 'reproduces gender biases because they pay insufficient attention to ways in which patriarchal power relations shape how particular bodies negotiate and access urban space'. For example, researchers addressing homelessness often ignore homeless women on account of the fact that they are less likely to occupy public space. Bronwyn McGovern (2013) in her excellent study on homelessness in Wellington, New Zealand does not focus on women, but she does draw out a range of insights on embodied subjectivity by focusing on the story of one homeless man – 'Brother' – who was an iconic figure on the inner city streets for many years.

CASE STUDY 5.2 'BEN' HANA A.K.A. 'BROTHER' AND 'BLANKET MAN'

In the late 2000s PhD student Bronwyn McGovern carried out ethnographic research becoming part of the life of Bernett Hana ('Brother'), a well-known street dweller and local identity who lived everyday life on a street corner in Wellington, New Zealand (see McGovern 2013). Using the core concepts of space, body and social interaction McGovern examined how broader social structures impact upon local settings to 'make, remake, and unmake' people. Brother, who passed away in 2012, was also commonly known by members of the public as 'Blanket Man' on account of him wearing nothing but a loin cloth and draping a blanket over himself when cold (which it frequently is in the windy city of Wellington). Brother was constantly monitored by the authorities. He preferred to live on the inner city streets rather than be housed, a decision that perplexed authorities who carried out psychological testing and medical procedures to probe Brother's mental health. In 2010, despite tests finding Brother to be in sound mental health, a judge ruled to have him committed to Wellington Hospital's psychiatric unit. McGovern focuses in her project on various scales including the body. She explains: 'By specifically examining the homeless body … I am able to elaborate on how [it] is made up and categorized through spatial practices' (McGovern 2013: 173). She describes Brother's semi-nakedness, dirtiness, odour, horizontalness (he often lay on the street), identity as Maori, and his humour (Brother could at times be sharp with people, but also had a playfulness that won him a great many friends). McGovern also describes how Brother or Blanket Man used his blanket not just as cushioning and warmth for his near-naked body, but as a wall of defence on the street, a kind of 'bodily armour' (p. 181) that helped him defend against unwanted intrusions from others. By laying out the blanket he was able to demarcate personal territory on the public streets. McGovern's rich study of Bernett Hana illustrates how it is possible to study cities, in this case in relation to homelessness, using multiple scales including the micro scale of the body to develop more multidimensional understandings of social practices and spatial urban formations (see Figure 5.2).

Other studies pay attention not just to multiple and intersecting subjectivities but also to multiple and intersecting scales. Dina Vaiou (2013: 65) examines elder care in Athens, Greece. She says:

> thinking about care at all these scales [bodies, households, urban neighbourhoods, nation states and transnationally] yields more complex understandings of recent migrations and of the gender aspects of those migrations, beyond often unfounded generalisations. This multi-scalar approach to care also opens possibilities for different understandings of the urban which bring to the foreground of enquiry the politics of the everyday.

FIGURE 5.2   'Brother' and researcher Bronwyn McGovern at northern entrance to Cuba Mall in Wellington, New Zealand (Source: Belinda Brown Photography, 2008).

The elderly have been one of the many 'victims' of Greece's austerity measures instigated to help 'manage' the economic crisis. Vaiou (2013: 56) approaches 'movement' (read: transnationalism, mobility, migration) not as 'an analytical abstraction, but rather as a set of bodily practices which shape, among other things, urban life'.

Phil Hubbard (2012) also pays attention to multiple scales and dimensions of embodied subjectivity although his focus is the sexual body. He reflects: 'Put simply, cities have long been recognized as spaces of sexual encounter, as sites where bodies come together, mix and mingle' (Hubbard 2012: xiii). One area Hubbard examines in-depth are 'impolite bodies' (p. 93). These, he argues, are bodies which refuse to conform to codes of public decency. They refuse to be 'normal', civil, well-mannered, proper, are deemed to be 'dangerous' and in need of reform, discipline and/or containment. Hubbard refers, as an example, to unmarried mothers. While some commentators on bodies and the urban, as previously mentioned, focus on representations and discursive constructions of embodiment, shying away from more material considerations, this is not the case for Hubbard. He refers to 'Saliva, shit, sperm, blood and urine' (p. 95) explaining that 'pollution taboos' around these bodily fluids 'have resulted in sex acts being surrounded by complex cultural conventions about when and where sex acts can happen' (p. 95), which in short, is out of public view.

Ayona Datta (2012) also discusses bodily acts in relation to private and public space analysing the lives of those in a squatter camp in New Delhi, India, since 2000, when in the aftermath of a series of court rulings about 'illegal' settlements, there was a wave of demolitions of 'unauthorized' structures across Delhi and other

Indian cities. This forced many of the urban poor to have to defecate on public land in the open. Punitive measures were then taken by the state to try and stop this.

Beyond human bodies

Alongside the previously mentioned research on body-space relations conducted in the 1990s, 2000s and 2010s, has been another corpus of work (although much smaller) which focuses on animal (non-human) bodies and the relationship between humans and animals. Humans and animals interact in a variety of ways and in a variety of spaces, including urban spaces. Binaries between human/animal, machine/organism, culture/nature and urban/rural and have in recent years been 'troubled'. Philo and Wilbert as early as 2000 edited a series of essays which address people's embodied relations with animals, that is, how humans place different animals into a range of different categories, according to notions of species, usefulness, domesticity or wildness. As a result animals are consigned to particular places and spaces.

One particular place and space to which animals are often consigned is pet or human companion. Heidi Nast (2006) explains that many city dwellers have a relationship with pets, the relationships involving affection, love, companionship, domination, family sociality and, increasingly, commodification. Humans have long pampered their pets in many societies around the world but in post-industrial societies over the past 20 years these 'investments' – financial, emotional and cultural – have escalated. Pet products and services are numerous and widely available. In discussing the bodies of people and pets Nast refers to de-clawing, 'pooping on a carpet', 'spraying the house', dressing dogs up as boys or girls, dancing dogs, dog yoga, euthanizing and cloning pets, dogs as fashion accessories for humans, and 'furries' (people interested in fictional anthropomorphic animal characteristics with human traits). Nast adds a visceral thread to a highly complex argument about 'post-industrial forms of hypercommodification' (p. 304) in the neoliberal city.

Conclusion

Overarching understandings of cities, their global networks and transnational connections are important but so too are understandings of how people in very embodied ways actually inhabit homes, schools, work places, recreational spaces, shopping malls, parks, and streets. As Nast and Pile (1998: 48) note 'different cities, different sociocultural environments actively produce the bodies of their inhabitants as particular and distinctive types of bodies, as bodies with particular physiologies, affective lives, and concrete behaviors'. There is not necessarily one ideal environment for the body in relation to its health and well-being. Bodies are not natural but are constructed, at least in part, by the built environments they inhabit.

Urban theorists for three decades have been researching bodies and their relationship with cities. Not all have chosen to work using this scale of analysis but some have sought to address 'How bodies move across urban spaces, which bodies are allowed in which places, [and] how bodies are placed in or out of the polity' (Peake and Rieker 2013: 13). Much of the research on bodies and urban space is underpinned by the notion that the power relations surrounding and inhabiting particular bodies differ markedly (Mitchell 2003). The rights of urban inhabitants to physically and socially access various spaces in the city are filtered through subjectivities such as gender, sexuality, ethnicity, race, class, body size, (dis)ability and age.

Cities and bodies co-produce each other; that is, relations between bodies and cities are constitutive and mutually defining. People, with particular bodily capacities, skin colour, shapes and sizes that are inscribed by class and so on, make cities. People plan, build, work, play, love and fight in cities. We react, consciously and unconsciously, to these urban places we create. In order to explore this constitutive and complex relationship between cities and bodies this chapter has presented a number of examples produced over a period of three decades including pregnant women's feelings towards a shopping mall, a man who makes inner city streets his home, 'cultural imperialism', violence and fear, eldercare, sex and the city and people's willingness to love and spend money on pets. Cities and bodies are inseparable. Perhaps it is not surprising therefore that over the past three decades urban theorists have come to more-explicitly consider the importance of the bodies in their analyses of urban space. One of the most significant dimensions of this research is that bodies are a locus of identity with some being privileged and valued over 'others'. This is valuable, meaning that in future years it is likely that we will continue to see more studies that examine the embodied nature of urban life.

Key reading

Datta, A. (2012) *The Illegal City: Space, Law and Gender in a Delhi Squatter Settlement*, Aldershot: Ashgate.

Ayona Datta examines the lives of those in a squatter camp in New Delhi paying careful attention to the (gendered) body. She addresses topics such as 'soiling' of the urban forest, discourses of 'dirt' and 'filth', and bodies that are constructed as 'polluting' and 'untouchable'. The everyday spaces of homes, lanes, streets, roads, and public areas, and the bodies that inhabit these spaces are subject to the 'force of the law', but they also find ways to contest and reconfigure their legal citizenship.

Hubbard, P. (2012) *Cities and Sexuality*, London: Routledge.

Phil Hubbard's book brings together urban theory and sexuality studies in useful ways. For example, it illustrates how the spaces of the city shape sexed bodies – especially those identified as deviant or capable of 'corrupting' others, whose bodies are often contained in disciplinary spaces such as treatments rooms and clinics.

Peake, L. and Rieker, M. (Eds.) (2013) *Rethinking Feminist Interventions into the Urban*, London: Routledge.

This is a collection of ten essays from authors working in a variety of geographical locations including Taiwan, Canada, the USA, the United Kingdom, Egypt and Greece. The overall argument is that knowledges about cities, including feminist knowledges, have tended to be produced through a problematic division between the Global North and the Global South. Many of the authors in this collection draw on 'bodies' to address issues of power, place and politics.

Young, I. (1990) *Justice and the Politics of Difference*, Princeton, NJ: Princeton University Press.

In Chapter 5, 'The Scaling of Bodies and the Politics of Identity', Young develops the notion of 'cultural imperialism'. A great deal of the oppressive experience of cultural imperialism, she explains, happens in mundane contexts such as on the street and involves bodily interactions (e.g. gestures, tone of voice, eye contact). Some groups are constructed as having 'ugly bodies' (p. 123) on account of their race, gender, sex, age, and/or (dis) ability invoking, in those who belong to culturally imperialist groups, feelings of aversion and abjection.

6

COMMONS

Andre Pusey and Paul Chatterton

Introduction

This chapter introduces the fascinating, long established yet also fast-growing interest in the idea of the urban commons. It is an idea that is being increasingly used in urban theory to explore and illuminate dynamics in the contemporary city. In this chapter we suggest that cities can be regarded as the ultimate commons. Related to this we suggest that the idea of commoning allows us to take a new look at the rich patterns of social life within the city and potential to deepen the commons. The implication is that the urban commons and its commoners open up new political imaginaries essential for pressing crises and tackling injustice and transforming urban life beyond capitalism. The commons therefore provides new insights and resources into the nature and potential of contemporary urban social struggles.

Commons and commoning

The commons have emerged as an alternative political keyword of our times. Much of the use of the concept of the commons dates back to the enclosures of English land, and dispossession of peasants from that land, just before the onset of the Industrial Revolution in the 1760s. Dispossessions from poor, peasant and indigenous peoples of vital resources and attacks on their livelihoods, then, have underpinned efforts to defend the commons for many centuries (Linebaugh 2014).

The commons consists of a shared interest or value that is produced through communal relations. It can form an alternative political ethics when common 'wealth' (for example land, water, seeds, air, food, biodiversity, cultural practices) that provides direct social and physical wellbeing, are faced with 'enclosure' in the form of the destruction of physical environments and the privatisation of resources and genetic stocks (Gibson–Graham 2006: 95–97). Protecting this 'commonwealth', and the communal relations that underpin them, from enclosure and dispossession is central to generating new forms of antagonism and solidarity (Hardt and Negri 2009). Mobilizing around the common are productive moments that build new group identity, shared understandings, and repertoires of tactics that can offer alternatives to capitalism (De Angelis 2003; Linebaugh 2008). As De Angelis (2003:1) notes:

> Commons suggest alternative, non-commodified means to fulfil social needs, e.g. to obtain social wealth and to organise social production. Commons are necessarily created and sustained by 'communities' i.e. by social networks of mutual aid, solidarity, and practices of human exchange that are not reduced to the market form.

The commons refers to much-more-than simple bounded territories, also encompassing physical attributes of air, water, soil and plants as well as socially reproduced goods and knowledge, languages, codes, information. The shared attribute is that they are collectively owned or shared between and among populations.

But it is also important to look beyond these basic physical attributes and see commons as complex social and political entities underpinned by particular social practices and relationships and forms of governance. The commons, then, can also be considered as a social relationship of the commoners who build, defend and reproduce the commons. The commons is made real through the practice of commoning, which does not simply reflect a set of bounded, defensive or highly localised spaces. Rather than a simple monolithic entity, the commons is complex and relational – it is produced and reproduced across networks that are linked and relate to each other across different times, spaces and struggles. Thus, we should not position the common as something always defensively on 'the back foot' to the more dynamic (and stronger) practices of capital accumulation. The commons is full of productive moments of resistance that create new vocabularies, solidarities, social and spatial practices and relations and repertoires of resistance.

The commons, then, is not a static entity. It is a dynamic, generative entity that is reproduced through these practices of commoning. Interrogating how the commons and commoning are created in and through space helps us understand how alternative political strategies are formed (see also Vasudevan et al 2008). As we explore below, Hardt and Negri (2009) see the common as a form of 'bio-political production' which has the potential to generate new relations between people and things. The commons, then, creates new ways of talking and acting, which citizens and activists are creatively using to challenge complex problems such as climate change, spatial injustice or urban displacement. Importantly, commoning can create a political imaginary which can, at the same time, be anti (against), despite (in) and post (beyond) life under capitalism (see Gibson–Graham 2006; Holloway 2010).

Understanding the commons is important given that contemporary patterns of capital accumulation seek to dispossess the poorest and most marginal groups in society of vital resources they depend upon and attack their livelihoods, as well as advance on the basis of the enclosure, appropriation and dispossession of land, resources and life-worlds. What we are witnessing in the contemporary moment of capitalist development globally is a particularly virulent form of primitive accumulation ushering in forms of enclosure akin to those seen in Europe during the eighteenth and nineteenth centuries, which is expanding into a whole host of new areas such as the Internet, plant patents and, most recently, the carbon cycle (see Shiva 1997; Dyer-Witheford 2001; De Angelis 2007). The commons consists of shared interests or values that form the potential base of communities that come together in the face of such losses and encroachments. The common, then, can be a bulwark against the excesses of contemporary capital encroachment and expansion. It is an essential defensive and productive act against enclosures and oppressions.

What does all this mean for the study of cities?

The first aspect relates to how we view the city. Cities can be regarded as the ultimate contemporary common. As the urban condition becomes the hallmark for the majority of humanity across the planet so too the city becomes thoroughly characterized by both the powerful forces of capital accumulation and the practices and potentials of the common. Antonio Negri and other autonomous Marxists have long expressed how the struggle against capital has far exceeded the gates of the traditional factory into wider society, or what they call the social factory. And indeed now, we could call the city the new form of the social factory. It is in this urban social factory that a new historical figure of social change, the multitude, as Hardt and Negri (2009) called it, becomes visible.

The multitude remains an ill-defined collective subject, but nevertheless carries with it the potential for revolutionary change that can build the urban commons. In the context of our increasingly urbanizing world, Hardt and Negri (2009: 250) insightfully comment that 'the city is to the multitude what the factory was to the industrial working class'. Hence, just as the whole city has become a potential site of economic production, as the working day and its practices spread across offices, cafés and everyday encounters in the street, so too does it become a potential site for resistance, struggle and the creation of commons. The urban multitude might point towards the new urban commoners.

Following on from this, the second implication for the study of cities is how commoning allows us to look again at the rich everyday life and dense patterns of social life in the city and their potential to deepen commoning. For Hardt and Negri (2009: 250) there is huge potential in the rich encounters and activities that make up the metropolis: 'the metropolis is the site of biopolitical production because it is the space of the common, of people living together, sharing resources, communicating, exchanging goods and ideas'. It is this everyday vitality found most notably in cities that gives them their potential. These everyday practices of the common within the city highlight the potential to decommodify a whole host of areas of urban life. Indeed, the urban commons suggests how alternative, non-commodified means can fulfill basic needs through social networks based on mutual aid, solidarity, and practices of human exchange that are not reduced to the market form (see also De Angelis 2007).

The contemporary city is where we can see glimpses of commons in terms of new radical spaces of democracy, ways of organizing, and non-commodified social innovations in a range of areas such as housing, education, food and energy (our case study on the Lilac housing project offers some pointers here). Hence, the city is both the ultimate focal point in the organization of neoliberal capital, but so too it is the ultimate site for resistance and struggle against this, creating alternatives through the productive endeavors of the collective multitude.

CASE STUDY 6.1 LILAC COHOUSING: CREATING A COOPERATIVE URBAN COMMONS

Lilac is a cooperative cohousing project in Leeds in the North of England, which comprises 20 homes based around the central common house, all constructed from straw bales and timber. Lilac can give us insights into the functioning of the urban commons on three levels – the institutional, the interpersonal and the spatial. At the institutional level, Lilac is legally a cooperative society, which exists for the benefit of its members. This kind of legal form embeds the idea of mutualism. Mutualism is a rich historical tradition enmeshed in the nineteenth-century philosophical debate that states that the association that emerges from interdependence can be beneficial and increase wellbeing. As a doctrine, it outlines how people can conduct relationships based on free and equal contracts of reciprocal exchange. It is based on a passionate desire for people to govern themselves and not have authority imposed upon them. From the nineteenth century onwards through a strong cooperative movement, mutualism provided a strong intellectual bulwark against the rampant individualism of the fast-expanding free-market capitalist economy.

For Lilac, this legal cooperative framework creates fertile ground for creating practices of commoning and identities as commoners. One particularly notable strand is the use of deliberative democracy to ensure that everyone has a voice and decisions are made equally between members. Consensually made 'community agreements' are used in areas of community life in Lilac such as use of shared spaces, food and pets. Members contribute to the self-management of Lilac through various task teams ranging from landscape and food to finance maintenance, while a governing board made up of voluntary participating members oversees the legal and financial aspects.

The second level that we can explore Lilac as an urban commons is through the kinds of interpersonal relations that it promotes. A significant focus of Lilac is on building a strong sense of community and interpersonal ties. In Lilac, residents have a different relationship to their housing tenure. Rather than being owner-occupiers of private property, residents in Lilac are members of a cooperative society and lease their homes after paying a member charge set at one third of their net monthly income. Through this regular payment, members can accrue equity in their housing. But the key difference is that the value of this equity is linked to national earnings rather than average house prices. Therefore, housing in Lilac is not a speculative commodity that can be bought and sold, according to the vagaries of market conditions. Instead, it remains affordable in perpetuity for future generations. This is a significant shift, as it adds up to a housing commons with increased stability in local housing markets and reduces tendencies towards volatile casino-like local economies. While money certainly does still circulate within Lilac and the project does depend on much debt financing, it has attempted to embed less-marketized forms of financial and social interactions.

The third aspect of Lilac as an urban commons is the physical layout. What we see in Lilac is a fascinating interplay between private, public and common spaces. One of the key principles of cohousing projects is to combine private self-contained homes with shared spaces. Residents have to continually negotiate the boundaries between the private and the shared as they navigate

FIGURE 6.1   Lilac cooperative cohousing project, Leeds, UK (Source: Paul Chatterton).

through their daily lives. One aspect of this negotiation relates to openness and availability in public spaces. The site has been designed to increase natural surveillance and neighbourly encounters, and therefore residents have to set their own boundaries and tactics for moderating levels of interaction with residents and others who might visit the project. Moreover, the boundary of the site represents the gateway to the broader public realm and here access with the general public has to be mediated. While the grounds of Lilac are private, the general public are not discouraged from entering, which blurs a traditional boundary between public and private, and sets it apart from the rapid growth of privatized housing estates. While Lilac has not fully resolved this issue, there is a greater desire to allow the housing community to be open to the public to reduce concerns about becoming a gated community. What this raises is a key question of a further exploration in terms of how each commons mediates its own boundaries between public and private and shared.

The third implication for the study of cities relates to new political imaginaries that a commons approach opens up. Tackling injustice requires not just successful attempts to mobilize against oppression, hierarchy and exploitation, although these are of course crucial. It also requires developing and advocating for new imaginaries and political vocabularies (Bonefeld 2008). This is not an imaginary which relies on old or established political tools and formulas. In trying to build the urban commons we find a political project that 'cuts diagonally across these false solutions – neither private nor public, neither capitalist nor socialist – and opens a new space for politics' (Hardt and Negri 2009: ix). Building an urban common also involves much more than capturing land and assets, although this is essential. It also requires the ability to control and imagine governance in new ways. Our example from the Real Open University (ROU) is one example of attempts to create new political imaginaries (see Case study 6.2).

The late Colin Ward, tireless advocate for an anarchist social policy and theory, stated '... the city is the common property of its inhabitants. It is, in the economic sense, a public good' (Ward 1989: 1). As a public good, the city can also be understood as a common that is governed by and for its citizens to maximize internal democracy, well-being and flourishing (see de Angelis 2007; Linebaugh 2008). The common is not a simple project of welfarism or nationalization. The city common does not just involve land and assets as common goods, but also its governance mechanisms. Policy and management options cannot be known or determined in advance of a commitment to participation. They emerge from this commitment.

Finally, the commons offers us new insights into the nature and potential futures of urban struggles. The city, then, is not just a site of diversity or cosmopolitan encounter, but is also a site of contemporary social change through its potential to create commons that can organize against capitalism. This productive moment of commoning and the social relations that produce and maintain it is a vital but

CASE STUDY 6.2 THE REALLY OPEN UNIVERSITY: NEW POLITICAL IMAGINARIES OF THE COMMONS

The Really Open University (ROU) was a radical education project based in Leeds. The ROU were engaged in an experiment in academic commoning, and in this section we examine three examples of commons-based activity that the ROU were involved with. The ROU was formed in January 2010 as a means to both protest against university budget cuts and the increase in tuition fees, but also against the further instrumentalization and neoliberalization of Higher Education more broadly. However, the ROU was not interested in romanticizing a golden era of the public university, or in simply defending the existing system, as clearly articulated in an early proclamation of the group that stated: 'we don't want to defend the university, we want to transform it!' The ROU's byline 'strike, occupy, transform!' embodied the group's desire to merge a praxis based on political antagonism and resistance with a transformative and affirmative politics of the common.

The ROU was a forerunner to the UK student protests that erupted in the autumn/winter of 2010, and the group participated in this emergent movement. The group's activities included: constructing a papier mâché costume depicting Marx's concept of the 'general intellect' and storming a live television debate about the tripling of student fees; the production of an irregular free newsletter called the 'Sausage Factory', taking its name from Marx's Capital; a three-day conference of varied talks, workshops and other activities, around the theme of 'reimagining the university', timed to coincide with a large demonstration against the Browne Review, which saw the occupation of a lecture theatre on the University of Leeds campus; and, lastly, the establishment of a six-month initiative called the 'Space Project', which was a city centre-based autonomous education space.

The ROU had an ongoing self-consciousness about producing commons – for example one idea explored at length in meetings, but which never materialized, was a plan for a 'Knowledge Commons' website, whereby the commonwealth of academic knowledge could be freely shared instead of trapped behind prohibitively expensive paywalls. Part of the intention of this was to expose the tensions and contradictions around academic labour, the production of the academic commons and their capture/enclosure by capital.

The ROU was also engaged in the production of commons in the form of spaces that were collectively managed, where it fostered a horizontal and collaborative environment. These spaces were non-profit and free to participate in. But the ROU was also about more than creating physical spaces and self-run courses; the process of creating these spaces also formed a community of commoners. Here the classroom (or department) is refashioned as the spaces of pedagogy created by the ROU, for example through its discursive 'concept meetings' or the six-month autonomous space: the Space Project.

REIMAGINE THE UNIVERSITY

JOIN THE REALLY OPEN UNIVERSITY FOR 3 DAYS OF CREATIVE
RESISTANCE. LOOKOUT FOR WORKSHOPS, LECTURES, DISCUSSIONS,
FILMS, OR OTHER 'INTERVENTIONS',
WE ARE CALLING ON STUDENTS, LECTURERS, STAFF AND OTHER NON
STUDENTS TO COME TOGETHER AND SHOW ANOTHER UNIVERSITY IS
POSSIBLE.

NOVEMBER 24/25/26TH

FIND OUT MORE AT: WWW.REALLYOPENUNIVERSITY.ORG

FIGURE 6.2   The Really Open University (Source: Paul Chatterton).

In operating within, against and on the edge of the university (Noterman and Pusey 2012), the ROU acted as a form of 'undercommons' (Harney and Moten 2013). This undercommons is populated by what have been termed 'para-academics' who mimic academic practices but refuse the instrumentalization of the academy (Wardrop and Withers 2014). This is a form of intellectual production that refuses measure and aspires to be 'doing' rather than abstract labour (Holloway 2010). The majority of the ROU were situated within the university, and operated at a subterranean level – not entirely off the radar, but not entirely on it. There was an imperceptibility. What, or who were the ROU? What was it about? Was it a protest or a seminar? There was a misfitting. This was an attempt to be in-but-not-of the university, of 'hacking the university' (Winn 2014). Over the two years that the ROU existed it contributed much-needed debate within the broader struggles around Higher Education, and began to ask important questions about the relationship between capitalism, universities and the common(s). Importantly these questions arose through prefigurative experiments in producing new forms of edu-commons: inside, outside and on the edge of the institution.

under-articulated component in our understanding of the contemporary city. The planet is riven by struggles in defence and promotion of the commons – be they in terms of resources and territory. Struggles such as the land occupations of the Movimento dos Trabalhadores Rurais Sem Terra [Movement of Landless Rural Workers, or MST] in Brazil, the Zapatista Autonomous Municipalities of Chiapas, Mexico, the South African Shack Dwellers movement, the Bangladesh Krishok (peasant) Federation (BKF), and the anti-fracking and anti-fossil fuel movements are indicative of attempts to defend and obtain social wealth and collectively organize social (re)production through antagonistic politics that directly challenge resource dispossessions of the poor (see Routledge 2011). Moreover, the wealth of temporary encampments and caravans that have characterized the anti-globalisation movement in places such as Gleneagles, Nice, Cancun, Durban, Adelaide and Edinburgh are all moments of experiments and building up expertise and capacity for urban commoning practices.

The interesting point to note is that many of the struggles are indeed rural or peasant-based. But, nevertheless, they have resonance for struggles in cities given that we are dealing with the globalization of rebellion that links the urban and the rural. Drawing on these diverse examples, it is important to see the common as a central demand/practice of translocal political networks, rather than as something which is necessarily bounded to specific places (Gilroy 2010). Therefore, the task of commoning is not just to (re)create locally controlled commons, especially for the most marginalized (although this is a crucial task), but also to mount a connected geopolitical challenge to move the present balance of power away from ever more powerful coalitions of multinational institutions and to strengthen a globally connected grassroots movement.

A commons approach allows us to see the potential for many rebellions in the city, both large and small. Not just rebellions in the city, but also against the city; struggles which are subversive and oppositional, transformative and prefigurative of possible, as yet unknown, urban worlds. We can see these struggles based around urban commoning through a range of examples, from more spectacular interventions such as urban gardening or subvertizing and adbusting, to more mundane everyday acts of kindness, social care and togetherness. Activists accept that commoning will be contradictory and will engage in practices and values that will sometimes feel embedded or trapped in capitalist ways of doing things, and at other times will be more creative or antagonistic. This is not to say that building commons are easy or free of internal power struggles. What is crucial is that they are prefigurative (they practice the future that they wish to see), open, experimental and have the potential to generate solidarities. Bringing the idea of the common into play in the city allows us to sharpen our analysis of the task at hand – the decoupling of life in the contemporary city from the logic of capital and capitalist work. Finding new ways to produce urban space can begin to form the bedrock of challenging capitalism as it is reproduced at the everyday level.

A number of different threads within contemporary urban studies draw on the idea of commoning. Borch and Kornberger (2015) present a useful review of many of these debates. First is the use of deeply engaged methodologies such as participatory action research and militant ethnography. These kinds of methodologies allow researchers to engage and work closely with urban social movements and promote their practices and aims, building solidarity and knowledge commons in the process (see Autonomous Geographies Collective 2010; Colectivo Situaciones 2007; Bresnihan and Byrne 2015). We encourage students and researchers who are interested in furthering the conceptual and practical application of the urban commons to explore the potential of highly engaged in radical methodologies in their work.

Second, there is also a significant body of work which takes the processes of enclosure, be they of knowledge or land resources, as a starting point of struggle to understand the emergence of commons. For example Hodkinson (2012) suggests that 'urban commoners' might contest what he calls the new urban enclosures by finding common cause around visions and practices to build a 'new urban commons'. Moreover, Bresnihan and Byrne (2015) outline creative practices of resistance against enclosure in the city of Dublin while Lee and Webster (2006) explore the enclosure of spaces of collective and public consumption and especially the emergence of gated communities in China. Drawing on process of neoliberal enclosure in Mexico, Jones and Ward (1998) explore how the ancient form of *ejido* common rights have been undermined by more recent neoliberal reforms of urban development.

Third, the political meaning and significance of the urban commons is a topic of some discussion in urban studies. Gidwani and Baviskar (2011) outline the need to protect the urban commons is a key democratic and ecosystem resource for communities in the future, while Amin (2008) points the key role of well resourced public spaces for nurturing democratic interaction and conviviality within an urban commons. Interestingly, Newman (2013), drawing on the example of how in Paris middle-class groups are becoming ever more vigilant citizens in order to manage and monitor the urban commons in a way that indeed might serve the needs of the middle-classes, and reduce diversity and conflict.

Fourth, there is an identifiable strand of work within urban studies on particular aspects of the urban commons. One significant thread relates to the potential of a housing commons. Under conditions of contemporary neoliberalism, struggles around housing have been regarded as a new frontier of enclosure and therefore ripe terrains for struggles to defend and enact new forms of commons (see Hodkinson 2011, 2012). Another set of recent work, especially by Tornaghi (2012, 2014), explores the role of urban agriculture in both defending against urban enclosure and promoting an urban commons. A further strand explores issues associated with struggles around climate change and extractive fossil fuel and industries (see Pusey *et al* 2012; Chatterton *et al* 2012; Russell 2014). Finally, there are also debates around independent and autonomous spaces as prototype commons within cities. Many of these spaces are emerging from radical anti-capitalist movements and embed commenting practices in self-managed spaces (see Pusey 2010; Chatterton 2010; Hodkinson and Chatterton 2006; Bresnihan and Byrne 2015).

Conclusion

What we have explored in this chapter is the diverse nature of the commons and communing, applying them to the city and urban studies and drawing on our own active engagement in two very different examples of commoning. We have demonstrated that the idea of the commons makes several important contributions to urban studies. First, the city is a key site of the commons, through the possibility of resistance and struggles and production of alternatives in urban life. Second, the rich patterns of social life in the contemporary city offer new potential for extending the commons and the production of alternatives. Third, the commons can create new political imaginaries of urban life based on the spread of rebellion and cooperation. Finally, the commons offers new potential for exploring urban struggles as a means of creating lasting alternatives to life under urban capitalism.

We strongly urge furthering the practical and conceptual use of the commons in urban studies. It could yield some exciting new research directions, novel avenues of enquiry for teaching, as well as opening up new lines of solidarity and engagement with those resisting the ongoing onslaught of urban neoliberalization. Given the ever-intensifying problems that cities face globally, ranging from fossil fuel dependency, biodiversity loss and climate change to democratic breakdowns, financial austerity and social unrest, we urgently need more intellectual and practical resources to build an urban commons to help resist enclosure and build a progressive and just urban future beyond capitalism.

Key reading

De Angelis, M. (2007) *The Beginning of History*, London: Pluto.

In this book De Angelis explores not only the processes of capitalist enclosure and commodification we have become increasingly accustomed to through the well-used capitalist mantra of 'the end of history', but more importantly the creation of commons and 'other values' which clash with capitalist value practices.

Hardt, M and Negri, A. (2009) *Commonwealth*, Harvard: Harvard University Press.

Hardt and Negri introduce the idea of the common as distinct from the commons and discuss its increasing importance in relation to what they term 'biopolitical production'.

Linebaugh, P. (2008) *The Magna Carta Manifesto: Liberties and Commons for All*. London: Verso.

In this book Linebaugh not only explores the history of the commons embedded within the development of capitalism and enclosure, but also introduces the ideas of 'commoning' to describe the active and collective doing of the commons, which radically transforms our understanding of the commons, not as static resources to be collectively governed, but socially produced and struggled over.

7

COMMUNITY

Deborah G. Martin

Introduction

Community captures the idea of social interaction, connection, and mutual reliance among a group of people. It may seem mostly a sociological concept – that is, 'community' identifies social ties and engagement. Yet in being social, it is necessarily also geographical, because communities occur in and across space, in particular ways that are important for understanding social life. Communities may express group identity based on a shared set of concerns or practices – the relationship of a factory manager to workers, where to send kids to run around, or how best to worship together – and thus expresses both social and geographical dimensions. 'Community' has been at the root of over a century of urban theorizing, from concerns expressed by philosophers such as Georg Simmel (1903) and Ferdinand Tönnies (1887) that city life created radically different forms of community from agrarian towns, to debates of the last decades about the possibilities for inclusive urban politics based on community (Young 1990; Green 1999; Joseph 2002). In this chapter, I trace some of the ways urban and political theorists have defined and imagined community for urban studies and urban life, pointing to its conflation with neighbourhood and the need to uncover the multiple socio-spatialities inherent in both terms.

Community as a foundational urban idea

In early urban theories such as those of the Chicago School of Sociology at the University of Chicago (the 'Chicago School'; see Case study 7.1) in the 1910s and 1920s, community was the core conceptual basis for understanding daily life in the city, the relations between social groups, and between these groups and economic processes (Burgess 1923; Park 1915; Park, Burgess and McKenzie 1925). This

CASE STUDY 7.1 THE CHICAGO SCHOOL OF SOCIOLOGY

The Chicago School established, in many ways, the scholarly discipline of urban studies. Based in the sociology department at the University of Chicago in the early twentieth century, scholars including Robert Park and Ernest Burgess sought to make sense of the incredible diversity and density of a Chicago teeming with immigrants from many different countries. In addition, Chicago was a major destination for the 'Great Migration', starting in the 1910s, of black Americans leaving the increasingly mechanized farms of the southern US for the factory jobs of the north and Midwest, in the manufacturing belt. In Chicago, this tremendous diversity of people came together in a city aggressively rebuilding after the 'Great Fire' of 1871. Chicago in the period from about 1870 to 1920 established itself as an industrial powerhouse, anchored by the meat-packing industry on the south side of the city, a gateway for working-class immigrants. In the poverty and chaos of these working-class neighbourhoods, Park and his associates developed a model for urban scholarship that focused in-depth research on the community and organizational networks that undergirded the immigrant experience in the city (Park 1915). A major aspect of their analysis was to try and understand how social mores were expressed and changed as cities grew and become more diverse. A major conceptual framework for the Chicago School was the interplay of mobility, as seen in the fact of migration itself, and the divisions of labour characterizing the industrial city. Chicago School sociologists expected to find regularities and social laws in their in-depth examination of the forces shaping vice, social organization, and mobility in the city.

work represented a departure of sorts from predominant sociological theorizing of the time, which argued that urban life was fundamentally different from rural and town life, in ways that both freed individuals from the constraining expectations of society, and forced new, socially alienating forms of organization on urban dwellers (Durkheim 1893 [1964]; Simmel 1903). The Chicago School sociologists argued that urban residents were very much interconnected in communities, and that there was an organizational logic to urban communities.

These scholars identified neighbourhoods, or residential districts, as the physical and morphological expression of urban communities. Community in their work was represented by social groups organized primarily around common countries of origin, language, and religion of the many recent immigrant groups to American cities, particularly their empirical focal point, Chicago. This work was based on observations of the explosive growth of the city in the period from the 1870s to the 1920s, a time of significant industrialization and immigration to the United States, including Chicago, which grew from 298,977 people in 1870 to over 2.7 million people in 1920 (United States Census 2014). The fact of common immigrant groups clustered together in certain locations of the city reinforced the idea that people organized themselves in cities based on their commonalities with one another, particularly in terms of religion, language, and cultural traditions.

The Chicago School sociologists argued that mutual support and common values were very much part of the dynamic of city life in immigrant neighbourhoods. Such arguments were extremely important for recognizing the underlying structure of these communities, despite the seeming chaos and distinct poverty and struggles within them. At the same time, however, these portrayals tended to represent urban communities as more internally coherent and homogeneous than they actually were. While certainly Chicago and other cities, then and now, have distinct geographical districts such as 'Little Italy' or 'Chinatown', which express culture through stores and restaurants, language, modes of work and recreation, Hirsch (1983) has argued that so-called distinct cultural areas of many cities have a significant amount of diversity within them, and may not even contain a majority of a group's own members. 'Little Italy' might have more non-Italian residents than those who claim direct or descendant heritage from Italy. Similarly, Italian immigrants, while visible in a district such as Little Italy, may also live in many other, less visibly 'Italian' neighbourhoods. The same can be said for immigrants of the 1880s to 1920s in the United States, and today in cities around the world. Indeed, Kay Anderson (1991) demonstrated that immigrant districts evolve and are shaped and defined through dynamic interplay of a designated immigrant group within the dominant society, especially local elites such as politicians. As she demonstrates, the persistence and symbolic landscape-marketing of districts like a Chinatown or a Little Italy are part community support and immigrant identity expression, and part strategic identity production for broader urban economic processes such as tourism (see Figure 7.1).

FIGURE 7.1 Community identity in the landscape: Worcester, Massachusetts, USA (Source: Alexander Scherr).

Although ideas of a cohesive community identity in cities is a powerful one and remains influential today, the sources of community identity are diverse, and come from external processes and attitudes as well as internal relationships.

Underlying meanings

Scholars have debated just what community connotes and denotes, and how that connects to urban life and politics. Iris Marion Young's (1990) critique of community drew out of her concerns about justice and difference; she cited ideologies of 'community' as problematic, arguing that people working to build communities always exclude as they seek to define and support those in a given community. For Young, city life is a locus of social difference, and thus a site for concerns about social interaction, distribution, and what Nancy Fraser (1997) and others call 'recognition' of communities of all sorts. Young argued that the emphasis on community, because of its reliance on boundaries and exclusions in order to establish cohesion, ends up undermining the ability of people to recognize, support, and engage one another. Further, the members of a community are supposed to recognize and understand one another. Yet in any community, internal differences abide: women activists for equal rights in the 1960s, 70s, and 80s in the United States, for example, were divided by class, race, and age, differences which wrenched the movement's cohesion and spurred questions (such as Young's) about the viability of 'community' as an organizing concept (Moraga and Anzaldua 1981; Frankenberg 1993; Young 1990). Young's solution to the problems of community relied upon a notion of 'unassimilated otherness' in cities, offering a celebratory approach to difference that advocated for diversity rather than seeking the social cohesion or boundary-making exclusions of 'community'.

Young's undeniably hopeful attitude to urban life sought to disconnect the ideal of community from notions of urban democracy, but ideas of community in cities continue to be strong in popular imaginaries. Indeed, the continuing significance of community development, and the emergence of community policing in the 1990s, point to the ongoing popularity of notions of 'community' in urban policies. Community policing offers an interesting example of such policies. It describes a strategy of urban policing that engages residents of cities in dialogue and mutual information-sharing and problem-solving with police forces. Steve Herbert's (2006) account of such policing in Seattle examines the underlying values and meanings of 'community'. Herbert notes that meanings of community include the classic sociological idea of deep communal ties that are threatened by industrial and post-industrial urban relations (as in Tönnies 1887); social relations that evolve from participatory engagement in local governance; or, finally, what Herbert (2006) terms a 'thin' community based on sporadic connections among people sharing temporary goals. He argues that community policing relies upon various aspects of all three of these views of urban community, but cannot hold or maintain them. Indeed, he argues that urban community itself is partly produced by these ideals, including through state-based programs such as community policing itself. In other words, government itself generates community through assumptions that the community can and does exist in ways that will enable citizen-police interactions that will foster better public safety. After meticulous ethnographic research with police and residents of neighbourhoods in part of Seattle, Washington, Herbert concludes that community is an 'unbearably light' concept for policy.

Fundamentally, Herbert's analysis highlights a critical tension in urban analyses or uses of the term, 'community.' Of the residents interviewed for his case study of community policing in Seattle, Herbert notes (2006: 33): '[T]here is an invariant realization that the reality of propinquity breeds a desire for a lasting and predictable connection.' In other words, the fact of living near others fosters hopes for some sort of social relationship among neighbours. Community policing tries to enact these social relations, or build upon ones that are presumed to exist. These social relations of common location and hoped-for connection are most often conceptualized by urban scholars as 'community', yet it might be better understood as 'neighbourhood'. The key operative term in Herbert's observation about the desire for connection may actually be 'propinquity', that is, by virtue of proximity, people could, or should, or want to have connections. The word 'community', with its connotations of deep social relations and mutual support, may inadequately characterize much in urban social relations. An alternative for conceptualizing urban community may be to compare it with the term 'neighbourhood'.

Geographical community: neighbourhood

'Community' may articulate a multiplicity of socio-spatial dynamics, only some of which occur in urban residential districts, or neighbourhoods. The term 'neighbourhood' highlights propinquity as the primary dimension of urban social relations. Interactions and connections among neighbours can take a wide variety of forms, from regular and sustained social interactions and mutual support over time (Cox 1982), to little more than occasional waves, to open hostility and suspicion. Indeed, these myriad relations can exist among the same people over time! Urban community that is based on physical propinquity is best described by the term *neighbourhood*.

In research in a few different cities in the United States over the last several years, I have found that residential community is motivated by and through two major elements; the opportunity to participate in local governance and physical improvements to shared spaces (Martin 2003b), and threats to residents' perceptions of their neighbourhood character (Martin 2003a; 2013). In the first case, that of participation in local policy, the meaning and significance of urban community is very much 'generated' – as Herbert (2006) noted with community policing – by the fact of opportunities for input in local government decisions. In some US cities, local participation is an important component in planning input for everything from recycling programs to park renovation design, economic development, and land-use zoning (Sandercock 1998; Martin 2003b; Fainstein 2010; Carr 2012). These participatory planning programs offer a means for urban residents to produce the sociality and engagement that can foster a sense of local community (see Case study 7.2). Indeed such participatory initiatives are based on

CASE STUDY 7.2 PARTICIPATORY PLANNING

Participatory planning offers a mode of decision-making in which elected and appointed government officials formally consult with the people affected by a given land-use or policy decision, be they nearby residents, property or business owners (Arnstein 1969; Carr 2012). Planning for much of the first half of the twentieth century was based on the idea of rational, expert planners, who could make the best judgment based on their technical knowledge (Sandercock 1998). Starting in the 1960s and 1970s, this top-down approach evolved to recognize the importance of planning as a decision-making process, which was laden with politics rather than merely technical decisions (Davidoff 1965; Sandercock 1998). Participatory planning has, implicitly if not explicitly, the idea of 'community' at its core as the group that participates. In particular, planning based on a geographical area relies upon input of people who live or work in that physical space (as in Martin 2003b). As many scholars have noted, however, the ability to act in ways that foster community and real input in governance is very much reliant upon resources that many residents may lack, such as the time to attend meetings. Furthermore, the idea behind 'participation' in decision-making often assumes a kind of sustained engagement with other community members, and with city officials, which aims to downplay and diffuse conflict in the name of 'consensus' (Sandercock 1998; Fainstein 2010; Carr 2012). Urban neighbourhood communities that are most able to engage with local officials to achieve their goals are not necessarily those with the most needs (Heynen 2003; Carr 2012).

an underlying expectation that urban community exists in the arena of residential life. Thus a paradox is created: local public input programs seek to take advantage of a presumed neighbourhood community, and in doing so these programs foster activities – community meetings, formal input structures – by which community is produced, made, or, as Herbert suggests, 'generated' in order for residents to interact in ways which give them connections to each other, and to policymakers.

In neighbourhoods I have studied that were seeking more input in local policymaking, core themes that were articulated in community meetings or through media accounts were those emphasizing the neighbourhood's *character*, focusing on landscape elements like single family homes, parks or community-tended places, and the family orientation of households in the area (Martin 2003b). These images and descriptions of the neighbourhood can form the basis of shared 'place-frames' (Martin 2003b) that help residents to define what they like, share, and wish to maintain or enhance in their neighbourhood. Such place-framing can help to establish community by defining common interests and values in place, moving from the fact of propinquity to values or goals that are known to be shared among a group of people who are involved in community organizing or formal input opportunities with local government agencies and officials.

In areas where residents have objected to a new land use – in one case, expansion of a local hospital; in another, social service housing – the discourses about the neighbourhoods were strikingly similar to areas where people were consciously organizing to address common goals or problems with local officials. The themes about the neighbourhood were also mostly about family character, housing types, maintaining property values, sidewalks and the social interactions fostered by them. In an interview with a resident of a neighbourhood from several years ago, where residents were concerned about the expansion of a local hospital into their neighbourhood, the resident described the area as follows:

> We have neighborhood [social] things now, in the summer or at the holidays, and literally, 40 or 50 kids are there. … And they're out, they're playing in yards. So there's a lot of, it's just family at this point. It's very little rental property. It's almost all families. A lot of them have kids, or are child-bearing years. People stop and talk, people sit on the porch. It's just got a very nice feel to it.

Similarly, in a totally different case and city, a resident gave a description of a neighbourhood where some residents were organizing against a land-use change that would bring social service agency clients into the area as residents:

> The neighborhood has a lot of older homes… It has a lot of older multi family homes. … By and large it is residential … and there are some rentals. The vast majority of rentals are homes… For example my neighbor, she rents out the 2nd floor and lives on the bottom. It is built as a two family home… A lot of them are built that way. It is also a very tightly packed neighborhood. It is also school centric and religious.

Regardless of the particular reasons for organizing within the community, the descriptions of the neighbourhoods in these accounts have several common elements, emphasizing the residential land uses and physical spaces, such as yards or porches, which enable social interactions (Martin 2003a; 2013). While these portrayals emphasize physical features and the social aspects of families, another, less positive tone of some community discourses in urban neighbourhoods involves the articulation of potential new uses and residents – 'others'- who might threaten the cohesion and character of the neighbourhood. For example, in the case of institutional expansion in a neighbourhood, a resident questioned, 'Why does everything have to do, you know, bleed back into this neighborhood, of attractive houses? It could be nicer if there wasn't this looming threat' (see Figure 7.2). In this excerpt, it is clearly the physical change in the neighbourhood that affects its character. The threat is to the infrastructure, which could, as the resident detailed in other aspects of the interview, affect the form of social intercourse, such as the

FIGURE 7. 2   Traffic is one concern for community activists protesting land-use changes: Worcester, Massachusetts, USA (Source: Alexander Scherr).

informal gatherings and kids playing in yards. Other types of community activism in relation to land-use change are much more explicit in defining boundaries that relate to social categories, as was quoted in a media story about opposition to social services in a neighbourhood in my research: '[Opposed residents] said the home for 15 girls will exacerbate traffic and parking problems, lower their property values and pose a threat to neighborhood safety' (Williamson 2005).

The concern about people in the neighbourhood who might somehow threaten its character, its economic (and social) value, and the ways of life are the heavy burdens of community that worried Young (1990) (see Case study 7.3). Any community defines insiders and outsiders. For neighbourhoods, the community identity that may be expressed and shared by some will rarely be shared by all. Thus, outsiders are part of the same place, but may not be within the community. By the same token, communities that are not tied to a particular geography will also produce boundaries of in and out, and these boundaries will change over time. Fundamentally, community may not be synonymous with any particular place or locality, but it will, none the less, have spatiality, because of the geographies of its members, and the specific individual and social spaces they use and engage with as they interact with one another.

CASE STUDY 7.3 THE 'COMMUNITY' OF FERGUSON, USA, IN 2014

A contemporary illustration of the burdens, mischaracterizations, and dilemmas of 'community' can be found in the painful case of Ferguson, Missouri, USA, in 2014. On 9 August 2014, Michael Brown was fatally shot by a police officer after a confrontation in a street in Ferguson. After the shooting, some of the details of which remain contested to this day, protests broke out and riots occurred for several days in Ferguson. The core issues in the protests were the use of deadly force, and racialized perceptions between police and black residents in Ferguson, and yet, also, in the USA as a whole. The fact that the white police officer, Darren Wilson, apparently believed he was in danger in the confrontation with the teenage (and black) Mr Brown, and used deadly force by shooting Mr Brown, highlights the vast perceptual gulf between whites and blacks in the USA, particularly between police officers and black Americans. Yet, unpacking these relations is complex, and loaded with notions of community; media portrayals sometimes refer to 'the Black community' or 'the Police' without acknowledgement of the complex and internally contradictory dynamics embedded in these categories. There is no singular 'black' community in the USA, even as there is a widespread understanding that black people in the USA encounter police differently than white people do. The very fact that black teenagers – from Trayvon Martin in Florida in 2012 to Michael Brown in Missouri in 2014 – can be shot by self-appointed neighbourhood watch members, or by police with little or no legal repercussions, points to a deeply racialized understanding of gender and race that produces fear and uncertainty, and – when combined with guns – deadly outcomes. While American blacks share a fundamentally suspicious relationship with police, neither blacks nor police are singular and fixed identities, and indeed the two identities can merge in one body (see Chapter 5, this volume). So too in Ferguson, a nominal 'community', where the local ('community') police force donned riot gear in order to respond to its own residents' frustrations. The idea of a 'community' – whether it be the municipality of Ferguson, the police force, or a set of residents defined by racialized language of skin colour – is hopelessly inadequate for describing, or confronting, any of the multiple identities, understandings, and material relations at play.

Conclusion

Community has a long history in urban scholarship. It has been used to identify distinctive racial–ethnic commercial and residential districts of cities, and has fostered particular forms of governance through participatory planning. Community as social interaction, engagement, and shared values among a group is itself an ongoing production through a variety of processes: mobility of people, topography of locations, socio-economic and cultural differences among people, political processes of citizenship and state socio-spatial organizing and policymaking. Thus community cannot be taken-for-granted as a social fact, but needs investigation as socio-spatial outcomes of multiple processes. In most urban scholarship, community is conflated with the local, through the concept of neighbourhood. Through this conflation, both community and neighbourhood symbolize one another, such that neighbourhoods – really no more than residential districts – carry a burden of social connection, and community is presumed to develop primarily from proximity. Despite the challenges inherent in using these words as descriptive or analytic categories in urban studies, both are none the less useful: for the imaginaries they enact and motivate, and the ideals of urban social intercourse they connote. An enduring form of urban analysis will thus continue to be examination and investigation of the multiple meanings, processes, and ideologies that underlie urban interactions and policies.

Key reading

Herbert, S. (2006) *Citizens, Cops, and Power: Recognizing the Limits of Community*, Chicago, IL: University of Chicago Press.

In this book, the author theoretically and empirically explores notions of community. He offers a review and analysis of classic texts and theories about community, and then deploys these understandings in a case study of community policing practices in Seattle. An excellent illustration of urban ethnographic methods.

Martin, D, G. (2003) 'Enacting neighborhood', *Urban Geography*, 24(5): 361–385.

This article posits that the term 'neighbourhood' derives meaning primarily through practice, or social relations. It extends notions of community to geography by materially grounding the concept in 'place' via the concept of 'neighbourhood', and offers a post-structuralist emphasis on discourse, with attention to practice.

Park, R. E., Burgess, E. W. and McKenzie, R. D. (1925) *The City*, Chicago: University of Chicago Press.

Classic articulation of the 'Chicago School of Sociology' that forms the basis of urban theories of a social division of labour in cities – which corresponds to specific land uses and racial/ethnic districts. In this theory, urban growth occurs via in-migration of immigrants, and subsequent assimilation and outward mobility of residential locations towards the urban fringe. This work established 'urban ecology', which likened social groups (defined by race/ethnicity) to competitive plant species. The influence (and uncomfortable notions of race for today's reader) of Darwinian approaches to social understanding is evident.

Tönnies, F. (1887) *Community and Society (Gemeinschaft and Gesellschaft)*, London: Routledge and Kegan Paul.

One of the first sociological explorations of the impacts of industrialization and associated urbanization on social relations and social organization in Western Europe in the late nineteenth century.

8

COMPARISON

Jennifer Robinson

Introduction

Urban studies generates theory through engaging with a diversity of urban experiences and outcomes. Whether considering the world of 'cities' as specific urban settlements, or the range of urbanization processes which shape these outcomes and which often stretch way beyond any specific urban territory, urban theory must build its understandings across diverse and often highly divergent outcomes. The conventional method for approaching this challenge is comparison. This chapter will set out the benefits of comparative thinking for urban studies, drawing on some classic examples of such research. It will then rehearse the limits of such traditional methods for producing knowledge about cities when the goal is to enable any city to contribute to theorization – thus, when the goal is a more global urban studies. The chapter will encourage readers to question the foundational assumptions of comparison, and using case studies will offer some pointers towards a reformulated comparative practice, where scholars can relish destabilizing existing concepts as they stretch across different cases, and where the openings for generating new concepts can emerge in any city.

Comparative urbanism might be thought of as 'thinking cities through elsewhere', and has been accomplished through reflecting critically on wider theoretical analyses from specific locations; or reading widely the work of scholars writing about other cities to learn from a wider array of urban experiences; or through composing original research on two, several or many cities within the same research project. By exploring what comparativists in urban studies have actually done by way of comparisons, we are able to assess the pragmatic methodological conventions of working with differences amongst cities which have shaped the field – as opposed to the formal, quasi-scientific methodological guidelines which can significantly limit the contribution of comparative methods to the global field of urban studies (Robinson

2011). We can also trace how the processes of urbanization have themselves inspired new comparative methods – tracing the multiplicity of connections amongst cities has led to the development of relational approaches to comparison, for example, which work with these links to provoke insights.

What do we need comparison for?

Demands for urban studies to adopt a more global approach to understanding the urban are arriving from a number of different directions. Many dynamic urban processes cut across the kinds of divisions that framed the analysis of twentieth century cities. Developmental differences now not only distinguish wealthier and poorer cities from one another along lines of inequality, poverty and the services available to enable urban life, but they also tie cities together through shared ambitions to succeed in certain arenas of global visibility, or to ensure collaboration in addressing urban problems, as well as through the multiplicity of circuits of travel, trade, investment and imagination which shape new developments and remake urban inheritances. There can be no *a priori* analytical separation between cities on the basis of measures such as GDP or income, for example.

Similarly, while regional differences in growth and opportunities certainly persist – not least in the differential insertion of cities in Africa and Asia into the global economy through the early years of the twenty-first century – the interconnections and flows amongst cities across the world make these divisions at times unhelpful to understanding emerging and historical urban conditions. The extraordinary rise of Chinese influence in cities in Africa, for example (Ndjio 2009), and the circulation of speculative developments on urban peripheries across the African continent driven by Asian (and Russian) capital investments (de Boeck 2011), mean that cities in these two continents are significantly inter-referenced (Roy and Ong 2011) and mutually influencing each other. Urban theory needs to respond to and build on this connectedness in developing analyses and concepts.

National repertoires do remain important for understanding processes of urbanization: the challenges of developmental needs in growing cities, cultural and political attitudes to urban life and the deeply contested local politics of urban economic development and globalization. But scholars in many different cities around the world have for some time now been drawn to think through a diverse array of wider circulating processes to understand the cities they are working in, such as transnational labour migrations, speculative financing, urban design, suburban or ex-urban developments, urban redevelopment, neoliberal policies.

This explosive exteriority (extraversion) of the contemporary urban condition can be aligned with a broader post-colonial critique of the foundational influence of wealthier and western cities in urban theory to press towards a more global scholarly practice (Robinson 2006; Roy 2009; Parnell *et al* 2009; Simone 2011). The expectation of such a critique is that the experiences of all cities around the world are considered relevant to understanding the nature of the urban. The call for a more global urban studies is based on the simple but generative assumption that all cities are cities; thus the formulation of 'ordinary cities', which insists that all cities properly belong to the realm of theoretical reflection on the nature of the urban (Robinson 2006). The term 'ordinary' is of course not to detract from what might at times seem like the extraordinary and impressive nature of some aspects of cities (Taylor 2013) or to argue against their differentiation in respect of many features. Rather, the ordinary cities manoeuvre seeks to carve an analytical pathway across the divisive categorizations that plagued twentieth century urban theory (developmental, regional, continental, analytical – see Robinson 2006), and to insist that there is no privileged starting point for constructing theorizations of the urban. This has established grounds for new practices of theoretical reflection appropriate to building understandings of the urban across the great diversity of urban outcomes around the world.

Such opening towards a more global urban theoretical practice has been immensely productive, with many initiatives to recast urban studies, for example, from the perspective of cities 'beyond the west' (Edensor and Jayne 2012) and from secondary cities (Chen and Kanna 2012); by revisiting western-centric concepts, such as gentrification (Lees et al 2015) or city-regions (Vogel et al 2010) in a more global register; by asserting the value of a 'Southern' perspective' (Watson 2009) or an Africa-wide contribution (Myers 2011); and not least by proposing a revitalized comparativism as a basis for generating new theoretical insights across a diversity of urban contexts (Nijman 2007; Ward 2010; McFarlane 2010; Robinson 2006, 2011).

Comparative analysis in urban studies has conventionally played only a limited role in bringing very different (as opposed to fairly similar) cities into analytical conversations and formal quasi-scientific comparative methods can at times be an active hindrance to such wider theoretical conversations (Robinson 2011). However, the core elements of a comparative imagination also hold out much promise for inspiring new kinds of tactics and practices suitable for opening theoretical engagements to a wide array of cities. This can be expressed most generally as 'thinking cities through elsewhere' (Robinson 2016): building comparisons through putting *case studies* into wider conversations; *composing* bespoke comparisons across diverse outcomes or repeated instances; *tracing* connections amongst cities to inform understandings of different outcomes or to compare the wider interconnections and extended urbanization processes themselves; and *launching* distinctive analyses from specific urban contexts or regions into wider theoretical conversations.

Urban comparative vernacular: thinking with difference across shared features

The comparative gesture

Urban studies has an easy familiarity with a comparative imagination – frequently writers invoke a comparative gesture to quite different contexts in order to isolate historical processes which seem most relevant to their analysis, or to signify the wide range of possibilities associated with a particular urban process. Bagnasco and Le Galès (2000: 5), for example, in their collaborative comparative analysis of European cities, reach out to urban India and rural Japan to set the extremes of the density of settlements, such an important feature of their understanding of the distinctive urban qualities of the cities they are exploring.

With the growing insistence on the need for a more global urban imagination, the comparative repertoires of urbanists are expanding, and the geography of such comparative gestures will inevitably become richer as scholars and practitioners read more widely on urban circumstances around the world and are able to draw in examples to support or set limits to their analysis from a much wider array of places. Thus, perhaps the most transformative agenda for global urban studies would be to create new reading practices (Jazeel 2012) and establish new guidelines for what constitutes the relevant literature to consult in developing new insights, as well as establishing higher expectations for urban theory to engage with research from a diversity of urban contexts.

Composing comparisons

Light gestures of comparison, unsupported by careful assessment of empirical evidence, have their dangers, though, as Loïc Wacquant's (2008) important comparative analysis of 'advanced marginality' in US and French cities highlights most effectively. Popular and scientific discourse in European cities developed a comparative analysis of the ghettoization ('Americanization') of European cities through the 1980s and 1990s, seen to be reinforcing stigmatization of poorer areas and hardening the racialization of political discourse. Wacquant composes an ethnographic and sociological comparison of advanced marginality in Chicago and Paris to establish the quite different functional location, spatial form and social relations associated with the French *banlieues* (suburban housing estates), which are based in the first instance on class location (and so are heterogeneous ethnically and in terms of nationality), and still drawn into both social and political relations with the wider city and welfare state. This is in stark contrast to the US hyperghettos, which have developed as a result of conscious state abandonment, foundational racial discrimination and which are deeply segregated from the wider city, both socially and spatially.

The carefully composed comparison across these two superficially similar cases allows Wacquant to identify some of the shared trends actively producing what he calls 'advanced marginality' in many post-industrial economies: disconnection of global economic cycles from low-paid employment; reconstruction of the wage relation through the growth of precarious, contract and less-protected jobs; spatial stigmatization; loss of support networks and decline of quality of living environments; and the absence of a political identity associated with the class decomposition, splintering and fragmentation of the 'precariat' (Wacquant 2008: 244). However, in order to oppose the authoritarian policing and work-fare responses which the 'ghettoization' thesis supports as policy responses, Wacquant energetically uses his nuanced comparative analysis to emphasize the range of political choices which states have: to invest in neighbourhoods, to cultivate citizenship, and to draw residents of poorer neighbourhoods into social, economic and welfare relations – as opposed to seeking a draconian penal solution to marginality as in the USA.

Wacquant's study opens up for us some of the methodological tactics that can make variation-finding comparative research conceptually productive. He is very insistent that sustained '*comparative* ethnography, based on parallel fieldwork in two sites chosen to throw light upon theoretically relevant invariants and variations' remains a vital research tactic – as opposed to what he sees as the 'currently fashionable "multi-sited fieldwork" (Wacquant 2008: 10) more committed to tracing connections across different places (discussed in the following section). 'Composing' comparisons across more than one city remains, in his view, an essential component of explaining the specificity of urban trajectories – in his study, the historical processes responsible for specific experiences of spatial concentration and exclusion in different contexts. But for him, this kind of comparison is also an important way to generate wider conceptualizations that might be useful in other contexts. Here he is clear this is not about providing 'lively illustrations of theories elaborated outside sustained contact with the prosaic reality', but about 'enrolling ethnographic observation as a necessary instrument and moment of theoretical construction' (Wacquant 2008: 9–10).

At the same time Wacquant himself makes wide-ranging use of concepts and insights from other urban contexts, both advanced economies and places where the production, for example, of an 'excess reserve army of labour' through mass unemployment is far advanced (following Fernando Henrique Cardoso and Enzi Faletto writing on South America). South American writers, such as Janice Perlman, are also well placed to provide insights on the spatial relegation of the poor to marginal neighbourhoods, and their spatial stigmatisation. However, Wacquant is clear that the specific arrangements of marginality in France, for example, are not hangovers of a colonial past, but an effect of the transformations of the *most advanced sectors* of Western economies, 'as they bear on the lower fractions of the recomposing working class and subordinate ethnic categories' (Wacquant 2008: 232). In drawing comparisons with South America, he excavates the future of contemporary Western capitalist economies not in any simplistic sense that these are directly following trends in other contexts, or importing colonial inheritances, but to illuminate similar processes at work across these contexts, especially in relation to the new and specific production of 'vertiginous inequalities' within the USA and European economies and the selection of growth paths that permanently exclude large sections of the population from opportunities for secure wage labour.

For Wacquant, as with many authors turning to comparative analysis, an important goal is the ambition to develop useful concepts to be able to describe and delineate the processes at work, and more broadly to contribute to theorization and wider policy debates. This is an essential feature of comparative practice, drawing writers from many different contexts to present their findings in relation to the limits or potential of existing theorizations, or to seek to build relevant concepts through exploring different cases.

Reinventing variation-finding comparisons

If the comparative gesture at the heart of urban studies is an inspiration towards a more free-ranging comparative imagination, methodological conventions which lead scholars to compose bespoke comparisons in order to develop wider conceptual insights (what Tilly (1984) calls 'variation-finding comparisons') have, in principle, established some relatively firm procedures and limits to thinking across different urban contexts. Holding certain broad variables constant, such as national GDP, national political forms, or city size, is seen as a way to limit variation and allow focus on specific explanatory processes, such as local level governance arrangements (Kantor and Savitch 2005). None the less, the huge variety of features potentially held in common or varying across the selected cities and the small-n nature of these studies means that these strategies in fact barely reduce variability in the system and do not address significant issues of endogeneity (where supposedly independent cases already have shared influences) caused by the significant interlinkages and co-causality within and across cities. These conventions of case selection which mean that only relatively similar cities end up being compared have existed partly because researchers build on unexamined assumptions shaping urban studies more generally – such as, that cities with different average levels of development are categorically different and not able to be compared at all (Robinson 2011). Conventionally, then, relatively stern limits have been set to which cases can be compared through the ways in which the grounds for comparison have been established. The politics of the 'third term', the variable which operates as the comparator, that is, the basis on which different cases are drawn together, is very much what is at stake in exploring the possibility for expanding comparative practice from analysing most (or more) similar contexts towards being able to put quite different urban contexts into analytical conversation (Jacobs 2012a).

However, if we look in detail at examples of comparison in urban studies it is not clear that it is these relatively restricting formal criteria for the selection of cases which have enabled conceptual generativity. Consequently, we could afford to take a much looser approach to identifying shared features across cases as the basis for comparison; this would support comparisons ranging more widely across different cities. Case study 8.1 presents the example of variation-finding comparison by Susan Clarke (1995) to illustrate this point. Her analysis of how restructuring affected eight US cities was carefully composed according to strict formal criteria for comparison – but her key insights emerged out of thinking with the variety of outcomes for variables which had not been subject to any methodological controls. Her study demonstrates well how thinking across several different urban outcomes can generate innovative conceptualizations (theorizations) which have the potential to be put to work elsewhere – Raewyn Connell (2009) defines 'theory' as the potential to speak beyond the individual case. The focus of conventional urban comparison is indeed on generative conceptual work – putting different contexts into a designed analytical arrangement to allow for conceptual reflection and theoretical innovation. In fact they seek to directly interrogate existing concepts across the different observed outcomes. This is essential work for a more global urban studies, if it is to be able to extend or subvert existing theorizations and to practically open up understandings of the urban to a greater diversity of urban outcomes.

CASE STUDY 8.1 VARIATION-FINDING COMPARISON IN PRACTICE (CLARKE 1995)

In an exemplary comparative study, Susan Clarke sets out to determine how 'restructuring' is articulated in local urban institutions in the US. In an effort to move beyond some of the limits of regime theory she uses 'new institutional approaches' to draw attention to the complex ways in which local political institutions respond to and are transformed by the changes brought by widespread fiscal and economic transformations, focussing on local economic development policy. Her conclusions indicate that whether more democratic or more market-centred approaches to local economic development result depends on the balance of power of local community and business organization, as well as whether (more or less integrated) local bureaucratic procedures are in a position to dominate decision-making rather than turn to more market-oriented logics with flexible, external institutions and funding mechanisms. Her institution-focussed analysis is alert to the many different influences on local economic development policy – interactions with federal programmes, engagement with wider circulating policy ideas, formal bureaucratic and democratic structures, the invention of new institutional and financing vehicles, or informal relations with business.

Clarke thoughtfully composes a comparison across eight smaller US cities, with two cities selected from four states, and tries to balance variation with holding important variables constant. Thus she includes both declining and growing manufacturing areas to be open to different kinds of responsiveness to economic change, but tries very hard to ensure limits in variability:

> First, each pair is located within the same state to minimize comparability problems arising from interstate differences in state authorized spending, service responsibilities, legal municipal requirements and state policies and regulations. Second, each pair was active in federal redevelopment programmes and suffered comparable cuts in federal aid during the 1980s. Third, in varying ways, each pair of cities has previously incorporated minority and neighbourhood interests in federal redevelopment projects.
>
> (Clarke 1995: 520)

The comparative approach allowed a subtle analysis of how institutional diversity shapes outcomes in local economic development policy and also brought significant insights into how particular institutional arrangements might shape the emergence of specific urban regimes, and lead to their transformation. This diversity of institutional response, though, which was the main conceptual achievement of the work, did not play a part in the research design: the criteria for selection established the shared relevant features which enabled the identification and analysis of this variety – i.e. all cities had experienced participation and then cuts in federal programmes; all had some history of community participation in these programmes. But importantly, variation in the variables that emerged as the key features explaining different outcomes (bureaucracy, participation) were distributed randomly across the study, did not form a basis for case selection, and emerged only in the course of the analysis.

Thus in practice, the criterion for case selection which matters – the comparator, the third term, or the grounds for comparing the cases – is the presence of the shared features which the author wishes to explore, namely, a changing federal funding context; participation in governance; a response to cuts. Within the bounds of these shared features there is significant variation along many axes (economic growth, social divisions and inequality, social mobilization, state-level funding and legislation, local institutional cultures) across the different cities, some of which are relevant and interesting to consider in relation to how they help to delineate different instances of (responses to) the shared phenomenon. Following this model, stretching comparisons beyond the national boundary or across cities that might otherwise seem very different, based on shared transnational processes or other experiences, is feasible.

Globalisation and cities: towards relational comparisons

Cities are not discrete territorially bounded entities. Thus as urban studies has engaged with the ways in which urbanization is a process with a complex, intertwined geography of both connections and territorializations (McCann and Ward 2010) – or as Brenner and Schmid (2014) style it, characterized by both extended and concentrated urbanization – new and relatively inclusive lines of analytical connection across different urban experiences have been opened up by following the spatial form of the urban. The variation-finding comparative imagination has been complemented by emerging comparative practices, which work with the multitude of interconnections that draw different urban contexts into the same empirical and analytical field.

One important consequence of this approach has been the expansion of the range of cities brought into comparative juxtaposition. Global and world cities analyses played an important role in opening up the possibility for thinking across different kinds of cities because of their participation in shared processes of globalization, and indeed provided strong grounds for placing cities from different regions and (to some extent) different levels of economic development together in the same analytical category (Friedmann and Wolff 1982; Sassen 1994; Taylor 2004). The comparative imagination has been definitively stretched by these empirical and analytical developments.

The focus on the globalizing networks which shape urban outcomes has also opened up the possibility of deterritorializing the comparative imagination, to include comparing the wider networks themselves. Thus Kris Olds' path-breaking (2001) study explored Vancouver and Shanghai together through analysing the different networks which were drawn on in the 'megaprojects' of 1 Canada Water and Pudong Island. The comparative tactic here was novel – to compare the different networks of a family firm of Hong Kong-based property developers investing in Vancouver and drawing on and forging close ties to generate trust and embedding localized commitments, and of a group of architects (he focuses on Richard Rogers) invited to contribute to a design exercise for Shanghai's mega-project developments, whose lack of engagement with local issues saw them produce proposals with little purchase on local histories and imaginations. The two cities are treated quite equally, and both are placed within the category of 'global city', caught up in the same design and investment circuits.

More generally, a critical post-colonialism (King 1990; Robinson 2006; Roy 2009; McFarlane 2010) has established the potential for cities everywhere to be drawn into wider theoretical conversations. As theoretical statements about urbanization are exposed for their locatedness, the expectation is that urban studies would be informed by a great diversity of experiences, articulated by 'new subjects of urban theory' (Roy 2011a) and supported by attending to the diverse processes of globalization that shape differentiated outcomes in cities around the world (Brenner et al 2009; Ong 2011; Simone 2011). Urban studies has moved far beyond the relatively functionalist approach placing cities in relation to the totality of the world economy pioneered in early analyses of world cities. Notably, accounts of neoliberalization of urban governance and policymaking (Brenner et al 2009) and wider analyses of policy mobilities (McCann 2011) highlight the multiplicity of links that tie cities around the world into shared processes of imagination and intervention. Similarly, the expansion of the kinds of global processes shaping the production of different cities – including, for example finance, architecture, migration, international agencies, property developers, urban forms (high-rise, urban mega-projects, satellite cities) – means that the grounds for bringing cities together on the basis of the shared processes (with of course differentiated outcomes) have proliferated.

A new form of 'relational' comparative method has therefore been proposed, by amongst others Gill Hart (2008) and Kevin Ward (2010), in which the relational nature of urban processes – notably their production through relationships with processes, places and actors elsewhere – informs comparative exploration. The methodological practices, or experiments in comparison, which have been developed in the wake of these and other initiatives to renew urban comparison (McFarlane 2010; Robinson 2011; 2016) provide a robust foundation for developing more-inclusive urban analyses.

Tactics for a more global urban studies: experimental comparisons

The analytical stage has been set for the proliferation of experimental comparative tactics. Thus moving from the variation-finding assumption that shared experiences across relatively similar cities would form the proper basis for analysis, comparative urbanism today faces the challenge of working across a greater diversity of cities, which share and contribute to a multiplicity of interconnecting processes shaping urban outcomes. And following the planetary urbanization hypothesis, these outcomes will not only be territorially concentrated in urban regions but the operational landscapes and influences of the urban are potentially stretched out across the planet, posing strong challenges to the territorial imagination of urban studies' vernacular comparative practice (Brenner and Schmid 2014).

In response to these empirical and theoretical developments, how might a reformatted comparative practice proceed? Any comparative analysis for understanding the global diversity of the twenty-first century city will need to be able to trace and work with the multiplicity of connections which exist amongst cities around the world, bringing many different urban contexts into closer conceptual proximity. With post-colonial critiques of the field in mind, such a reinvigorated comparative practice will also need to address the demand that theoretical insights from cities beyond the west – in fact, cities anywhere – be understood as possible starting points for distinctive theoretical conversations. New methodological initiatives therefore seek to stretch comparisons to support opening theory to insights from any city, and indeed seek to use a comparative imagination to stretch existing urban theories to their breaking point.

Beginning, then, with the specific interconnected processes creating different urban outcomes, Jane Jacobs' (2006) analysis of the different fates of residential high rise in London and Singapore (see Case study 8.2) provides some important pointers towards a new comparative imagination. Rather than starting with pre-defined territorialized cases, in which particular outcomes are imagined as resulting from wider processes (such as neoliberalization, or restructuring) hitting the ground differently, which tends to preserve intact these wider explanatory features, comparative practices could engage with urban outcomes through tracing their *genesis* – their production by means of specific connections, influences, actions, compositions, alliances, experiences, across the full array of possible elements of urban life: material–social–lived–imaginative–institutional.

CASE STUDY 8.2 REPEATED INSTANCES (JACOBS 2006)

Jane Jacob's (2006) discussion of the serial production of the modernist residential high-rise building across the globe has made an important contribution to reimagining comparison. Here she focuses on the distinctive achievement of each almost-the-same repetition through globalizing circulations and specific assembling of diverse elements to produce, differently, each building. Across the world, then, the achievement of a global sense of urban modernity, visible in the repetitive architecture of international modernism, emerges from the relatively unpredictable multiplicity of circulations and manifold elements able to be assembled into each construction – buildings which are both repeated and yet produced as original objects, with an equally original yet partly repeated and interconnected set of meanings crafted locally, differently each time (King 2004): 'the making of repetition – or more precisely, repeated instances in many different contexts - requires variance, different assemblages of allies in different settings' (Jacobs 2006: 22). Her examples evoke the strikingly different histories and fates of residential high rise: in London, poor construction and low valuation of residential high-rise as a solution to working-class housing-need led to the alarming collapse of a prominent apartment block; in Singapore, early resistance to high rise living and the social difficulties of this (including suicides and harm to pedestrians from falling objects) was overcome and the vast majority of the population now live in well-maintained and relatively successful high rise apartments.

For Jacobs, together the many different instances of modernist residential high-rise *produce* the global *effect* of international modernism: each instance is a singularity (a specific and distinctive outcome), emergent from an array of interconnected practices, ideas and relationships, and not an example of an already given global process coming to ground to shape local phenomena (Jacobs 2012b). Her interventions direct us to reconsider how we might approach comparison through tracing the shared genesis of phenomena, focussing on how they are produced, or assembled, rather than presuming that certain pre-given wider processes are at work shaping particular outcomes, whose interpretation is then already determined by the pre-given explanatory framework. And she also invites us to explore how strong grounds for comparison might be found in the 'repeated instance', which is such a significant feature of the urban landscape – from architectural design to gated communities; from circulating brand images to low income housing finances; from strategic visioning to market stalls; and numerous other examples.

Forging comparative insights across a wider range of urban contexts can be grounded by following the tracks of globalizing processes, thereby expanding the directionality and reference points for theorizing – this is the manoeuvre offered by relational comparison, for example (Hart 2008; Ward 2010). A shared connection – for Hart (2008) this was the interlinked process of regional industrialization in Taiwan and South Africa as Taiwanese investors took up incentives for rural industrialization offered by the apartheid government – reveals differences across cases. In her example of the shared process what became apparent was how land redistribution and dispossession shaped urbanization and conditions of labour in both cases, thus strengthening insight and analysis of processes in each context. Similarly Theodore (2007) considers how personal migration histories and consequent links with labour and revolutionary movements in South America have informed the mobilization of informal workers in US cities, shaping urban outcomes there. And Jacobs (2006) deftly draws attention to the ways in which repeated instances (in her case the residential high-rise) are produced from highly specific assemblages of practices, ideas and materials, to generate a global effect rather than being the outcome of a 'process' of globalization.

Constructing comparators through tracing connections can draw on the great diversity of the globalized linked processes influencing many cities. This produces the potential for a much more decentred theorization, bypassing cities conventionally placed at the core of urban theoretical developments through too-strong a focus on only selected elements of globalization, such as advanced producer services in global cities analyses, or the transnationalization of formal economic activities. Arcs of connection can be traced to bring quite different urban contexts into analytical conversation. In Simone's *City Life: From Dakar to Jakarta*, for example, African circuits of trade, migration and investment linking different cities there with Asia ground a series of ethnographic engagements, drawing urban experiences across these two contexts into wider reflections of urbanization. The result is an analysis of urban economic and political dynamics which foregrounds transient, ephemeral and emergent processes. In another influential example, the collection of essays in *Worlding Cities* (Roy and Ong 2011) trace strong circuits of inter-referencing amongst Asian cities, effectively ignoring the dated reference points of Western urbanisms, producing an understanding of new centres of innovations in urbanism and also signposting new centres of urban theoretical production. The referencing becomes, if anything, inverted as increasingly the styles of urbanism which articulate the idea of being a global city in the west (and elsewhere) become citations of these emerging Asian urban forms; the tracks of design and investment which shape urban developments in many regions of the world are already closely intertwined with Asian-centric circuits.

In a further important step, some researchers have demonstrated that understanding contemporary urbanization requires that the wider connections shaping urban outcomes and the extended processes of urbanization are themselves subject to comparative interrogation. Here Peck and Theodore's (2012: 24) discussion of the extended case method explores how this generates a 'methodological problematic' which:

> is not simply one of accounting for transactions or transfers, but encompasses the origination and reproduction of multisite policy networks or fields,

which as 'transnational policy communities' may become social worlds in their own right.

The methodological field can be opened up to incorporate the value of considering the differential global connections of cities – as in Söderström's (2014) comparison of the different relationality of Ougadougou (in Burkina Faso, Africa) and Hanoi (Vietnam, Southeast Asia), reconnecting to the wider world after a period of internal development, the one through more political and developmental connections shaping cities across Africa; the other through the readily available economic connections mobilized in the Asian context. Not only, then, is the potential for comparative methods strengthened and stretched across a greater diversity of urban contexts by building comparators through shared transnational experiences, but the great diversity of these shared processes and the variability of such processes across space and time makes it important to build understanding across different kinds of transnational connections and mobilities.

Generative conceptual insights can also be supported through inventing analytical proximities across different contexts, connected or not, substantially reinvigorating the grounds for comparison proposed by formal variation finding methods and practice. And here, again, the geographies of comparison are being both expanded and decentred. Oren Yiftachel (2009a and 2009b) proposes a comparative analytical grounding tracing an arc through a 'South-East' urbanism (bringing Southern urban experiences into conversation with the Middle East), supporting conversations about informality ('grey spaces') and working generatively with the comparison across Palestine-Israel and South African cities, as the shared political technologies and spatial architecture of racialized rule bears reflection across the two cases (Yiftachel 2000). Bringing insights from one context to another can also be diversified across a wider variety of cities. Nijman's (2007) interesting exercise in 'multiple individualizing comparisons' with four other cities – to explain 'different pieces of the puzzle' of Miami – places the contemporary experience of Miami in conversation with a number of different cities, from different periods in their present and past, to enhance understanding of Miami as a refuge (Amsterdam), transcultural capital (Hong Kong), capital of vice (Shanghai), and ambitious world city (Dublin).

Also innovative in terms of the direction of comparative learning is Roy's (2003) paper, which brings her research on squatter movements in Kolkata to bear on homeless peoples' organizations in California. Here, the differential form of substantive citizenship claims highlights how in the USA a restricted form of 'propertied citizenship' limits the range of claims homeless people can make on the state, compared to the full demands for redress which feature in Indian squatters' movements. In this way, perhaps 'unexpected comparisons' (Myers 2011) might emerge, drawing cities into comparative reflection in surprising and perhaps unpredictable ways. New lines of decentring comparison are also emerging, as scholars draw insights across a range of different contexts – as Shatkin (2014) sets out to carve an analytic across Chinese and Indian urban experiences, or Waley

(2012) proposes regionally contextual comparisons between Japanese and Chinese cities. The generativity of such comparisons lies as much in the potential of thinking across some measure of shared histories and contemporary interconnections in regional globalizing economies as it does in the invitations which diverse urban outcomes present to scholars to think again, to draw insights by following new routes for thinking, and along the way decentring the western experiences which have so long weighed heavily on urban scholarship in different contexts.

Given the great diversity of transnational processes shaping cities and the multiplicity of urban outcomes to consider around the world, the potential for insights from comparative reflection is significant. Expanding and proliferating the potential for conversations across different urban contexts supports the ambition to sharpen conceptual insights, generate new analyses of processes, or contribute to conceptual innovation. Insights and concepts, then, might emerge in any urban context, and be launched into wider conversations. For example, the extraordinarily productive engagement with the concepts of 'informality' (see Chapter 1, this volume) to catch the emergent, ephemeral processes shaping many cities (Simone 2001; AlSayyad and Roy 2004) has been demonstrated to have the potential to bring stronger insights to studies of wealthier urban contexts, such as the limits of governing all large complex cities (Le Galès 1998) or the centrality of informal networks to financial and other economic clusters (Storper and Venables 2004).

Conclusion

Reformatting comparative tactics for twenty-first century cities needs to work with a complex understanding of the spatialities of urbanization. Emerging imaginations of the urban call for comparative tactics, which are inspired by the ways in which the urban can come to be known as it is assembled transnationally, across a multiplicity of specific but strongly interrelated outcomes. In this chapter I have explored the opportunities for *building comparisons across shared features*, which traditional comparisons have relied on to generate conceptual insights. I explored relational and experimental *comparisons that work with interconnections to trace the genesis of urban outcomes*, focusing on transnational urbanization processes. I have also stressed the post-colonial expectation to *theorize or launch concepts from any urban context*. Together these support the possibility for proliferating insights across a world of cities, based on the extraordinary array of interconnected processes generating urban outcomes around the globe.

The 'actually existing' work of urban scholars has demonstrated both the great value and the changing nature of the comparative imagination and comparative methods for building analyses of cities in a world of cities. Further consideration of comparative methods requires attending to all aspects of the methodological and ontological imagination subtending this work (for a broader philosophical discussion see Robinson 2016). In this regard, here I have considered what, for example, constitutes a case – both territories and connections, places we might recognize as cities as well as those forming part of extended urbanization; and what are the grounds for comparability – interconnected processes and shared features both offer potential for surprising insights and abstract conceptual development. Further comparative experiments and methodological innovations are awaited, as are the analytical payoffs in which understandings of the urban are potentially open to insights from all cities.

Key reading

Jacobs, J. (2006) 'A geography of big things', *Cultural Geographies,* 13: 1–27.

 Jane Jacob's article demonstrates how to build comparative analyses across interconnected repeated instances, in this case the residential high rise. She treats each case, Singapore and London, as distinctive, but uses the different outcomes in each to illuminate the other. Her analysis is also closely influenced by Bruno Latour, and this focuses on the ways in which assembling each development relies on creating alliances of actors and materials (also see Chapter 4, this volume).

Kantor, P. and Savitch, H.V. (2005) 'How to study comparative urban development politics: a research note', *International Journal of Urban and Regional Research,* 29: 135–151.

 This article from two of the most prominent comparative urbanists outlines some of the potentials and pitfalls of comparative methods. It is an excellent resource for anyone embarking on new comparative research, and treats many of the conundrums of designing a comparative project. While a little too quickly dismissive of including 'Third World' cities in comparisons with 'more developed' cities, their closing section encourages more research to explore that possibility.

Simone, A. (2011) *City Life from Jakarta to Dakar: Movements at the Crossroads,* London: Routledge.

 This is an accessibly written text with many lively examples, which demonstrates how urban theory can be decentred. It builds analyses of urban life through case studies of urbanization processes and urban life in Asia and Africa, and through the various links between these two contexts, as well as exploring some links through European situations. It gives a good insight into the thinking of this highly original urban scholar. AbdouMaliq Simone demonstrates how careful ethnographic studies in distinctive urban contexts can contribute to original theoretical analyses that can be productively used in different situations.

Söderström, O. (2014) *Cities in Relations: Trajectories of Urban Development in Hanoi and Ougadougou,* Oxford: Wiley-Blackwell.

 This is an excellent example of how to do comparisons that take seriously both the distinctive territorial politics of cities, as well as their extended connections to other places. Most innovative here is that Ola Söderström, a Swiss scholar, compares the different kinds of connections woven out of two cities emerging from relative isolation. This book should stimulate methodological innovation for those setting out to do comparative research.

Ward, K. (2010) 'Towards a relational comparative approach to the study of cities', *Progress in Human Geography* 34: 471–487.

 Kevin Ward's article explores a number of the key issues that have inspired urbanists to revisit comparative methods in new ways. He offers some thoughtful suggestions for different ways to think about urban comparison. The paper is also an accessible and very useful introduction to relevant literature to date.

9

CONSUMPTION

Steve Miles

Consumption lies at the very heart of the contemporary city. As Pimlott (2007: 9) puts it, the city appears to have become an object of the ideology of consumer capitalism, 'a distended ... scene of consumption'. The significance of consumption in the urban realm lies not solely with its symbolic value, but also in how it is predicated upon new forms of material organisation and the ways in which we as citizens of contemporary society move through the urban landscape in light of those new forms (Kärrholm 2012). But in many ways urban theory's engagement with consumption in the city lies at a crossroads.

There is a rich tradition of studying consumption in an urban context. During the latter eighteenth and early nineteenth century Veblen (1899) was concerned with the conspicuous display of discretionary economic power and how it was that urban residents used consumption to provoke the envy of others. Similarly, Simmel (1950) identified how certain social structures prescribe relationships in the metropolis: city life as the product of a world of over-stimulating consumption. Benjamin (1999), meanwhile, described the flâneur, the urban explorer whose

---

CASE STUDY 9.1 DEFINING CONSUMPTION

At its most basic level consumption is about the 'the selection, purchase, use, maintenance and repair and disposal of any product and service' (Campbell 1995: 102). However contemporary scholars of consumption are more interested in the ways in which symbolic and material resources are mediated through the markets and the implications this has for the relationship between the individual and society, and indeed, for the individual's experience of city life (Slater 1997).

CASE STUDY 9.2 BENJAMIN AND THE *FLÂNEUR*

One of Walter Benjamin's key concerns were the arcades of nineteenth-century Paris and it's here that he locates the shift to Modernity through the commodification of things. The experience of Modernity is thus represented in Benjamin's work through the figure of the *flâneur* who is, effectively, the personification of the alienated modern city. Benjamin's work conveys the historical manifestations of consumer culture while bringing to mind the shopping malls of today.

---

demise came with the 'democratic' excesses of consumerism. More recently, it has been fashionable to condemn the consumer city as a neo-liberal aberration (see Chapter 18, this volume): as a demonstration of everything that has gone wrong with a society determined by its relationship to the market. Arcades, department stores, shopping malls, and the public spaces of the city are by their very definition spaces of consumption and as such the city constitutes the very realm within which the individual's relationship with modernity and/or postmodernity is played out on an everyday basis. In this chapter then, I consider some of the key characteristics of the city of consumption, 'good' and 'bad', and what they mean for how social scientists might conceive of the city of today, as well as that of the future.

The post-industrial city

The contemporary city is the product of a process of industrial decline and of reinvention. It is the product of a particular set of historical circumstances, in terms of both the decline of the industrial model and the apparent failure of Modernity to recreate the world in a human guise. In this chapter I want therefore to consider the suggestion that the contemporary city has forfeited its commitment to progress in favour of a less stable, less optimistic apparently temporal existence. It's important of course, as Jayne (2005) points out, not to over-simplify by drawing an essentialized picture of the consumer city as nothing more than the culmination of the relationship between modern production and 'post-modern' consumer lifestyles – as expressed through questions of the social life of production, social administration, the role of leisure and the aestheticization of lifestyles and the like. Indeed, such an approach would be anathema to many geographical notions of place. By alluding to such a dichotomy the unique and complex characteristics of particular cities are, furthermore, liable to be lost in the crush. But having said that there are, none the less, certain key dimensions of social change that have had a profound impact on city life and there is a case for arguing, I would suggest, that consumption is the primary driver of these.

The city that has emerged from the process of economic decline and uncertainty appears to be, in the broadest of terms, permanently scathed and disoriented. With the old certainties of industrial production long gone, a gaping socio-economic hole was left that desperately needed filling. All cities have had to address this problem. They have done so in a number of ways and in doing so they have apparently sought to assert their unique identities. But the point is that the tools they can call upon to do so are necessarily and conversely constraining. And furthermore, as Sharon Zukin (1998) points out, the more contemporary cities have sought to distinguish themselves from one another the more they have in fact ended up looking and feeling the same.

The re-envisioning of the city as a site of consumption can be understood as part of a broader process: a new urban consensus around what Arantes *et al* (2000) have called the 'city of single thought' that emerged in the aftermath of de-industrialisation. The consensus concerned was one that coalesced around entrepreneurialism (see Chapter 11, this volume). This created a situation in which cities were obliged to compete in order to secure a larger slice of the entrepreneurial cake. Of course, some commentators, such as Jones and Evans (2008) and Krugman (1996) have questioned the extent to which a territorial entity can itself be competitive, but what is certain is that in the shadow of globalization there has emerged a knowledge-based economy so that in effect the city has become, or indeed has had to become, the spatial manifestation of global change: an incubator of innovation. Thus cities came to compete for a place on the economic high table, or if not, to fight for some of the scraps left behind. The city became the new unit of business: it came to be determined by an aspirational notion of entrepreneurialism so that cities, regardless of the protests of geographers referred to above, have indeed effectively become reified into active agents rather than mere 'things' (Harvey 1989a; Miles 2010). The problem here, and something worth thinking about when reflecting on the impact of the consumer city, is this is in fact a very dangerous process, as it may lead those responsible for strategizing cities to forget that the city is constituted by real people with real needs, desires and motivations.

Place branding and the commodification of the city

One example of the way in which the contemporary city has effectively become both commodified and homogenised is that of culture-led regeneration which was perceived, at least in some somewhat optimistic quarters towards the end of the twentieth century, to represent a genuine answer to the travails of the post-industrial city. A particular kind of discourse thus emerged around the creative city. Authors such as Landry (2006) and Florida (2002) offered up creativity and, more broadly, culture as a prescription: as a means of reinventing the city in a post-industrial guise. But in a sense this process wasn't about establishing the creative city at all. It was about maximizing the potential of the city as a site for consumption.

The broader context in which the above process took place was that of place branding. The creative city or at least the aspiration that the creative city implied was born out of a propensity to see the city as a marketable commodity. Place marketing is a reductionist practice that almost entirely neglects the everyday realities of the place it depicts. The rationale for place branding lies in its role in helping a city to construct a competitive identity. It not only serves the purpose of communicating key messages about a place, but it also stimulates and executes a range of creative ideas that add to our perception of that place. This reflects a process in which symbols and representations have become essential components of how we perceive the city. For Anholt (2003) this is all about equating a place with a degree of 'rootedness' or distinction that appeals to the consumer. The clear danger here then is that place marketers have a tendency to impose a universal vocabulary depleting a city of the meanings that made it what it was in the first place. As such, it has been argued that the creation of 'city myths' is part of a broader process of commodification in which idealised images are relentlessly reproduced until they become more real than the reality itself (Goodwin 1997). For Florida (2002), a city thrives primarily because creative people want to live there. A city succeeds if it can provide an offer of amenities and experiences that allow these creative people to validate their identities. In this sense the 'consumer city' masquerades as the 'creative city'.

The role of consumption in the city is therefore at least in part to do with a process of de-territorialization. Urry (1998) sees this as a reflection of a broader historical process in which societies have effectively been 'hollowed out' insofar as global flows and communications are so essential to how we relate to the city that our experience of time and space has been fundamentally altered. In effect, the city becomes a transitory space. As individuals we are not determined by the spaces we inhabit, rather we construct a biography or a narrative according to how it is we choose to interact with that space. The point here is that the transitory nature of this space is arguably maintained through our relationship to consumption. If we accept Crewe's (2001: 280) definition of consumption as being more about an 'ongoing process' than it is a 'momentary act of purchase' we can begin to perceive of consumption as a key intervening factor in how human beings engage with the city. Consumption is effectively the stage upon which the city is constructed. The city has arguably become a space for the instrumental purchase of commodities.

It almost goes without saying that every city is different, that no two cities are ever exactly the same. However, the main result of the intervention of consumption into the city has arguably been to make such a statement less authoritative than it would have been in the past. Many commentators have described a process in which the cities of the world have become increasingly homogenised. In order to demonstrate the nature of this process I want to consider the work of two particular contributors to the debate, Pimlott (2007) and de Châtel and Hunt (2003).

Pimlott's work on the USA is geographically intriguing as far as he posits that a consumer society is characterised by a process in which places and specifically, the interior of places come to resemble each other regardless of their actual location. From this point of view the city of consumption has not emerged naturally, but is the product of an instrumental imposition of a programme or indeed, of an ideology. The formula for the construction of such places is thus apparently deliberately reiterated so that people are funnelled towards the exchange of consumption which emerges in its own right, as a kind of adhesive logic. The consumer city, fantasised and spectacular as it is, becomes naturalized despite its essential artificiality. Another way of putting this is that the contemporary city becomes a home for a series of dreamworlds, 'and the continuing role of fantasy in the construction of the urbanised public's psyche' (Pimlott 2007: 300). One of the implications of this process and of the dominance of consumption in how we have come to perceive the city is that the public of the city has from this point of view effectively been transmuted from the street to the interior: the city becomes defined by the captive experience that the consumer society constructs for us:

> the abundance of … signs, as shop-fronts, advertisements, kiosks and discrete publicity-entertainment environments, creates a highly artificial atmosphere of spectacle and interiority, wherein consumption is 'naturalised', unavoidable and inevitable: the consumer is obliged to be part of the environment's *life*… Interiority is reinforced, and artificiality prevails.
>
> *(Pimlott 2007: 296)*

For Pimlott the problem here is that the implied emphasis of the above on private ownership, designed to achieve maximum financial returns, creates a situation in which the achievement or imposition of a system 'supersedes the making of a *place*' (2007: 275). You could therefore argue that from this perspective residents of (and indeed visitors to) the city become nothing more than a mass of potential consumers. Their relationship to the city is effectively defined by their status as consumers.

The above argument is taken yet further in the work of de Châtel and Hunt (2003: 4) who argue that retail principles have become an extension of the way we live. To put that another way, 'retailisation is partly about its implicit requirement to stealth bomb our acquisitive genes, about the way it seeks immersion in our lives, wherever we are, to create the *ability and the desire* for us to buy things'. For de Châtel and Hunt then retailization is about the colonization of space – urban and otherwise – a colonization in which our relationship to the world around us is determined by our ability to buy what it is that the world has on offer. Retail has thus become one of the few means by which we experience public life in the city. As Kärrholm (2012: 7) puts it: 'Shopping has outgrown the role of being an important urban function and become a necessary condition for urbanity itself… The shopping mall wants to become a city, the city wants to become a shopping mall.' In contemplating this dilemma Kärrholm points out that on the one hand consumption could be seen to be controlling and manipulative in the way it has taken hold of the city spaces, while in other ways it is something we actively engage in as citizens of a consumer society. The end result of this process, however, is a set of spaces that simply package the act of consumption: a bland, predictable set of spaces that force retailers, in particular, to reimagine the mundane activity of buying products into some kind of an experience: a memorable experience that gives the product being experienced 'added value'.

Authenticity

A key element lying at the heart of this debate then is the notion of authenticity. Is the city that is produced by the consumer society an authentic one? Zukin (2009) thinks not, arguing that the end-product of this process is a 'hegemonic global urbanism': a socio-spatial regime that raises local property values and submerges local character in the process. This is further reflected in the work of King (2004), who sees the production of architecture as being the primary material and visual realm by which competing symbolic and spatial orders are played out in the city. King considers that global urbanism has prioritized an altered version of universality: a process of homogenization through commodity consumption, though he ultimately argues that despite the power of such processes they are never immune from the vagaries of individual agency.

The role of architecture in reproducing the contemporary city is also considered in the work of Williams (2004) who discusses the way in which the city has increasingly been constructed in the form of a visual tableau to be touristically consumed. Williams (2004: 229–230) argues that the contemporary city is not so much materialised, as staged. For Williams, urban theatre is the cornerstone of a city designed to attract investment and tourism: 'the architecture of the recent past in England has been preoccupied with making stages, on which the user or inhabitant becomes a performer, whose actions can be viewed as theatre'. Williams thus describes the process by which the city has become a theatrical spectacle for consumption by tourists. From this point of view the city is effectively and self-consciously presented to the consumer through the lens that that performance provides. But the problem here as Williams (2004) goes on to point out is that this focus on the spectacular constitutes a limited and limiting reading of the city. What is more it highlights the fact that the contemporary city is constructed around the needs of a privileged audience.

Williams' argument then, is that as a result of the above process the city has been theatricalized rather than having been built up organically. What has become important is not so much what lies beneath the surface of the lived city, but how that city looks; as such, regeneration represents something of a panic driven by the misconception that design solutions can resolve decades of urban problems. The contemporary city is aestheticized. It has become a series of stage sets that only fully fledged consumers can fully experience. The consumer city is, thus, a neurotic city (Williams 2004).

One of the key problems here is that the public discourse associated with notions of gentrification and regeneration claims to be all-encompassing (Zukin 2009). It implies that the benefits to be had from a consumer society are universal, but such an approach neglects the culturally specific ways in which such processes are translated and negotiated. A further concern is that the understanding of the city that such a position promotes tends in turn to see the city as being accessed by the individual by right: a city is a site of consumption insofar as it provides a set of resources which the individual citizen can consume. In this context, Zukin refers to the work of Muschamp (1995) who argues that a city defined by business no longer 'recognizes the difference between creating and consuming'. Indeed, the consuming city can perhaps best be understood as a city of images and tastes:

> It is clear that media images and consumer tastes grease the wheels of global urbanism, anchoring the power of both capital and the state in the spaces of our individual desires, persuading us that consuming the authentic city has everything to do with aesthetics and nothing to do with power.
>
> *(Zukin 2009: 551)*

Zukin (2009: 552) describes a situation in which the aesthetic appeal of a city defined by consumption imparts a pacifying effect so that any notion of authenticity is reduced to the experience of consumption, rather than that of the construction of a just or diverse city, and as such 'we must politicize the meaning of authenticity to include the right to put down roots, a moral right to live and work in space, not just to consume it'. But the issue here is that the contemporary city does not actually need to be authentic, its task is to *invoke* authenticity and it does so through spaces that frame authenticity as something to be consumed, in the form of shopping opportunities, waterfronts, museums, galleries and themed environments (Prentice 2001).

This apparent obligation to consume is further illustrated in the work of Pendlebury *et al* (2009) who point out that there is an immense pressure to present heritage locations in ways deemed appropriate by the tourism industry. That, in other words, heritage sites and specifically designated World Heritage Sites have to present a commodified version of themselves, but in doing so raise question marks over the authenticity of what it is they offer. One version of events then is that the consumer city is essentially inauthentic. But a contrary position is that retail, in particular, has always been a key means of integrating city life. From this point of view there is no doubt that consumption has deeply affected the way in which the public domain is developed, and perhaps it is doing so in ever more profound ways. But ultimately this profundity is expressed through a more 'effectively controlled and functionally homogenous' city (Kärrholm 2012: 134).

The insecurities of the consumer city

It is one thing to acknowledge the nature of consumption's impact on the city, but it is also important to comprehend the theoretical significance of this process. The geographer Robert David Sack (1992) has done this particularly effectively in arguing that the world in which we live is characterized by the complex interplay that exists between homogeneity and variety. That, in other words, in deciphering the impact of consumption upon the contemporary city it is important to recognize the fact that while on the one hand, at the global level, diversity appears to have diminished (after all, the more we consume it, the less diverse the city effectively becomes), at another, individual level, the variety that the individual experiences is increased. What Sack is describing here is a world, like I suggested above, that is less secure than it was in the past: a world in which from the perspective of the consumer anything is possible, but in a thoroughly disorienting way. The city is effectively a setting for the production and display of commodities – and thus becomes as much about our imagined relationship with place than it does any kind of a stable reality in which that place exists. In short, the solidity of place has been undermined by a set of circumstances in which the market has delivered on what appears to be a post-industrial obligation to offer the city up to be consumed. Augé (1995: 103) puts this another way when he argues that in 'supermodernity' the individual consciousness is subjected to an entirely new kind of consciousness associated with the proliferation of what he calls 'non-places': the way in which the anonymity of places frees up or liberates the individual who is 'subjected to a gentle form of possession'. This allows him or her the passive joys of identity-loss or as Augé (1995: 105) suggests, a sense of 'an unending history in the present'. This is place defined economically, in which the utopian needs of the organic society are long forgotten.

Critiquing consumption and the city

It's important to remember, above all perhaps, that the relationship between consumption and city is a contradictory one. The consumer city gives and it takes. But this degree of complexity is not always fully explored in the literature. The reason for this is that consumption and the free market world that it implies are so easily objected to on moral and political grounds. There has indeed been a tradition of left-wing approaches to the contemporary city that have condemned it on the basis of its neo-liberal intent and consequences.

Authors such as Goldberger (1996) and Hayward (2004) have pointed out that the virtues traditionally associated with the public realm have effectively been corrupted and sanitized, and that a simulacrum of civic public space is the result (see also Swyngedouw 1989) of a city so tied up to the requirements of the market. Other authors have lamented the rise of the 'private city', and draw attention to the potential social problems caused by the so-called neoliberal city (Goldberger 1996: 27). Thus, Brenner and Theodore (2002) refer to the 'creative destruction' inherent in the geographically uneven spatial changes that have occurred as a result of neoliberal economic change and the belief that unfettered competition represents the ideal mechanism for economic development, when in reality it constitutes a dramatic intensification of coercive forms of state intervention. They go on to highlight the extent to which neoliberalism creates new forms of urban inequality according to whether or not an individual or social group fits the eligibility criteria of the consumer society as part of the imposition of a neoliberalized urban authoritarianism. As such, the argument here is that the main aim of neoliberal urban policy is to mobilise the city as an arena for market-centred growth and for elite forms of consumption (Brenner and Theodore 2002). Indeed, the way in which such processes have served to create an unequal city has become a particular preoccupation for urban scholars (Hubbard 2004: 666) who are particularly interested in the reinvention of city centres as 'corporate landscapes of leisure' from which marginal groups are displaced.

There is indeed a lot about urban consumption that *should* make us feel uncomfortable. Such a city is inevitably divided. But the condemnation of the neoliberal city comes at a price. There is a perennial danger that by condemning such a city on political grounds we cut ourselves off from understanding what it actually means to live in such a place and operate according to, and at times in opposition to, its norms. Perhaps herein lies the key theoretical challenge for those social scientists seeking to understand the relationship between the city and consumption. The picture I have drawn above is one of a critique that questions the ability of a city, determined by consumption, to provide economic and social benefits to a broad constituency. These approaches have coalesced around the dangers implied by a public life dictated to us through our experience as citizens of a consumer society. This notion of consumption as a source of citizenship has been developed elsewhere (Garcia Canclini 2001). But such a prospect remains anathema to urban scholars who have tended to rally against the city of consumption as a symbol of a market-based model of social and economic change that has apparently favoured the gains of some at the expense of the considerable losses of the many.

Research into the role of consumption could be said to have reached something of an impasse. The time has come to rethink the hegemonic impact of consumption as a driver of the competitive city ands to seek to understand it from within. In other words, there is a profound need to understand why consumers engage so enthusiastically with the apparently inauthentic pleasures that a city driven by consumption appears to offer. Only when scholars can fully understand what it means to consume and how it feels, and only when sufficiently pragmatic methodologies are developed to critically investigate consumption as an entity that can both give and takes, will genuine progress be made.

Conclusion

Consumption is ideological and the city is the primary space within which the implications of this ideology are played out. There is no immediate sign that consumption will come to have any less of a role to play in how we engage with the city. Notions of the genuinely sustainable city, at least as far as consumption is concerned, remain on the fringes of our everyday experience of the city. The city is from this point of view, essentially unsustainable. It is in this context that studies of consumption continue to enrich urban theory. They do so by providing a counter-balance to more productivist approaches and by highlighting the fact that the way we experience the city as consumers tells us something very significant about how the relationship between structure and agency is framed within the space that the city provides.

Key reading

Harvey, D. (1989) 'From managerialism to entrepreneurialism: the transformation of urban governance in late capitalism', *Geografiska Annaler. Series B, Human Geography*, 71 (1): 3–17.

This seminal piece established how it is that the image of prosperity, regardless of the reality that lies beneath the surface, must be seen to win out. Harvey is thus concerned with how it is that the contemporary city is stylized in such a way as to elongate its economic value.

Jayne, M. (2006) *Cities and Consumption*, London: Routledge.

Jayne is concerned with how cities are moulded by consumption and vice versa. His was one of the first and most accessible efforts to bring debates around how consumption frames the urban experience together into one volume, and is particularly useful in helping us to understand the relationship between consumption, the city and notions of identity.

Williams (2004) *The Anxious City: English Urbanism in the Late Twentieth Century*, London: Routledge.

William's book talks about the aestheticization of regeneration: the process by which cities are regenerated in such a way as to revitalize the lifestyles of the consuming classes while making no fundamental difference to the urban communities that existed beforehand.

10

ENCOUNTER

Helen F. Wilson

Introduction

It is a Wednesday afternoon in Istanbul's Grand Bazaar. Crowds of people negotiate their way through the narrow walkways of the city's main market – dubbed one of its best attractions. Residents, tourists, shoppers and traders are forced into tight spaces and intimate proximity. Market traders call to passers-by. Prices are haggled and information is exchanged in multiple languages against a backdrop of low-level chatter. Along with their economic significance, it is this form of sociality and casual exchange that has made the urban marketplace a significant site of encounter and academic inquiry. In recognizing how they bring strangers together, along with differences of class, religion, ethnicity and culture, they have become commonplace sites for the study of urban diversity. Whilst the engagement facilitated between people can be as little as a quick glance or as much as a conversation (Watson 2009: 1581), such spaces are thought to have the capacity to shape how we view those around us and thus to shape our behaviour (Wood and Landry 2008). The key question for urban scholars is how.

Markets are not just about encounters with people, but about encounters with objects, ideas and economies. In Istanbul's Grand Bazaar, carpets are neatly stacked, shelves are loaded with Turkish ceramics and walls are layered with hanging fabrics. Bejewelled lamps vie for attention, while shoes are laid out for people to rifle through. Spices, dried fruit and sweets are piled high, and handbags, Turkish gold and beads are lined up in rows from floor to ceiling. These materials are encountered through touch, smell, taste and sound. They are also shaped by market regulations, formal and informal economies, gender politics and urban imaginaries, to name just a few factors that we might consider.

Attending to all of these elements and asking questions about how they become significant in moments of contact is just one of the challenges that theories of

FIGURE 10.1 Istanbul's Grand Bazaar (Source: Helen. F. Wilson).

encounter have tried to address. With this in mind, this chapter outlines how social contact has been theorized in relation to the organization and experience of public life and space. It begins with early writing on the city as a site for encountering strangers and then outlines how urban scholars have called for a better understanding of the links between encounter and social relations. With these foundations laid, the remainder of the chapter details how work on encounter contributes to theorizations of urban diversity, urban politics and urban space, before finishing with another ordinary, yet important, site of urban encounter to bring these debates and reflections together.

Cities, encounters and the organization of public life

Henri Lefebvre (1996) described the urban as a 'field of encounters'. Indeed, for many scholars encounters are what produce the city and its distinct character as a site of possibility, continuous flux and disequilibrium (see also Merrifield 2013; Stevens 2007). Cities are diverse, incomplete, congested and disorderly (Amin 2008) and are routinely described as sites of 'throwntogetherness' (Massey 2005), where different, previously unrelated objects and people are flung together (Amin

CASE STUDY 10.1 'THROWNTOGETHERNESS'

Doreen Massey's (2005) concept of 'throwntogetherness' is regularly cited by scholars interested in urban encounter. Taken from her work *For Space*, the concept is used to describe how space and place are formed by the 'coming together' of different, previously unrelated things. For Massey, places are not fixed but are rather 'events' that are produced by the temporary and happenstance constellations of objects, people, histories, ideas and many more things beside. According to Massey, 'throwntogetherness', is what makes place special, for it means that place is never static, but is rather temporary and unpredictable. This unpredictability is what interests urban scholars, as it confronts us with the 'unavoidable challenge' of urban life, which is the constant need to negotiate 'multiplicity' (ibid: 141). This, Massey argues, is an important responsibility.

Demant and Landolt (2014: 175) use the notion of 'throwntogetherness' in their work on youth drinking in public spaces in Zurich, Switzerland. On a Friday or Saturday evening, Katzenplatz Square, located in a middle-class neighbourhood of the city, is a space where you can usually encounter teenagers who have congregated to meet with others, to buy drinks, change public transport or 'frolic'. The development of this drinking space for young people is as a result of the 'throwntogetherness' of disparate factors such as proximity to home, availability of alcohol, informal social networks and regulation' (ibid: 174). Those congregating are usually young men, from a range of socio-economic, national and educational backgrounds, for which the square provides a relatively informal and private space to drink, with little chance of being seen by parents. At the same time, the square is also considered a site of interruption as it is a site where different 'expectations, uses, and 'interpretations' (ibid: 176) of the square collide. This leads to sometimes-fraught encounters with police, social workers, local business owners and residents. As such, this space of encounter cannot be understood without attending to the constellations that make it possible.

and Thrift 2002). On this basis, encounters become a point of interest for scholars who want to understand what happens in these unexpected, fleeting or casual moments of 'throwntogetherness' and how they might produce and shape the city (Darling and Wilson 2016).

There is a long tradition of writing that has focused on the opportunities for encounter that the city provides. For example, early writing by prominent sociologist Georg Simmel focused on the rapid expansion of modern cities at the turn of the twentieth century, to document its effect on urban life. As people flooded into cities and industrialization changed the rhythms of everyday life, the likelihood of encountering unknown others grew, leading Simmel to claim that 'the stranger' – a figure that was simultaneously proximate and distant – defined the metropolis and changed how people related to others. He also documented the development of a 'blasé attitude' in urban populations and speculated that the general approach of urban dwellers towards one another was 'one of reserve' (Simmel 1903: 15). According to Simmel, this reserve exhibited a 'slight aversion' towards others that had the capacity to break out into conflict at any given moment.

In contrast, for other prominent scholars the increased opportunity for encountering strangers or 'unknown others' was something to be celebrated and for the purpose of this chapter I want to highlight three such theorists. First, Jane Jacobs' influential text on *The Death and Life of Great American Cities* (1961) included a focus on the 'liveliness' of public space and the improvisations that urban dwellers made as they negotiated crowded sidewalks. Jacobs not only likened these negotiations to a form of urban 'ballet' but argued that the ability to engage in this form of trivial contact was essential to the development of trust, respect and the more formal organization of public life. This interest in trivial contact was developed further by Iris Marion Young (1990) who argued that the intermingling of strangers was not only a necessity but an ethical concern. For Young, accepting strangers was central to a wider politics of cultural recognition that resisted the tendency to suppress difference. In Young's terms, cities that fostered the intermingling of differences without driving for homogeneity were also cities that were more likely to foster a more democratic politics.

Finally, Richard Sennett's (1971) work has demonstrated a similar enthusiasm for the possibilities that cities provide for encountering others. Indeed, Sennett argued that a decline in opportunities for contact with unknown others would result in the death of public space and an increased desire for the insularity of private life. Whilst keen to retain the opportunities that cities provide for pleasurable, unexpected encounters with people we may never know intimately (see Sennett 1978), Sennett has advocated for more 'meaningful interaction' between strangers and has emphasized the important role that urban design can play in developing it. As an example, this could include the development of strategies for moving people through the city in such a way as to encourage gathering, stopping and mixing with others, perhaps though the strategic placement of benches or the widening or narrowing of streets.

Critiques of social contact

Taken together, these readings of urban contact have been central to the development of work on urban encounter and have been noted for their celebratory tone and investment in the possibilities that might arise from social encounter with unknown others. These optimistic readings have also been reflected in urban policy, which has placed emphasis on designing public places that are capable of bringing different people together. The clearest example of this is the UK's community cohesion agenda (Cantle 2005).

CASE STUDY 10.2 COMMUNITY COHESION

The Community Cohesion agenda in the UK emerged in response to riots that took place in multi-ethnic Bradford, Burnley and Oldham in 2001, involving white and Asian youths in areas characterized by economic deprivation (see Amin 2002). Segregation was identified as a key factor in fuelling anxieties and conflict. In response, Community Cohesion policies identified different types of contact to outline the forms most desirable for developing meaningful exchange. For example, 'social incidental' contact described unplanned interaction arising from everyday activities such as shopping, while 'social organizational' contact depicted interaction arising from planned social activities such as sports (Wood and Landry 2008: 110). Identifying different forms of contact in this way was meant to provide local authorities with a means of assessing what forms of contact should be promoted in their area and what 'zones of encounter' might enable it to flourish. However, these policies were criticized for their poor address of the shifting and complex geographies of intergroup relations, as well as the structural inequalities that shaped them.

Despite these celebratory accounts there have been numerous critiques, which have laid the grounds for the proliferation of work on urban encounter that has emerged over the last decade. Amin (2008) has emphasized just how unpredictable the dynamics of 'mingling' are, forwarding the argument that proximity alone is not a guarantor for transformation. In short, encounters can produce conflict and anxiety as much as they can alleviate it (see also Vertovec 2015). Valentine (2008) has contributed to these concerns by critiquing the assumption that encounters with others might positively change attitudes or reduce prejudice. She has argued that celebratory accounts of public space have failed to explain just exactly *how* positive change occurs as a result of contact. Drawing on Allport's (1954) 'contact hypothesis', Valentine has cautioned scholars not to be too quick to assume that positive or civil encounters in public space necessarily translate into a change in attitudes (see Case study 10.3). For instance, Valentine and Waite (2012: 20) offer the example of 'heterosexual people of faith' who might be willing to 'accommodate LGBT people' in public space based on politeness or tolerance, but are not willing to change their deeper beliefs on homosexuality. This challenges the assumption that bringing people together in public space will necessarily have an effect on how people view others.

At the same time, it has been noted that many of the celebratory accounts of urban encounter focus on a narrow range of public spaces, which as Amin and Thrift (2002) argue, risks overlooking the other sites, connections and experiences that are significant to the formation of our identities. As an attempt to address this issue, Watson (2006: 14) called for research into more 'irregular, haphazard and ordinary' urban spaces and for a better investment in the more 'complex and textured' understandings of the people and places that are evoked by these debates.

As such, in contrast with the 'grand piazzas [and] endlessly rehearsed shopping malls' (ibid) of the kind that are often prioritized by grand architectural projects and planners, Watson's call was to address spaces that are random, marginal, contingent, imagined, both invisible and visible – such as allotments, public baths and street markets.

CASE STUDY 10.3 THE CONTACT HYPOTHESIS

The development of the 'contact hypothesis' has largely been accredited to the psychologist Gordon W. Allport (1954), who argued that interpersonal contact between different groups and individuals could help reduce conflict, prejudice and misunderstanding. According to Allport's theory, to diminish anxiety or prejudice, positive contact in a variety of social contexts is needed in order to develop familiarity and knowledge about others. One claim to emerge out of the contact thesis, is that positive contact between a majority individual and a minority individual, would likely lead to a 'scaling up' of positive feeling, whereby the expression of more positive feelings towards a minority individual would be extended to the minority group as a whole.

Some scholars, interested in the idea of urban encounter and questions of living with difference, have returned to Allport's work to consider whether it is possible to identify the conditions under which contact becomes positive (see Valentine 2008; Wessel 2009). However, some geographers have been cautious about the generalization of positive effects. Matejskova and Leitner (2011) undertook research with immigrant and non-immigrant residents on the outskirts of East Berlin in Marzahn, which has a high settlement of *Aussiedlers* (German minorities from the former Soviet Union). Focusing on different spaces of encounter, from spaces that facilitated only fleeting encounters (such as streets, shopping plazas and public transportation) to those that facilitated longer term engagement (such as immigrant integration projects, churches and workplaces), they argued that cities are full of different opportunities for contact, which have differing capacities to transform relations. Most contact in Marzahn was fleeting, which had little ability to alter prejudiced attitudes and in some cases, made anxiety about difference worse. Whilst there were instances where attitudes towards individual immigrants were improved through contact at some of the community centres where integration projects were held, it was noted that 'close encounters with individual immigrants were not generalized to the whole immigrant group' (ibid: 736). More importantly, they noted how unequal power relations and structural frameworks also reduced opportunities for contact, such as the development of Russian-language only spaces, the high unemployment rates amongst immigrant communities and shifting local political contexts. As such, their study calls for a more careful consideration of contact and its positive potential.

Contributions to urban theory

Taking these concerns forward, work on urban encounter has sought to better theorize and understand how contact takes place. This includes a greater concern for diversifying the types of public space considered, the power dynamics that might be involved, and the different structures and influences that shape encounters in different contexts. By detailing how urban encounters happen, the open-ended possibility and ambiguity of encounters is recognized, with efforts concentrated on developing a better understanding of the circumstances that give rise to different effects – whether this is new forms of coexistence and conviviality, or new sources of conflict or prejudice. On this basis, the remainder of the chapter outlines three contributions that work on encounter makes to urban theory. These relate to issues of: urban diversity; urban politics; and urban space.

Urban diversity

By definition, the term 'encounter' is often used to describe a meeting of opposites, or even adversaries, and is often defined by its unexpected or unplanned nature. Thus, as a specific form of social contact, encounters have been associated with conflict, risk, hostility and danger. Yet within the social sciences, an interest in encounter now reflects a more general interest in the meeting of differences more broadly (Darling and Wilson 2016). As such, work on encounter has addressed questions of urban diversity such as race and ethnicity (Wilson 2013), class (Schuermans 2013) and sexuality (Valentine and Waite 2012), along with other forms of social difference (e.g. homelessness, Cloke *et al* 2011). In so doing, scholars have considered how differences intersect but also how they are negotiated on an everyday basis, offering an understanding of how categories of identity and belonging are destabilized and drawn into question through different forms of contact. This work demonstrates that difference is not fixed, but is rather fluid, changeable and continuously made, remade and contested. On this basis, the city is not taken 'as a container' in which 'already formed differences' encounter one another, rather differences are *produced* 'in and through the city' and thus in and through encounters (Isin 2000: 233; Haldrup *et al* 2006).

Whilst early writing on the city tended to focus on bodily, face-to-face contact between strangers, recent work has developed a much wider account of how difference is encountered. Wise's (2010) notion of 'sensuous multiculture' recognizes how difference is encountered through smells, sounds, and taste, as much as it is through bodily contact and vision. For example, religious diversity might be encountered through the sonic landscape of the city, which might include calls for prayer and other forms of 'religious noisemaking' (de Witte 2016; also see Darling and Wilson 2016). Difference at Istanbul's Grand Bazaar, where this chapter began, can be encountered through the smell of burning oil or spices, whilst non-human difference might be encountered through the sound of scurrying feet or the traces left by elusive animals (Power 2009: 46; see Chapter 5, this volume). Swanton (2010) has also emphasized how important materials are to social interaction. Encounters with mundane objects and the material elements of the urban landscape are entangled with processes of differentiation. For instance, in documenting how race takes shape through encounters with things such as rucksacks, salwar kameez, road signs, food and security cameras, Swanton demonstrates how multicultural life is formed by encounters between people, objects and atmospheres (ibid).

Finally, whilst work on encounter has tended to focus on human diversity, a focus on human-animal encounters can offer important insight into constructions of urban nature and wildlife (Lorimer 2015; Wolch 2002). For example, Yeo and Neo (2010) focus on the case of long-tailed macaques (*Macaca fascicularis*), a small monkey species, in *Bukit Timah Nature Reserve*, a borderland area in urban Singapore where continuous residential development has encroached on natural habitats. This encroachment has produced a space of 'trans-species sharing' and encounter (ibid: 682), where encounters are often understood in terms of conflict, producing exclusionary geographies that mark macaques as 'out of place' (ibid: 696; for another example see Power 2009). Strategies to avoid aggressive encounters with macaques and to secure houses against them demonstrate how macaque encounters have shaped practices of homemaking. At the same time, non-verbal communication between human and macaque has also enabled more empathetic encounters, which have not only reduced trans-species conflict, but also problematized local efforts to cull the macaque population (ibid: 694). In short, human-macaque encounters continually question understandings of belonging, nature-society binaries and understanding of trans-species sociality.

FIGURE 10.2 'What not to do when you encounter wild monkeys': National Park guidelines, Singapore (Source: Helen F. Wilson).

Urban politics

This chapter has already demonstrated how links have been made between encounters and the development of democracy, cohesion and civic values (Young 1990). Because encounters have the ability to shape understandings of difference and to destabilize social boundaries, encounters can thus be instructive for theorizing urban politics (Darling and Wilson 2016; Merrifield 2013).

For example, the notion of 'encountering' has been deployed by academics interested in social movements and urban activism, who have focused on the urban conditions of proximity that have enabled diverse powers and actors to come together, resonate and form new relations, points of coordination, collaboration and contestation (Halvorsen 2015; Uitermark *et al* 2012). For these scholars, the city is a site through which political imaginaries and processes are imagined, enacted and contested as social activists, authorities, residents and workers are brought together. For example, the Occupy movement has been presented as a useful case for thinking through the 'radical potential' of urban encounter. The Occupy London camp has been described as 'an ensemble of people, things (for example, tents), and the built environment', which provided 'spaces of encounter in which power was negotiated and new social relations were produced' (Halvorsen 2015: 314). Here, city workers, tourists, members of St Paul's Cathedral, local police and social activists from across the country were brought together in unpredictable ways. Encounters under these circumstances enabled new connections, and the formation of new powers and political struggles from the bottom-up in order to stake claims to space and make demands of those in privileged positions (Halvorsen 2015).

Of course, it is important that an interest in the political potential of encounter is necessarily accompanied by recognition that encounters are unpredictable and so do not always have positive outcomes. Encounters can enchant, delight, surprise and thrill, but they can also produce anxieties, shocks and more violent effects. Some urban scholars have theorized such ambiguity by drawing upon Pratt's (1991) concept of the contact zone, which is useful for considering the unpredictability of encounter, particularly in instances of highly unequal relations of power (see Case study 10.4). Pratt's contact zone was originally deployed to describe spaces of colonial encounter where 'cultures met, clash[ed] and grapple[d] with each other' (ibid: 34). These zones were characterized by 'rage incomprehension, and pain' but also by 'exhilarating moments of wonder and revelation, mutual understanding, and new wisdom' (ibid: 39). As such, particular emphasis was placed on the capacity of encounters to facilitate social learning and it is this potential for learning that can make encounters politically productive through destabilizing certainties and provoking disruptions.

CASE STUDY 10.4 CONTACT ZONES

Mary Louise Pratt's (1991) concept of the 'contact zone' described the highly unequal spaces of colonial encounter in which different cultures met and struggled. For Pratt, the contact zone was full of 'ambiguous potential' and as a result was always laden with risk. Pratt emphasized the opportunity for learning that contact zones presented, describing them as having the potential for co-constitution, improvisation and interaction, despite the unequal power relations involved. Scholars deploying the concept of 'contact zones' are therefore concerned with understanding their possibility to facilitate learning.

Askins and Pain (2011) utilized the concept of the 'contact zone' when examining a participatory art project in northeast England that worked with young people of African and British heritage from one neighbourhood. The aim of the project was to bring together the two segregated groups to facilitate interaction and explore ideas of belonging. The children were invited to focus on their experiences of bullying and racism through sketches, discussions and cartoons and then on more positive representations of the northeast and African countries through paintings. In so doing, they utilized pens, papers, paints and glue, which had to be shared and negotiated. The art materials became part of the contact and the process through which the children started conversations and negotiated their relationships with each other. As such, the space of the art project became an example of a contact zone – a zone characterized by both conflict and potential, but a zone also capable of destabilizing categories of belonging and instigating new forms of learning.

Urban space

Finally, the focus on encounter contributes to our understanding of space. In order to move beyond the narrow range of public spaces engaged in earlier work, studies have attended to a multitude of sites, including; spaces of work and education, such as playgrounds and offices (Ellis *et al* 2004, Wilson 2013); food and consumption, including marketplaces and cafes (Laurier and Philo 2006; Watson 2006; Vertovec 2015); public transport (Wilson 2011); streets and plazas (Swanton 2010); community centres (Matejskova and Leitner 2011); and leisure spaces such as parks, gyms and hairdressers, all with a concern for detailing the local contexts in which difference is encountered (see Figure 10.3). However, a focus on spaces of encounter (Leitner 2012) is not only about the 'predicaments of those who find themselves thrown together' (ibid: 829). Instead, it is about how space shapes the unfolding of encounters and how space is simultaneously shaped *by* encounters.

As a means to better contextualize and illustrate how spaces of encounter have been engaged I want to provide one example by looking at the public bus as a space of encounter.

FIGURE 10.3   Grand Central Station, New York City: ordinary spaces of mass transit have been a popular focus of the study of 'throwntogetherness' (Source: Helen F. Wilson).

The bus

For many people, bus journeys are a necessary part of everyday routine and are often considered to be an unremarkable part of public life. However, buses are important spaces of encounter. They are sites of close proximity and enforced intimacy, where interaction with others is not only possible but often unavoidable. At times of high volume, bodies are thrust up against each other and personal space is re-negotiated and shaped by codes of conduct, behaviour rules and material objects such as handlebars, fixed seats, luggage and shopping bags, which restrict and direct movement or enforce stasis. It is because of this close bodily contact that buses have been described as sites where social boundaries such as class are crossed (Hutchinson 2000). It is often split-second decisions made about fellow passengers that inform seat selection and because people have little to go on, difference is often read off the body. Once on the bus, difference is also encountered through sounds, smells and materials, all of which are accentuated by the enclosed nature of the space. For example, a city's diversity might be encountered through the mingling of foreign languages (Back 2007), whilst the unwelcome presence of other people might be sensed through the nose or even the unwelcome touch of another body, all of which are difficult to avoid.

Rather than simply a means to get from A to B then, buses can be spaces of conversation, avoidance, musical performance, work, xenophobic outburst, community organization and daydreaming, and this can be shaped by many factors, including the time of day. As such, encounters can be convivial, hostile, fearful, or indifferent. YouTube clips of violent encounters in spaces of mass transit in both UK and Australian cities, have led Paul Gilroy (2012) to describe sites of mass transit as the spaces where 'lived, unruly multiculture' is tested. The filming of violent encounters by fellow passengers has allowed them to be viewed by millions across the world, triggering debates about misogyny and xenophobia. This demonstrates that whilst such encounters may be fleeting, they can be used as a 'prism' through which to consider aspects of the contemporary urban politics of race, immigration and belonging.

The bus is also a site of routine, and whilst for some, it might be used only on rare occasions, for others it is a key part of daily life where the negotiation of difference becomes a familiar occurrence. Some writers have even suggested that widespread use of public transport can make city populations more open to diversity and more comfortable in its presence, but this is a link that has yet to be effectively researched (Kennedy 2004).

Finally, it is important to ask questions about what shapes the possibilities for encounter with others. For example, in some cities, buses are most commonly associated with the young, the elderly and with people on low incomes, as buses do not carry the same connotations of autonomy, security and prestige that come with other forms of mobility. They can also be associated with anti-social behaviour, such as drug-taking, violence and vandalism, which has a profound affect on who chooses to use the bus. At the same time, thanks to the advance of technology and the use of hand-held smart devices, passengers can tune-out, switch-off or take part in workplace activities whilst on the move, changing how people interact with the environment and people around them. Precisely because of the intimacy that buses require, it is also important to consider the long history of policies that have sought to segregate passengers (for example, by race) in order to prevent or restrict 'undesirable' encounters (see Alderman *et al* 2013).

Conclusion

This chapter had three aims. First, to outline the lineage of work on the relationship between social contact, urban proximity and the organization of public life. Second, to draw out the renewed interest in urban encounter and the more ambivalent, contested and textured nature of urban space and diversity. Third, to outline the lines of inquiry pursued by scholars working on urban encounter and their contributions to theories of urban diversity, urban politics and urban space. In outlining these contributions I have also made connections to issues of power, temporality, sensuous geographies, identity and the more-than-human, as key themes that are brought together by the varied work on urban encounter.

The chapter purposefully finishes in the same way it began, with an ordinary space of encounter. The example of the bus demonstrates the contextual understandings of space that are necessarily developed to better think through questions of encounter. It demonstrates how foundational theories of urban contact (whether Jacobs' urban ballet, or Simmel's indifference to the proximate stranger) can inform our understanding of bus encounters. Yet, it is also vital to recognize that the bus is a space profoundly shaped by rules of conduct, materialities (for example, fixed seating), routines, demographics, sounds, smells, atmospheres and security systems that make this site of encounter unique and that trouble any attempts to develop a universal, or automatically celebratory account of contact.

I want to finish the chapter with a quote from Kurt Iveson (2007: 46) who argues that:

> rather than demanding that urban inhabitants be open to 'encounters with strangers', we need to learn more about the circumstances in which particular people have taken the risks associated with these strange encounters and transformed their cities in the process.

For urban scholars interested in encounters and their potential it is to these circumstances that they have turned their attention. However, in drawing out the key contributions that work on encounter can make to understandings of difference and diversity, I would finally emphasize the scope that theories of urban encounter hold for thinking beyond 'particular people' and for including all those other agents, objects, and animals that constitute the city and urban life.

Key reading

Amin, A. and Thrift, N. (2002) *Cities: Reimagining the Urban*, Cambridge: Polity Press.
   This text outlines the importance of encounters to the building of urban life and for understanding the complexities of the city. It is notable for its move away from face-to-face encounters between people to consider other forms of relation and more-than-human geographies.
Darling, J. and Wilson, H. F. (Eds.) (2016) *Encountering the City: Urban Encounters from Accra to New York*, Aldershot: Ashgate.
   This edited collection brings together contributions from across the social sciences to showcase recent work on urban encounter. The collection demonstrates how this work contributes to theorizations of urban diversity, temporalities, politics and more-than-human geographies.
Merrifield, A. (2013) *The Politics of the Encounter: Urban Theory and Protest Under Planetary Urbanisation*, Athens, GA: University of Georgia Press.
   This text focuses on the value of 'encounter' for theorizing urban politics and social movements.
Valentine, G. (2008) 'Living with difference: reflections on geographies of encounters', *Progress in Human Geography,* 32 (3): 323–337.
   This intervention is useful for its critique of celebratory accounts of urban contact and encounter and for asking how contact might change social values and prejudice.
Watson, S. (2006) *City Publics. The (Dis)enchantments of Urban Encounters*, London: Routledge.
   This is a compelling, ethnographic account of how difference is negotiated and contested in ordinary spaces of encounter. It also offers valuable reflections on theories of urban public space and the problems of universal accounts.

11

ENTREPRENEURIALISM

David Wilson

Introduction

Large cities in the global north currently experience a redevelopment surge steeped in a feverish drive to be more entrepreneurial. Amidst the destruction to people and neighborhoods wrought by austerity policies and the mortgage foreclosure fiasco, development alliances (local governments, real-estate interests, business interests) strive to cultivate a richly entrepreneurial city that refines their previous (pre-2006) 'go-entrepreneurial' city project. From London to Madrid to Berlin to Chicago, the drive is to more deeply 'business-up' downtowns, re-make investment climates, foster neighborhood gentrification, and implant new cultural, social, and economic facilities to invigorate the engines of economic growth. The built environment, the supposed spine of the project, is served up as the all-important stage and setting. Out of this, as now chronicled, has emerged a new technocratic vocabulary – smart growth, city sustainability, urban innovators, the creative class, creative city re-birth – that emphasises a market-rooted, go-global development agenda.

My chapter deepens our understanding of this drive to go entrepreneurial by suggesting the reality of and theorizing a new, most recent phase. I build on a tradition of updating our understanding of this rise in these cities as realities continue to inexorably evolve. I illustrate a central thesis: these cities have increasingly (post-2006) experienced a shift from a 1980s-inaugurated entrepreneurial developing (Harvey's (1989a) stage of the entrepreneurial city) to what I term a hyper-entrepreneurial developing. I chronicle, on the one hand, that some characteristics in these stages have remained the same – i.e. this drive to re-entrepreneurialise continues to rely on use of a complex, multiple-scaler discourse and a politics of redistributing wealth and resources. I suggest, on the other hand, that these cities now experience development under altered alliance goals, modes of conduct, discursive strategies, and discourse-enabling content (in addition, the intensity of redevelopment has changed). An important point emerges from this: while we continue to be in blatantly neoliberal development times, I suggest, we must recognise this entrepreneurial project as ever-changing and in constant need of being tracked.

The entrepreneurial city: theoretical debates

Current studies of the present drive in global north cities to entrepreneurialise have generated a number of contentious debates. Resolving these debates will help us better understand the contemporary development scene here. To further understand them, we must first briefly note that there are three major points of agreement among urbanists to comprehending this decades-old drive. First, this drive pivots around a shift in emphasis from Keynesian to supply-side policies. City growth, it follows, less relies on public spending than privatism to function as the city's economic engine. Second, a politics of resource attraction has arisen as a major centrepiece to guide redevelopment plans and practices. Here programs and policies privilege a drive to bring in resources to the city (new investment, new jobs, new city 'energizing' populations, new companies and corporations) that will supposedly maintain urban economic and social stability. Third, a new institutional alliance has formed to generate entrepreneurialism, which potently connects the public and private sectors as business-driven partners. At the core of this, the local state moves beyond its traditional activities of land-use control and planning and providing for public services, to become a major player in urban land development. Yet, considerable disagreement also exists among urbanists in their attempts to understand this current city entrepreneurializing. Three points prove most contentious; let me elaborate.

First, scholars debate where dominant development alliances (amalgams of local government, business interests, land and property interests) 'reach' beyond their cities to forge growth. As urbanists recognise, all cities are engineered to 'grow' (an imperative in capitalist cities), and a core city strategy is to pursue some form of 'structured complementarity' by building favorable linkages to the wider economy (Jessop 2011). New investment must be secured, new developers must be appealed to, new jobs have to be obtained. The search beyond the city ultimately becomes a quest to forge alliances with governments, private institutions, companies, businesses, and investors. But what wider economies become the focus? The regional economy? National economy? International economy? Which scale prevails? All of these? Only one of these?

One position has dominated: development alliances are seen to privilege one link, the local-global (see Marcuse and Van Kempen 2008). Buoyed by the influential works of Harvey (1989c), Jessop (1998), and Hall and Hubbard (1996, 1998), these alliances are universally seen to cast their gaze at this level above all others. Out-competing global players and cities, it seems, is the holy-grail, the necessary act for maximum city possibility (Thornley and Newman 2011). The global, in assertion, enraptures these development alliances. These formations, seeking thriving cities, believe global businesses, global institutions, and global cities primarily must be duelled with. Multi-national players, and London, Tokyo, Shanghai, Madrid, and New York are the villainous competitors that must be bested. There is seemingly no exiting this reality, the new global competitors must be confronted. Thus, key economic alliances are sought, especially with multi-national operatives: no spaces on the earth are immune from this search for partnership. The ideal: to attract and display in a city globally networked financial institutions, businesses, social organizations, and cultural facilities.

Yet, an alternative position advanced by Smith (2011) and Taylor *et al* (2012) has attracted attention. It has two parts. First, this reach to re-entrepreneurialise is more complicated; global strategies to situate one's city may require an intricate engaging of other players and cities at smaller scales (the regional, the state, the national). True global strategies, here, need to also forge regional and sub-regional connections. The hysteria of new global realities, to Taylor *et al*, does not automatically translate into knee-jerk responses where city development alliances dive exclusively into international competition. Second, a 'multi-scaler' vision onto this question is essential, rendering the issue of one scale's relative importance over others problematic. Here every 'scaler reach' involves a reach to multiple scales, and economic operations across scales seamlessly entangle and embed. As Smith (2011) puts it, scales dynamically collide and interconnect, separating scales and rendering them discrete unities is one of the day's pernicious academic constructions. This alternative perspective thus posits development 'outreach' as a spatial stretch that always engages multiple scales which conjoin in a dynamic, meaningful way.

The second central debate comes out of a provocative point noted early on by Hall and Hubbard (1996, 1998) that is still unresolved: to what degree has this alliance-driving entrepreneurialism replaced its old focus on provision of services and resource redistribution? Harvey (1989a), taking the lead on this issue, identifies an entrepreneurial fervor that has quickly and dramatically supplanted the once crucial focus on resource redistribution and resource delivery. He notes a rapid and dramatic rise of an entrepreneurial ethos within these alliances that has overwhelmed resource delivery. Jessop (1996, 1998), similarly, identifies a near-complete supplanting of one for the other. His 'new competitive state' notion identifies a growth alliance that glories in and strikes to push history out of the way as it advertises itself as making the city economy solvent in the face of a new menace – heightened inter-city competition in new global times. Indeed, Hall and Hubbard (1996, 1998), after posing the question of how influential this drive to entrepreneurialism is, come around largely to agree with these responses.

But recent works by Peck (2014a), Merrifield (2014b), and the authors in a special forum assessing David Harvey's classic piece 'From managerialism to entrepreneurialism' suggest otherwise. Recognizing complex and multiple political projects that these alliances must be involved in, they identify the mutual co-existence of both a politics of resource attraction and a politics of resource redistribution in current neoliberal times. These alliances purportedly partake in both because each is still an essential undertaking. Here both approaches do not merely exist, both are significant political responses in these cities. On the one hand, alliances drive aggressively to use resources for growth and development: the lifeblood of city sustenance – jobs, investment, housing – must be secured. Also, avenues of profitability are relentlessly opened up for real-estate capital (builders, developers, realtors) as a dominant power broker. On the other hand, resource delivery continues: there are still people – the poor, the segregated, the subaltern – that need to be pacified and tended to. These cities are, as much as ever, turbulent places as inequalities fester and deepen. Thus, beneath the rhetoric of state retrenchment and the killing of the welfare state, alliance abandonment of 'the political reproduction project' is impossible.

The third major debate has focused on specifying the relationship between this re-entrepreneurializing and a powerful force in the world today: globalization. Globalization and this city 'businessing-up', all agree, are intricately connected. But what is the precise nature of this relationship? In a dominant position, Harvey (2005, 2012) and Sassen (2011) have led the way to identify the influence of a globalization that is fundamentally real and deepening. This notion of globalization, rooted in a casting of a historically unprecedented changing, presents a force that has speeded up economic transactions and makes businesses and industry hyper-mobile across the globe (in a 'time-space compression'). Now, economic units (plants, headquarters, businesses) in cities have a historically unparalleled mobility and can easily relocate to other cities, regions, nations and continents, and effectively hold cities hostage. Capital continues its restless and relentless search for profitability enabled by new technologies and expanded territorial playing fields. Cities, now, seemingly have no choice and respond in one predictable way: they strive to become more economically competitive. Finding ways to effectively play the global game – i.e. to attract and retain their lifeblood of jobs, investment, cultural institutions, and educational institutions – must be their new mantra.

But more recently, urbanists have highlighted another side of globalization purportedly neglected in accounts of its power: its discursive deployment. To Wilson (2007, 2014) and Carver (2010), a 'global trope' now wields tremendous influence in these cities. Here, globalization is served up as a potent imaginary – a scary thought – that is wielded like a cudgel to punish and discipline populations to adopt neoliberal (re-entrepreneurial) redevelopment (see Chapters 12 and 18, this volume). Global-speak, an ominous and frightening offer, now shoots through the circuits of placed-based communication in these cities. The realities of new global times are never denied in this work; at issue is the deliberate inflating, exaggeration and caricature of this reality. Invoked in this exaggerating, for all to digest, is a supposed new world order that is poised to wreak havoc on cities if they fail to fight back in meaningful ways. In this context, the script of the 'global trope' is made to feature an ominous cast of characters: new spaceless, place-detached entrepreneurs, teetering cities, a potentially city-annihilating globalization, and heroic, global-fighting neoliberal planners and business people. The outcome, ultimately, is tangible: publics often come to accept an 'entrepreneurializing' of cities as inevitable and logical.

Theorizing the updated entrepreneurial city

Even with this vibrant current literature, I believe that large cities in the global north have moved into a new phase of going entrepreneurial that has been largely unrecognised and non-theorised. This section illuminates this point by engaging the just-discussed debates. I show three key things about this new hyper-entrepreneurial phase in response to the aforementioned debates: it continues to crucially involve a politics of resource redistribution amid the fervent privileging of a politics of resource attraction, it operates through a multi-pronged spatial strategy (appealing to the local, region, national, and global scales), and it widely uses the rhetorical foundation of 'the global trope' to advance its projects. Offering these three notions provides an important update to how this current entrepreneurial project in these cities proceeds. To make these points, I interrogate this entrepreneurial drive's recent past and present. Turning the clock back very briefly is important: it enables me to identify the original thrust to entrepreneurialise with the current strategies to deepen this in the new 'hyper entrepreneurialism'. To chronicle my points, I draw on the recent experiences of two cities, Chicago and Glasgow, which have received much attention for their ambitious drives to become more entrepreneurial.

Initially, most northern cities jumped into Harvey's 'entrepreneurial growth' amidst the 1980s rise of neoliberal sensibilities. Mayors found these sensibilities – a deepening belief in private market primacy, social welfare retrenchment, use of government to propel filter-down growth, an atomistic vision of societal process and social causation – ideal for driving this development. Now the private sector could dominate the enterprise, multiple fractions of capital could be more easily serviced (real-estate capital, industrial capital, corporate capital, finance capital), and social welfarism and the politics of resource redistribution could be blamed for city woes and offered as things in need of being scaled back. The election of often prominent, charismatic mayors in the 1980s (e.g. Daley II in Chicago, Pasqual Mira of Barcelona, Robert Gray of Glasgow) aided the cause. This upsurge in mayor charisma was not accidental: such operatives in ascendant neoliberal times could now speak boldly and aggressively about the need for fundamental change in development policy.

Before long, these mayors and their planning appendages were unveiling a bold city vision: to make their places global, cosmopolitan cities. Now, this move was cast not as a luxury or simply one possible option, but rather a necessity in this day and age. Globalization was the recently identified new reality that encircled these cities – it was served up as an ominous force – and forging competitive urban environments rather than 'welfare-saturated' cities supposedly made sense. At the same time, in narrative, the tyranny of government over-reach and excessive social welfare had finally been identified: it was time to change the course of action. In America, these narratives frequently drew on a national discourse from nine years earlier: Ronald Reagan's 'new morning in America'; in the UK and across Western Europe Margaret Thatcher and Adolfo Suarez's 'spring-time in Europe'. Cities, now, would purportedly become energised by re-tooling development policy to confront their ills: the downtown's 'flagging soul and clout' and 'flabby, ineffectual city reinvestment climates'.

Nearly from day one, mayors and their development alliances struck out to physically upgrade especially downtowns and nearby neighborhoods. Gentrification had emerged in many of these cities, and this was aggressively built on. Also, economic cores were re-made to make them compatible with a supposed new economic reality: new post-industrial economic times. The major tools used were tax incentives, tax increment financing, public-private partnerships, historic preservation development, land banking and re-release of land for development, and forms of urban homesteading. In Chicago, this movement resulted in a fast, dramatic expanding of gentrification across the Loop and beyond that came to encompass the Old City, Lakeview, Bucktown, West Bucktown, and Wicker Park areas. At the same time, new culture districts, consumption facilities, and economic incubator zones appeared in the Loop. In Glasgow, its first international large-scale marketing campaign was begun, 'Glasgow's Miles Better' initiative, which inaugurated gentrification in its west end and downtown. Set in motion was not merely the metaphor of the ugly duckling turned beautiful white swan, but also the denial of a persistent past of industrial grit, grinding poverty, and gaping uneven development.

By 2006, the story became more complicated: many mayors across the global north, now emboldened by their successes, began to move their cities into a 'hyper-entrepreneurial' development stage. Most visibly, they led a push for an unprecedented upscaling of their downtowns and nearby neighborhoods. In Chicago, Mayor Daley and his development alliance initiated this change through engineering and tapping a global hysteria which was sweeping America, and dramatically hyped the need to create a second-to-none international city. This drive centered around nurturing key things: a robust post-industrial economy downtown, a richly international-symbolic city, and a lushly consumption oriented city. Between 2006 and 2010, the results reflected this: Loop gentrification increased by an estimated 26 percent, and general investment here by 21 percent. In Glasgow, too, this city drive deepened. Scotland's largest city began to advertise and re-build itself as a financial, commercial, and tourist hub. Not surprisingly, gentrification of the west end and downtown sprouted across the city, engulfing more than 20 once industrial blocks. Its long-neglected waterfront (the Clyde River), now being fully recognised for its upscaling possibilities, experienced the growth of more than 25 new redevelopment projects (see Case studies 11.1 and 11.2).

At the core of this now is a change in tactics and strategies. Most notably, the new (post-2006) vision is driven by something novel: declarations of an unprecedented hyper-competitive reality that ensnares these cities. Globalization is now cast, in Neil Smith's (2011) terms, as a viral scourge. Moreover, this theme is delivered by alliances now posturing as brave and blunt neoliberal truth-tellers that need to offer a kind of new shock treatment (the need to concentrate public and private resources in select spaces, demand that all residents be productive and civically contributory) while proclaiming themselves the new heroines of the

CASE STUDY 11.1 THE UNDERSIDE OF CHICAGO'S REDEVELOPMENT

A dimly recognised process marks Chicago's current state of development. In the shadows of the downtown remaking, poor African-American communities are relentlessly regenerated and repopulated. Institutional actions on the ground ceaselessly motor the process: the deed is done in the dullness of everyday real-estate workings. Thus, Chicago's 2004 revised zoning ordinance (four years in the making, spearheaded by the Mayor's Zoning Reform Commission) promotes tall, high-density development and affluent Loop tenants, and pushes 'lesser' housing and its associated kind of tenant elsewhere. In response, dense tower and medium-dense mid-rises flock to 'hot' sections of the Loop – Printers Row, West Loop, New Eastside – that now swell with white-collar 'creatives'. A seemingly innocuous ordinance this way speaks volumes about who are the desired, civic contributory subjects in the current Chicago, and who are not. At the same time, the city's twenty-six Central Area tax increment financing (TIF) districts, spearheaded by the Lasalle/Central and Randolph/Wells designations, power gentrification in and around the Loop and displace ascribed 'riff-raff' to elsewhere. In this vicinity, the TIF districts funnel tax revenues extracted from these districts back into these zones. Land is often quickly valorised, and districts made ripe for new investment. These TIFs, advertised as value-free development tools, also speak to who is to occupy prime Loop and near-Loop land, and who is not.

CASE STUDY 11.2 THE UNDERSIDE OF GLASGOW'S REDEVELOPMENT

Glasgow's current development scene also afflicts the city's poor. Hidden from the city centre and nearby build-up, deprived areas have become more-profound receptacles for the poor and the residual. Stigmatised people, the anointed 'visual trash' in the new big build-up, continue to be decisively cordoned off as blighting influences on the downtown and new play spaces. Glasgow's recently revised zoning ordinance, tailored to attract new housing development to the core, is ruthlessly efficient, enabling tall, high-density development and the 'right' kind of resident to occupy the city centre, while consigning the building of affordable housing and its kind of resident to elsewhere. Glasgow housing activist M. McLean (2014) calls this ordinance 'the building block of our most visible dilemma today – uneven development and class segregation in current Glasgow'. Moreover, companies in Glasgow are notorious for steering the poor to 'their own areas'. A handful of dominant companies, uncompromising in their efforts to flagrantly fortress the city centre from 'a poor invasion', are termed by local activist M. McLean (2014) 'the modern day slum makers and "segregation creators of Glasgow"'. Like Chicago, Glasgow today has a sordid development underbelly.

modern city. This theater of dire economic times and alliance self-aggrandisement stages a power and acuity that bolsters the integrity of their city re-making message (it also conceals the difficulties of their reality: they continue to struggle to negotiate shifting political ground, engage new possibilities and constraints, and grapple with new forms of contestation). To be sure, global-speak existed before 2006, but not in this form. Global speak here, prior to 2006, was a different and more rhetorically tame affair.

Also now, in bold proclamation, these cities are widely depicted as something new: fully threatened, teetering cities on the brink of disaster. Here once enclosed and confident containers of the economic had recently become dangerously porous and leaky landscapes rife with a potential for a dramatic economic hemorrhaging. Chicago, in Planner C. Smith's (2012) words, 'is now at a crossroads and has been for some time now … we stare at a new economic reality that hangs over us … Chicago has faced crises before, but this one is new and just beginning. City survival, now, supposedly needs a key ingredient: re-entrepreneurialization'. In this imperative, Chicago 'now must push to build attractive cultural spaces, efficient labor pools, and bring in the new people and jobs that will help us succeed … This stuff is, plain and simple, civic business' (Mayor Daley 2006). In Glasgow, too, planners and public officials speak starkly of the wrath of new global days. In proclamation, Glasgow has morphed from a once-confident, stable economic center to an increasingly open-ended, unstable economic venue with a potential for a dramatic economic eclipse. Glasgow, like Chicago, is said to stare at an uncertain economic future as a kind of accumulation disorder and uncertainty hangs over it.

But more newness is at work here. At the same time, alliance utterances transform the offer of the global menace, and in a perversely innovative way: by meaningfully changing globalization's most basic content (what this is). Globalization, of course, has historically been widely identified across global west cities in one dominant way: as a force of blunt, transparent place-engulfing. Soon after the concept's dramatic rise in the global north after 1990, national newspapers, magazines, and speeches commonly offered this as a fully enveloping, totalizing kind of reality. This version of globalization, of course, has been a dominant discursive artifact of its time, a force 'out there' that jolted and externally loomed at a moment when global-city connections across the west were just being realised. At the very moment that global connections were prospering, it was communicated, its underside of inducing hyper-mobility of jobs and investment had gained full steam.

Now, however, this global offer changes. Whether the end product of cold calculation or simply a malleable concept's gradual evolution, many of these city alliances offer something else: a furtive globalization of evanescent penetrations, uneven morphologic effects, and sporadic bursts. Now, globalization, in Mayor Daley's (2005) terms, 'moves', 'hits and runs', 'is ominously out there', and 'strikes but always keeps us guessing'. In Glasgow, city planners speak widely of 'a globalization that deepens but also remains unpredictable in where it will strike and who and what will be immediately affected' (McGintie 2014). This globalization is now not only alive-and-well and biting, it is also painstakingly elusive, intangible, and ambiguous. It has purportedly moved to the city's centre, but continues under a cloak of mystery and furtiveness. This change in globalization's content is meaningful. With this replacement, a once-calculable, brute globalization becomes something much more haunting: a murky, difficult-to-discern-and-predict globalization. Here is a political trope par excellence: a once-engulfing totality, in a refined stroke, becomes something more sinister, a moving, unpredictable attacker of districts and areas.

At the same time, engaging in a politics of resource distribution is still pronounced. Despite often bold pronouncements of alliance desires to retrench the local welfare state and peripheralise social welfare concerns, the realities suggest otherwise. On the one hand, these development alliances speak boldly about the need to retrench the welfare state and the politics of redistribution. To mayor after mayor, it is essential that cities put their priorities in order – harsh global realities dictate this – and the crucial task of attracting resources to the city must take precedence over the once-popular concern for redistributing resources. In theme, the once-flourishing era of 'poverty politics' can no longer be afforded. However, there is a reality that is not widely publicised: the poor still need to be tended to as potentially explosive agents of social change. In these cities, much of the poor have grown poorer, many more residents have become poor, and these populations are a potentially volatile lot that need to be carefully managed. Indeed, while this contradiction in alliances was identified as early as 2002 by urbanists, this imperative to assuage and manage the potentially incendiary poor deepens.

Glasgow's current re-making bears this out. On the one hand, bold oratory speaks of cultivating the new entrepreneurial development above anything else. In this context, Glasgow nurtures its new crowning jewel: the upscaling of the Clyde Waterfront. Scottish Enterprise and the Scottish government have invested more than 3.5 billion pounds in this massive project (Clyde Waterfront Committee 2008). 'London has its docklands, Baltimore has its harbor, Amsterdam has its canals', Glasgow planner McGintie (2014) noted to me, adding, 'Glasgow has found ... at long last ... its meal ticket into the future, its city-synergizing water body'. Currently, this 20-km stretch, running east-west from Glasgow Green to the heart of Glasgow, has more than 200 projects completed or under way. But this drive does not slow the local state's provision of key resources – affordable housing, halfway houses, subsidised job skill programs, welfare payments – which is used to pacify the city's 120,000 poor people (a growth of 36,000 in five years in a city of 602,000). In 2014, nearly 30 percent of Glasgow's working age population (110,000 people) were unemployed. In 2015, to Glasgow Councilperson Vox (2015), the local welfare state is not only alive and well, it is expanding to patch over increasingly glaring social ills and deprivation. A historically redistribution-conscious local state, despite pretensions to becoming more 'real' and neoliberal, continues down this path.

Moreover, the drive to re-competitivise these cities operates through a conscious multi-scaler planning strategy. Desire to out-compete far-away cities, it seems, is not by itself the Holy Grail, finding a place in regional and national economies is also sought. This multi-scaler economic goal is widely seen as a building-block enterprise where appeal to diverse scales adds up to a globally competitive development strategy. A successful global competition, it follows, is felt to require a spatially blanketing strategy of all scales. To miss one, it seems, is to dramatically compromise the project. Not surprisingly then, these redevelopment alliances welcome local and regional developers as well as international developers, regional capital as well as international capital, national companies as well as global companies. Out-maneuvering regional hubs as well as international hubs for new jobs is deemed crucial to them. City governments, not surprisingly, now fervently sell their neoliberal themes of rejuvenated investment climates and friendly business alliances to nearby suburbs and states as well as to far-flung, distant global cities.

This is the case, for example, in current Chicago. Its Loop, since 2006, has experienced an additional US $2 billion in new investment that has built the likes of Millennium Park, the new consumptive Pilsen, architecturally sparkling University Village, One Museum Park Condominiums, the Heritage at Millennium, and an array of other projects. The South East Loop, once an iffy investment zone for developers, now sees a flood of new chic housing projects and supportive retailing powered by the presence of mega-park Millennium Park. To alliance members, this successful global re-making is as much about attracting new resources and people from the region and across the Midwest as enticing their international counterparts. 'We know the reality of new global times', noted Chicago Planner Weeks (2013), 'we need to make a play to Chicagoland, the state of Illinois, and nearby states too. Only through being successful at these levels, we believe, can the go-global drive work.' Weeks notes a conscious decision to entrepreneurialise the city with these other scales in mind. His 'play to the domestic', the stuff of besting the Clevelands, Pittsburghs, and Baltimores for scarce resources, ensures a successful internationalizing project.

Conclusion

Many urbanists are wary of the current re-entrepreneurializing of global north cities. This process continues to be tinged by the shallow claims of neoliberal sensibilities, and deepens the highly uneven distribution of benefits and costs across space and across classes. In this vein, theorizing the current reality of these cities 'going global' is an urgent task. Driving alliances armed with neoliberal values and protocols today profoundly influence these cities. As public and private sectors dramatically meld into one taut formation, development has become more bold, streamlined and efficient. Development now has an unprecedented global ambition, cuts across vast corners of these cities, and is proclaimed the saviour to stabilise purportedly struggling cities. In rhetoric, this development is the modern engine of economic growth so vital to any city. Yet, much confusion exists about this reality, even as it continues to be currently studied, especially with respect to this dynamic's central debates (three continue to occupy centre stage).

At stake in this updated theorizing is the need to grasp crucial things about this development: how they deepen inequalities, what institutions and networks now lead this charge, and how this transformation is being imaginatively and materially engineered. These issues strike at the heart of identifying David Harvey's (2013) 'new inequality' in these cities. To Harvey, a deepening socio-spatial splintering further fragments urban space that fosters more glaring zones of affluence and more naked zones of neglect. Most of us know the image well: increasingly upscaled downtowns and expanding rows of gentrified neighborhoods concealing in their shadows festering and neglected poor communities. In these shadows, the poor are forgotten by government and the private sector and suffer. By persuasively theorizing these questions something important can result: progressive political intervention can tackle the growing socio-spatial divide between the haves and the have-nots. More informed human activities can upgrade qualities of life, particularly of those who have been most marginalised. Knowing the tactics, strategies, and discourse tropes of these alliances can enable effective counter-responses. Much conceptual work needs to be done here, and its benefits are potentially profound.

Development alliances here, this chapter has shown, now exhibit a clear shift from a 1980s entrepreneurial development to a hyper-entrepreneurial development. There is more than a simple change in the intensity of the restructuring here. As documented, these development alliances currently move ahead offering new goals (to hyper-entrepreneurialise the city, to build a sharply visual-centric city), altered discourse content (refining what globalization is, deepening globalization's imagined power to afflict cities and lives), and revised 'communicative tropes' (making and using deeper and more subtle fears for public consumption). At its core, a global trope peddles a supposed new reality – globalization – that is now way beyond civil society, current politics, and city control. There is seemingly no other option but to play the game of global competition effectively for city survival. Globalization, so wielded, is used to justify flagrant elite restructuring and abhorrent human marginalization. Under the guise of essentialness, cities are being re-made in routinised practices that afflict so many and in profound ways.

Yet, looking beyond the present, we must realise that the future of these cities is still up for grabs. Future patterns of alliance intervention and their outcomes are difficult to predict: much is fluid and variable. There are possibilities for a better future. For these cities continue as landscapes of experimentation: these alliances must continue to keenly read evolving political and economic trends (e.g. upsurges in resistance, new political formations) as they offer ideas, visions, discourses, and actions. Not to be forgotten, they continue to proceed ahead as frequently unstable, reactive, and ever-evolving social formations (Wilson 2004). They are not the end of history, but rather one agent of change in a present moment that are also seeds for future change. Development alliances, conveniently casting themselves as blunt, inevitable, and irreversible, are anything but these. Nothing is set, all is evolving as a mix of emulation, policy transfer and technocratic ethos – funneled through place specificities – guides this newest development foray. Clearly, change in the form and features of these alliances and how they will shape these cities is coming, but it is difficult to know what is next.

Key readings

The following four readings are foundational to understanding the rise and evolution of this drive to entrepreneurialise global west cities:

Harvey, D. (1989) 'From managerialism to entrepreneurialism: the transformation in urban governance in late capitalism', *Geografiska Annaler. Series B. Human Geography*, 71: 3–17. A key conceptual analysis of the rise of urban entrepreneurialism.

Jessop, B. (1998) 'The narrative of enterprise and the enterprise of narrative: place marketing and the entrepreneurial city', in T. Hall and P. Hubbard (Eds), *The Entrepreneurial City*, Chichester: Wiley. A foundational account of the centrality of narrative in the rise of the entrepreneurial city.

Merrifield, A. (2014) 'The entrepreneur's new clothes', *Geografiska Annaler: Series B, Human Geography*, 91 (4): 389–391. A timely and important essay that highlights the ever-changing rhetorical forms that urban entrepreneurs work through.

Peck, J. (2014) 'Entrepreneurial urbanism: between uncommon sense and dull compulsion', *Geografiska Annaler: Series B, Human Geography* 96 (4): 396–401. A nuanced and intricate essay about the pervasive survivability of current urban entrepreneurialism.

12

GENTRIFICATION

Loretta Lees

Introduction

Despite calls at the dawn of the twenty-first century to allow 'gentrification' to disintegrate under the burden of its reconceptualization (Bondi 1999), the term has not disintegrated; quite the opposite, it has become stronger and more internationally known. The literature on gentrification itself has grown, offering new understandings, and it remains at the centre of critical work in urban theory. Gentrification scholars have not been afraid to 'burden' it with new concepts.

Gentrification studies is trans-disciplinary and this is one of its strengths. Although dominated by scholars from geography, there has been significant cross-fertilization of ideas and concepts between sociology, planning, urban anthropology, urban economics, and policy studies. Perhaps the most significant concept that it has been burdened with recently is the challenge of 'comparative urbanism' (Lees 2012, 2014a, forthcoming; see Chapter 8, this volume) and this means increasing learning across disciplines from around the world. The concern was that gentrification studies in different parts of the world were using a term – 'gentrification' – that had been coined in the Global North, and theorizations/conceptualizations of it that were based on studies of cities in the Global North. Lees (2012) challenged gentrification scholars to move away from an 'imitative urbanism' – the idea that gentrification travelled to the Global South and was copied from the Global North, to a 'cosmopolitan urbanism' which could respond to the postcolonial challenge of how 'gentrification' was constructed in gentrification studies, and of dealing with new forms of urbanism around the globe.

That same year a Greek scholar Thomas Maloutas (2012) asked similarly whether the use of the Anglo-American term 'gentrification' facilitated or impeded understanding of processes of urban restructuring in different contexts. Ley and Teo (2014) were also concerned about the 'conceptual overreach' of 'gentrification' from the Anglo-American heartland to the cities of Asia Pacific and specifically Hong Kong, and explored the identification and naming (or absence of naming) of gentrification in Hong Kong. But they concluded that just because the word 'gentrification' is missing from public and academic discourse in Hong Kong it does not mean that it is not happening. Similarly, Lees *et al* (2015), in their global collection on gentrification, show that 'gentrification' with or without the label is affecting cities all over the globe, from Karachi to Abu Dhabi to Beijing. Meanwhile Latin American scholars have begun to assert a specifically 'Latin American style of gentrification' (Janoschka *et al* forthcoming) that could link Mexico, Brazil, Chile and Argentina, due to the colonial legacy of Spain and Portugal.

The term 'gentrification' was coined over 50 years ago now by the British sociologist Ruth Glass (see Lees *et al* 2008, 2010). Her coinage referred to the invasion of middle class people into previously disinvested inner city, working class neighbourhoods in London, where they renovated old properties and displaced working class communities. Half a century later and that same process has mutated beyond the renovation of old properties and beyond London. In what follows I outline, if briefly, what gentrification is today – that is 'planetary gentrification', followed by a discussion of resistance to this socially unjust process.

Planetary gentrification

In their subsequent book Lees *et al* (forthcoming) go one step further and use the term 'planetary gentrification' (see Case study 12.1 for a definition) to refer to gentrification around the globe today. Key to this process is (re)investment in the secondary circuit of capital – the built environment, real estate. Importantly, in some parts of the Global South this is happening at the same time as investment in the primary circuit of capital (industrial production), for example, in China; whereas in other places (re)investment in the secondary circuit is trumping (re) investment in the primary circuit, for example, Dubai. The book underlines the fact that the planet is living in a 'property moment'. Much of this state-led gentrification is taking place on formerly public land; and like with 'planetary urbanization': 'rural places and suburban spaces have become integral moments of neoindustrial production and financial speculation, getting absorbed and reconfigured into new world-regional zones of exploitation, into megalopolitan regional systems' (Merrifield 2013: 10). As such planetary gentrification is also destroying the old dichotomies between urban and rural gentrification, inner city and suburb, and so on. The process plays out in the form of 'accumulation by dispossession', Harvey's (2003) upgrading of 'primitive accumulation' for the twenty first century neoliberal context, in which urban (also suburban and rural) land is appropriated and value extracted from it to facilitate capital accumulation (see Chapter 18, this volume).

Twenty-first century gentrification in the Global North is continuing unabated. Indeed, even cities like Los Angeles infamous for their suburban sprawl are experiencing gentrification in their old core (Lees 2014). There is no singular type

---

CASE STUDY 12.1 GLOSSARY OF TYPES OF CONTEMPORARY GENTRIFICATION

Creative gentrification: a term coined by Peck (2010) in reference to the gentrification caused by creative city policy.

Hyper-gentrification: a heightened, almost frenzied gentrification caused by over-heated property markets, in London, San Francisco, etc.

Mega gentrification and *mega displacement*: the large scale redevelopment of urban land in mega projects and associated massive displacements, usually through mass demolition and eviction (see Lees *et al*, forthcoming).

New-build gentrification: a term coined by Davidson and Lees (2005) in reference to new-build developments marketed at gentrifier types.

Planetary gentrification: a term coined by Lees *et al* (forthcoming) that takes seriously Merrifield's (2014b) call for a 'reloaded urban studies' that removes centre-periphery binary thinking and acknowledges the emergence of multiple centralities (gentrifications) across space (the planet).

Rental gentrification: where the gentrified property/apartments are for rent, not to buy.

Slum gentrification: a term coined by Lees *et al* (2015) referring to either the state-led, developer, or individual gentrification of slums.

Super-gentrification: a term coined by Lees (2003) referring to the re-gentrification of already gentrified inner-city neighbourhoods by a new breed of much more wealthy gentrifier.

---

of gentrification – there are multiple types, which have distinctive if inter-related features. The classic, pioneer, first wave or 'sweat equity' gentrification undertaken by 'new' middle class individuals (poor in economic capital but rich in cultural capital) still happens in cities and even rural areas, but it is less significant now. The dominant type of gentrification these days is new-build, and it is state- or developer-led. The big change has been the sale of public land into gentrification – the state-led gentrification of large areas of public housing in London, and in US cities like Chicago, New Orleans, and New York. This 'new' urban renewal (Hyra 2008; Lees 2012) of public housing bears many similarities to the large-scale urban renewals happening elsewhere in the world, e.g. Shanghai and Mumbai (see Weinstein and Ren 2009). Drawing on Merrifield (2013), Lees *et al* (forthcoming) have called these mega-gentrifications 'neo-Haussmannization'.

For Baron Haussmann the plight of the poor was not his administrative responsibility; like European nineteenth-century philanthropists Haussmann assumed the continued poverty of the poor and privilege of the rich. That assumption continues in some parts of the world, but less so others. In the US and the UK, for example, the 'renewal' of public housing is being undertaken through the auspices of 'mixed communities policy' – social uplift of the poor via social and economic trickle down effects, even if the end result is gentrification by stealth (see Bridge *et al* 2011). By way of contrast, in China (see Zhang and Fang 2004) and South Korea (Shin 2009) urban renewal is sold to the public as necessary modernization. These differences deserve more attention in the gentrification literature, and there are some suggestions that gentrification in the Global South may be more visceral (see Lees *et al* 2015, forthcoming).

Such schemes can be characterized as slum gentrification in both the Global North and the Global South. Indeed, slum gentrification is taking off globally, and can be found in cities as far apart as Lisbon (Ascensao 2015) and Rio de Janeiro (Cummings 2015). Slum gentrification has also underlined the fact that gentrification scholars need to learn from atypical examples. Lemanski (forthcoming), for example, demonstrates the correlation between gentrification and 'downward raiding' in South African slums. Situated in two different literatures 'gentrification' and 'downward raiding' have been seen until now as two very different processes, but the sale of state-subsidised houses in South Africa has meant that downward raiding has become implicated in gentrification. Lemanski demonstrates how urban theory can learn from diverse empirical sites and warns against the dangers of narrowly, empirically embedded concepts like gentrification and downward raiding. Cummings (2015) talks about the 'favela chic' related to slum gentrification in Rio de Janeiro, and this trend is global, for Burnett (2014) talks about poor and marginalized residents in the Downtown Eastside of Vancouver, Canada, having become a competitive niche for the promotion of distinctive and authentic culinary adventures. These trends for poverty tourism in and around slums demonstrate the commodification of poverty and a more complex type of gentrification.

Planetary gentrification is causing and revealing expanded notions of displacement (see Ghertner 2014) and pushing 'displacement' beyond the Euro-American geographies within which it was developed. 'Displacement' is a term that can be traced back to early sociological studies on the rapidly urbanizing cities of Europe and work by the Chicago School into the destruction of identity and community in modernizing American cities. As such it has real resonance in terms of researching gentrification in the rapidly urbanizing cities of the Global South. Re-theorizing displacement with respect to planetary gentrification would seem to be an important goal. This means stepping back from the Euro-American conceptualizations of displacement that dominate gentrification studies – e.g. Marcuse (1985). It means unpacking assumptions about 'accumulation by dispossession' and asking what such an idea of displacement conceals, neglects, and disavows. The aim is not to move away from generalizations about displacement but to expand the theoretical and conceptual repertoire of it where possible, and to seek methodologically innovative ways of investigating it around the globe.

Other new types of gentrification are emerging too. Rental gentrification is taking off in cities like London, New York and San Francisco, where owner occupation is out of reach of even the middle classes, as developers realize yet another gentrification niche. As Northern cities have progressed well into post-industrialization, Richard Florida's (2004) 'creative city thesis' has been an attractive one for policymakers and city 'fathers' (sic) because it recasts urban competitiveness between cities as cultural and economic 'creativity'. Creativity is now seen to be the key to economic growth, in a policy sphere where there are few, if any, new ideas. In Europe, hardly any other city has relied on Richard Florida's creative city thesis as heavily as Hamburg. It all began when Hamburg's then science minister, Jörg Dräger, distributed Florida's books to his fellow Hamburg Senate members, claiming that they offered good ideas for Hamburg's future. The city of Hamburg then hired the management-consulting firm Roland Berger to examine how Florida's theory could be applied to Hamburg. The result was a policy document titled 'Hamburg, City of Talent'. The result in many cities has been what Peck (2010) has termed 'creative gentrification', which extends 1990s urban entrepreneurialism into a more fully neoliberal world based on competition, middle-class consumption and place marketing. And as Lees *et al* (forthcoming) show, this thesis has been taken on board in the Global South too, as creativity becomes the new buzzword in the economic redevelopment of East Asian cities, like Shenzen, and others (see Chapter 11, this volume).

It is evident now that the state (including policymakers) and developers are leading gentrification agendas; the middle classes are the consumers, rather than the producers of gentrification today. Indeed, planetary gentrification urges scholars to revisit the role of the middle classes in gentrification. The emergence of new middle classes in Asia, Africa and Latin America is not the same, the temporality of this emergence differs, as does its relationship to the state, urbanization and (de) industrialization; their politics and ideologies are different too (see Lees *et al* forthcoming). There is a new global middle class that is urban(e), but how it acts in different places and at different times can be quite different. And how they act remains important, less so now in terms of their role in gentrification, perhaps more in their role against gentrification.

It might be that in East Asia where the middle class has been nurtured by the state and where a politics of property has consolidated as the region has urbanized and (de)industrialized the middle classes are less likely to stand up against gentrification. But in other parts of the world, e.g. parts of Turkey, Latin America and South Asia, the middle classes are involved in intra-class battles over gentrification. It might be that the middle classes that do rise up against gentrification are those whose material interests have not been guaranteed. As Weinstein and Ren (2009) argue, any comparative analysis of housing rights needs to focus on the structural conditions shaping housing rights regimes, the fragmented forms of urban citizenship produced by different regimes, and the opportunities for housing activism and resistance emerging in globalizing urban contexts, like Shanghai and Mumbai which they studied (see also Shin 2013).

FIGURE 12.1 Anti-gentrification protests, Taksim Square/Gezi Park, Istanbul, 2013 (Source: Tolga Islam).

Planetary resistance

The fact that resistance to 'gentrification' is becoming more noticeable around the world underlines the notion of 'planetary gentrification'. Resistance to gentrification made the international headlines in Taksim Square in Istanbul, Turkey, in 2013 (see Figure 12.1; also Angell *et al* 2014). Not since the Tompkins Square Park riots in New York City (see Smith, 1996) had anti-gentrification resistance made international headline news. The protests began when a small group of activists tried to stop the demolition of Gezi Park – one of Istanbul's few remaining green spaces in the larger Taksim Square; however, the protesters were also unhappy with the rapid urban changes in Turkey's larger cities. They criticized Erdogan's government for its neoliberal promotion of gentrification in Istanbul, arguing that the 'urban regeneration' plans hit the neighbourhoods that house the poor, immigrants, Kurds, and the Roma hardest. The police responded with extreme violence and the protests grew, with millions of demonstrators on the street, and copycat protests were held in dozens of cities across Turkey. More than 8,000 people were injured; six demonstrators and one policeman were killed. The struggle for Gezi Park further intensified and diversified unscripted encounters, transforming a public park – through the collective work of anonymous urbanites – into a commons (cf. Harvey 2012: 73; Lees *et al* forthcoming) (see Chapter 6, this volume). Resistance to gentrification has been more active and even successful in cities of the Global South and activists in the Global North could learn from their stories. Two good examples are the case of barrio women in Chacao, Venezuela (see Case study 12.2) and of the Pakistan Mahigeer Tahreek in Karachi, Pakistan (see Case study 12.3).

CASE STUDY 12.2 *BARRIO* WOMEN IN CHACAO, VENEZUELA, RESIST GENTRIFICATION (VELÁSQUEZ ATEHORTÚA 2013)

As a result of the reforms enacted by a socialist government in Venezuela to contest neoliberalism, *barrio* women in Chacao, Caracas, were empowered and mobilized into an urban struggle against the gentrification of their *barrio*. Gentrification began in Caracas in the 1990s with neoliberal decentralization reforms that promoted widespread gentrification, fed by oil revenues that concentrated global capital in Caracas. There were redistribution policies that helped the poor somewhat, but the rich benefited more from luxury housing and shopping and from zero-tolerance policing models imported from New York City (courtesy of a visit from William Bratton, ex NYC Police Chief), and the police protected the interests of the wealthy. This revanchist urbanism began to be contested. Latin America of course has a rich history of political mobilization from the Movement of Chilean Settlers (Castells 1973) to the Brazilian National Union for Popular Housing and the Urguyan Federation of Self-help Housing Cooperatives. The Settlers Movement in Caracas had two goals: the inclusion of poor *barrios* in their cities, and to fight against the accumulation of urban land and gentrification. The resistance was peaceful but it was helped by the support of the Socialist majority in the National Assembly who then passed a series of reforms that supported the People's Power (El Poder Popular), in a top-down approach that started from below. The aim was to allow the underprivileged full access to participatory democracy. The result was that the bourgeois could remain and retain its privileges but it had to live with a New Communal Order of the underprivileged. In Caracas, the government allowed them to build a pioneer's camp (a Campamento de Pioneros) on the land, to begin the process of building a Socialist Community for 600 families. They were given 38 million Bolivares (about US $9 million) from the government. There was fierce push back from the bourgeoisie, forcing Hugo Chavez's government to get the People's Power to govern with presidential Missions. This became a new model of social policy development that involved people contesting neoliberalism through the marginalized being involved in executing government programmes. Unlike in the Global North where public participation in urban renaissance has been a post-political process of inclusion that silences (see Lees 2012, on London), in Caracas it has been more real. There were three institutional building waves in Caracas that allowed *barrio* women in particular to achieve growing agency – in 2002

Urban Land Committees (Comites de tierra Urbana) were formed to regularize urban land; in 2006 Communal Councils were formed in which 150–400 families would formulate, implement and monitor central government public policy, allowing cooperatives to emerge. Then in 2008 Socialist Communes were made from aggregated communal councils, the aim of which was to address extreme poverty and corruption. This created spaces of insurgent citizenship and urbanism (cf. Holston 2009; Miraftab 2009). In addition the Settlers also created their own organizations that worked in the spaces the government had created; these organizations defended housing as a human right and contested the privatization of land, and neoliberalism. In partnership with the Socialist state they could stand up to the power of real estate elites, bankers and developers etc. In so doing they successfully fought off an 'urban renewal' project that wished to gentrify the Old Market in Caracas. There are few such successes in the Global North.

CASE STUDY 12.3 THE PAKISTAN MAHIGEER TAHREEK IN KARACHI RESIST GENTRIFICATION (HASAN 2012)

In Pakistan, since 1999, there have been attempts to gentrify Karachi's coastline using global capital and Dubai- and Malaysian-based real estate companies (see Hasan 2015). In 1999 there was a military coup and General Musharraf came to power. He appointed a Chief Vice Executive of Citibank as Finance Minister, a senior economist from the World Bank as Governor of the State Bank of Pakistan, and another World Bank person as Minister for Finance, Planning and Investment. Devolution saw the election of mayors for Karachi the second of which had international links. Dubai-based companies negotiated with the government of Pakistan and 'projects' replaced 'planning'. One of these projects, Sugarland City, which was initiated in 2006, involved the privatization and development of a number of Karachi's public beaches (a massive 26,000 hectares of land) which were to be developed by Limitless, a Dubai-based company launched by Dubai World, into a new residential, commercial and recreational city at a cost of US $68 billion. An MoU was signed between the Minister of State and Privatization and Investment and the Dubai World chairman. Limitless was set up in 2005 with the key objective of globalizing Dubai's portfolio of leading development companies by leveraging the know-how and exposure gained by Dubai World's real estate initiatives. But there was opposition to the project from the Pakistan Mahigeer Tahreek (the Movement of the Indigenous Coastal Fisherfolk Communities of Pakistan). They produced a letter titled 'Development to Destroy Nature and Displace People' in 2007, the outcome of discussions between various stakeholders, but especially local communities. As well as outlining the destruction of nature –

from green turtles, to mangroves, to fish and birds, they also were clear that the proposed development would displace people – the fishing communities who had been living on the coast for centuries. The project, it was claimed, would impact their livelihoods, which were based on subsistence fishing and beach leisure activities. Incredibly, despite more than 100 villages being in the project area, their future was not mentioned at all in the project proposal. The letter also claimed that given that lower and lower middle-class Karachiites would not be able to go to the beach this would increase the divide between the rich and poor in society. The letter was followed up by public demonstrations and a press campaign. Meetings were held with the Chief Secretary along with prominent civil society individuals, and, because of opposition from all segments of society, Limitless backed out of the project in 2009.

As said there are few examples of successful resistance to gentrification in the Global North, but Hamburg's plans for a new creative city (mentioned above) came up against an urban social movement of artists (see Novy and Colomb 2013) who did not want creative gentrification. Resistance coalesced around an alliance of activists called 'Right to the City'. Made up of over 20 citizens' initiatives they were united in their fight against gentrification: the deliberate and politically expedient upgrading of poorer neighbourhoods, leading to the displacement of the existing population with more affluent new residents. A manifesto against the branding of the city titled 'Not in Our Name! (see NION 2010 and www.nionhh. wordpress.com) was published by Hamburg artists, musicians and social activists in October 2009. Here is an excerpt:

> Dear location politicians: we refuse to talk about this city in marketing categories. We don't want to 'position' local neighbourhoods as 'colourful, brash, eclectic' parts of town, nor will we think of Hamburg in terms of 'water, cosmopolitanism, internationality', or any other 'success modules of the brand Hamburg' that you chose to concoct. We are thinking about other things. About the million-plus square metres of empty office space, for example, or the fact that you continue to line the Elbe with premium glass teeth. We hereby state, that in the western city centre it is almost impossible to rent a room in a shared flat for less than 450 euro per month, or a flat for under 10 euro per square meter. That the amount of social housing will be slashed by half within ten years. That the poor, elderly and immigrant inhabitants are being driven to the edge of town by Hartz IV (welfare money) and cityhousing-distribution policies. We think that your 'growing city' is actually a segregated city of the 19th century: promenades for the wealthy, tenements for the rabble.

Interestingly the response from the City was not militant, rather they appeared to back down and the Senate decided to buy back the buildings from the Dutch investors they had sold them to, costing them €2.8 million more than the Dutch had paid for the buildings in the first place. NION stated that this extra €2.8 million represented the cost the city incurred for listening to Richard Florida! Novy and Colomb (2013) argue that in cities around the globe there is mounting evidence of growing mobilization by members of the so-called 'creative class' in urban social movements, defending particular urban spaces and influencing urban development. In Hamburg, artists have maintained a progressive role in mediating between the City and activists, and some experts, including Richard Florida, now see Hamburg (ironically) as a new model for socially sustainable creative cities.

To date the literature on resistance to gentrification has been very limited (Lees and Ferrerai forthcoming) and dominated by case studies from the Global North, which must now change. If we are to take the charge of comparative urbanism seriously then we must start to learn from the successful resistance to gentrification in cities of the Global South and resistance to new types of gentrification in the Global North too. Cross-planetary learnings on gentrification should be the goal of twenty-first century gentrification studies.

Conclusion

It is evident then that gentrification as a process has become more complex over time, the interests and actors have diversified and it has become the leading edge of urban policy globally (Smith 2002). Significantly the movement and diversification of gentrification around the globe has not been North to South, but South to South and even at times South to North (Lees *et al* 2015, forthcoming). The emerging work to carry the burden of comparative urbanism into gentrification studies is already teaching us a lot. So what might the key concepts be for a twenty-first century gentrification studies that takes comparative urbanism seriously? I conclude with four suggestions:

1. *Spatial capital*: it is becoming apparent that the concept of 'spatial capital' theorized in Swiss core cities by Rerat and Lees (2011) as the locational advantage that gentrifiers use as a mark of social distinction is increasingly relevant in Southern cities where new forms of spatial capital are being created in previously undervalued urban areas (see Cummings 2015, on the *Metrocables* in downtown Rio favelas; López-Morales (2010) on the *Efecto Metro* in Santiago, Chile, which has opened areas to real estate developers and housing consumers, enlarging rent gaps, and enabling new social classes to capture spatial capital, so generating indirect forms of displacement; and Blanco *et al* (2014) on gentrification and mobility in Argentina).

2. *Planetary rent gaps*: rent gaps in cities around the world are being created so as to set up the necessary conditions for gentrification. Lopez-Morales (2011) discusses 'ground rent dispossession' in reference to how the local government in Santiago, Chile, has liberalized local building regulations (seeking to increase potential ground rents) and deliberately underinvested in housing upgrading (seeking to

devalue the current ground rent). Shin (2009) has shown the same process to be operating in South Korea, and Lees (2012) has researched the state-led disinvestment of London's council estates leading to what Watt (2009) has called a 'state-induced rent gap'. Slater (forthcoming) argues quite forcefully that we need to confront these new geographies of structural violence.

3. *Planetary Displacement*: 'accumulation by dispossession' is global, but it plays out differently in different places (Doshi 2013; Lees *et al* forthcoming), and these differences are important, both politically and practically, in the fight against gentrification.

4. *Planetary resistance*: collating detailed stories of resistance to gentrification from around the world (old and new) and cross-fertilizing the lessons from them has not yet been done; but if, as gentrification scholars, we believe in social justice in cities then improved conceptualization of resistance to gentrification needs our attention.

This means returning to some old concepts in (Anglo-American) gentrification studies – locational advantages and mobility practices (Berry 1985); rent gap theory (Smith 1979, 1982); conceptualizations of displacement (Marcuse 1985); studies of resistance (Hartman 1974) – and re-theorizing them for a twenty-first century planetary gentrification. This might well mean that we reject these earlier concepts and develop new ones theorizing *from* the South; this would mean 'unlearning' (drawing on Spivak 1993) existing dominant literatures that continue to structure how we think about gentrification, its practices and ideologies (Lees 2012: 156). The future of gentrification studies will be determined by how effectively we can challenge our preconceptions about this mutating process.

Current work in gentrification studies is enriching urban theory by introducing new concepts (like slum gentrification, spatial capital, etc.), by refuting the utility of some old concepts (like the new middle class as producers of gentrification), and, importantly, it is listening to ideas from urban scholars around the world, doing the real comparative learning that is necessary, not just for the future of gentrification studies, but also for urban theory itself.

Key reading

Lees, L., Shin, H. and López-Morales, E. (forthcoming) *Planetary Gentrification*, Polity Press: Cambridge.
    The launch text of a new series by Polity Press on the future of cities, this monograph discusses gentrification around the world in terms of planetary urbanization. It introduces some new concepts, e.g. slum gentrification, mega gentrification and mega displacement.
Lees, L., Shin, H. and López-Morales, E. (Eds) (2015) *Global Gentrifications: Uneven Development and Displacement*, Policy Press: Bristol.
    This edited collection provides detailed case studies of gentrification from around the world, from cities as diverse as Lagos, Damascus and Athens. All the cases are from cities beyond the usual suspects in the Global North.

Lees, L., Slater, T. and Wyly, E. (2010 first edition; second edition forthcoming 2017) *The Gentrification Reader*, Routledge: London.

This reader provides students with the key classic essays written in Anglo-American gentrification studies to date.

Lees, L., Slater, T. and Wyly, E. (2008) *Gentrification*, Routledge: New York.

This is the first, and remains the only, textbook on gentrification to date, and provides an accessible introduction to gentrification studies.

13

GOVERNANCE

Mark Davidson

Introduction

In the 1980s and 1990s, urban scholars documented how 'government' was being replaced by 'governance' (Cochrane 1989, 1991; Harvey 1989a; Pickvance and Preteceille 1991). This process was principally concerned with the 'growth in non-elected and quango (quasi autonomous non-governmental agencies) local institutions, the increased participation of business actors in local decision making (Peck 1995; Peck and Tickell 1995; Strange 1997), and the introduction of new types of competition-based urban politics …' (Ward 2000: 169–170). The governing of cities therefore underwent a transformation that brought in new actors into city hall, created new institutions and gave preference to the interests of business. David Harvey (1989a: 4) described this transformation as a shift from managerialism to entrepreneurialism:

> Put simply, the 'managerial' approach so typical of the 1960s has steadily given way to initiatory and 'entrepreneurial' forms of action in the 1970s and 1980s. In recent years in particular, there seems to be a general consensus emerging throughout the advanced capitalist world that positive benefits are to be had by cities taking an entrepreneurial stance to economic development. What is remarkable, is that this consensus seems to hold across national boundaries and even across political parties and ideologies.

The shift from 'government' to 'governance' therefore involved the forms of, and motivations for, urban politics undergoing radical change.

Many have criticized the form of urban politics that emerged in the 1980s for its lack of concern for social welfare and regressive impacts on income distribution (for a recent example see Dorling 2014). Raco (1999: 271) claimed that 'heightened emphasis on competition between places for public- and private-sector investment has led to the marginalization of issues that are seen as 'negative' or harmful to the cause'. This emphasis on the perceived benefits of private ownership (i.e. privatization) and a declining role of the state in remedying social problems has, of course, not simply concerned cities. These ideas have driven a broader remaking of the state (Brenner 2004), where ideas of market intervention, social mitigation and welfarism have been replaced with an over-riding concern for economic growth, free markets and privatization (Harvey 2005).

Although the shift to neoliberal governance has been dramatic, particularly in places such as the United Kingdom, some have questioned the extent to which the shift from managerial to entrepreneurial governance can be described in dualistic – before and after – terms (Imrie and Raco 1999). For example, Ward (2000: 181) argued that conceptualizing this transfer from government to government in dualistic terms is only 'partially revealing'. Instead, it is argued: '[T]here are a hybridity of possible (and potential) forms, incorporating a range of institutional frameworks, and modes of policy development and implementation' (ibid 181). Put simply, the transition from urban government to urban governance has involved much continuity, as well as significant change. In attempting to conceptualize contemporary urban governance, it is therefore necessary to identify more specifically how government contrasts to governance. From this basis, our attention can then move to consider (i) why governance seems to have become the predominant conceptual approach used to understand how contemporary cities are regulated and controlled, and, (ii) understand what is particular about contemporary forms of governance. In terms of the latter, the end of the chapter will discuss how contemporary urban governance appears to have an uneasy relationship with notions of 'good governance' within entrepreneurial/neoliberal ideology.

Theorizing governance

The term governance is used in numerous ways. People talk of 'corporate governance', 'good governance' and 'organizational governance', amongst applications of the term. Often the 'governance' is used to suggest something broader than government. It concerns a wider set of regulatory mechanisms that elicit – or demand – certain modes of behaviour and organization (Foucault 2008). Urban governance can therefore be understood as those modes of behaviour and organization that can occur within or through the city. For example, we might talk about certain lifestyles as types of urban governance. Or, we might talk about particular types of policy initiatives that bring together private and public actors as being a form of urban governance. Given there are so many different ways that urban societies are organized and regulated, it is difficult to conceptualize them all as a whole. It is therefore necessary to identify how urban scholars have used the concept of governance.

Urban scholarship is often concerned with the city as a political entity. Yet this can be interpreted in a variety of ways. Some look specifically at the ways that (urban) society formalizes political decisions and generates laws and institutions to uphold them (i.e. government). This can be broadened by thinking about how the state develops and/or enables certain practices and logics that condition people in particular ways (i.e. governmentality). Beyond this, we can look at both governmental and broader forms of social organization as mechanisms that (re) produce cities (i.e. governance). Here we will focus on the last of these perspectives: the city as a site of social coordination and reproduction (see Rose and Miller (1992) on governing beyond the state).

In order to think about the ways in which an incredibly complex social arrangement like a city is coordinated and reproduced requires a powerful theoretical basis. We must think about those core processes that cut through the immense variety of communities, peoples and practices that are present in cities. In this chapter I will draw on two political philosophers who have attempted to develop such theories: Jacques Rancière and Slavoj Žižek. Whilst neither of these theorists has specifically developed a theory of governance, both are concerned with thinking about how society is structured, how it reproduces itself, and how we can change it. As such, many urban scholars are now attempting to use their theories to generate an understanding of how the contemporary city reproduces itself (e.g. Dikeç 2005; MacLeod 2011).

The following sections will draw on both of these theorists to develop an understanding of urban governance. We will begin with Rancière's theory of politics (Rancière 1999). This theory will enable us to think about what governance is and how it relates to questions of social change. We will then move onto Žižek's discussions of ideology (Žižek 1989). In particular we will examine how Žižek's theories help us understand the perverse logics of contemporary urban governance, where seemingly illogical choices continue to dominate. The chapter's conclusion will introduce some political questions that emerge from contemporary interpretations of urban governance.

Politics and governance in the city

> The triumph of the West, of the Western idea, is evident first of all in the total exhaustion of viable systematic alternatives to Western liberalism.
>
> *(Fukuyama 1989)*

Above is a very (in)famous quotation from Francis Fukuyama's *The End of History?* essay. In this piece Fukuyama claims that the end of the Soviet Union was a pivotal point in human history. For him it represented a point where debates about different types of social organization – what we might call forms of governance – had ended. The failed communist experiment – if that is what it was – had proven that only one type of government/governance was legitimate: liberal, democratic and capitalist. Since Fukuyama wrote this essay, he has been subject to all kinds of criticism. But it is hard to argue that he was wrong. Has the western mode of governance (i.e. liberal democratic capitalism) not become globally dominant? It is certainly difficult to find any mainstream debates about other legitimate forms of social organization. However, there is an important distinction many have made between the powerful nature of these ideas and their actual implementation into governance practices.

Jacques Rancière argues that democracy as it exists today certainly does not reflect the ideas it is founded on. To simplify Rancière's (1999) philosophy, democracy is founded on the idea of equality: one person, one vote. Other social arrangements do not have this founding principle. For example, you can have society organized and governed according to race (e.g. South Africa under apartheid), wealth (e.g. a plutocracy) or religion (e.g. religious states), amongst many other orders. In these examples, not everyone has the same right to govern; some are more powerful and influential than others. As such, they have different beliefs about legitimate government (e.g. the highest-ranking religious figures must decide since they are closer to god). What makes democracy unique is that nobody has a natural right to govern: democracy is 'founded on the absence of any title to govern' (Rancière 2006: 44). Rancière uses this understanding of democracy to theorize politics. He argues politics are moments of social change whereby democracy is reaffirmed. This means that politics occur when an inequality is corrected and democracy (re)produced. Given that democracy, as a mode of governance, is legitimated via the idea of equality, all governmental actions are to be assessed accordingly.

This interpretation of politics gives us a criterion with which we can assess contemporary urban governance. We can ask if cities are governed in ways that are based on democratic or hierarchical forms of governance. Rancière uses the terms 'policing' and 'distribution of the sensible' to describe these governance arrangements. The latter term tries to capture the ways that society assesses what is possible or legitimate. If something is deemed insensible, it is often removed from public consideration; and therefore beyond the realms of political discussion. This term therefore indicates how Rancière's political theory goes beyond the state in considering how society is governed. One example of the 'distribution of the sensible' might be the ways in which societal understandings of race have changed in the United States. In 1950 it would have been difficult to find public discussion about an African-American president. To propose such an idea would have been popularly illogical (i.e. insensible) due to racist tendencies within that society.

Using this basic outlining of Rancière's theory of politics, we can begin to assess governance in contemporary cities (see Davidson and Iveson 2014). In the last 30 years, the city has been governed according to the logics of entrepreneurialism (see Chapter 11, this volume). As an entrepreneurial city, governance must make a city more competitive. With cities fighting amongst each other for investments, middle class residents and tourists, cities, it is argued, must compete with other cities in order to maintain economic and social vibrancy (Harvey 1989a). Oftentimes this translates into slashing tax rates for corporations and developers, displacing undesirables and preferring some businesses over others. Those asked to direct such governance projects are those 'in the know'. They range from policy advisors (e.g. Richard Florida, Jan Gehl, Charles Landry), to business owners, and to political consultants.

CASE STUDY 13.1 THE RISE OF THE URBAN CONSULTANT

This shift from government to governance has been accompanied by the rise of the 'urban expert'. These people, who are usually white, privileged and well-educated men, include the likes of Richard Florida and Rem Koolhaas. The role for such experts within the everyday governing of cities is wide-ranging but includes architecture, planning, social reform and policymaking. The central theme running across the various roles of the urban expert is the idea that the efficient governing of the city is a matter of technical expertise, not necessary democratic choice. It therefore becomes important for the city to capture/attract the best experts to make their city efficient and, therefore, competitive (see McCann 2011 on the globalization of urban policy). In this context, the authority to govern the city does not simply come from its demos, but rather from the privileged knowledge of the globe-trotting urban expert (see Peck 2005 for critique). McNeill (2009) has compared the rise of the expert, globe-trotting architect to the celebrity: 'the signature architect is a form of celebrity which extends into commercial, professional, and even general public recognition' (ibid 154).

When cities take the advice of these 'experts' and respond to the demands of corporations, are they governing in a democratic manner? It is difficult to argue they are. It is clear some viewpoints are given preference over others. For example, the investor is prioritized over dissenting voices. The logics of entrepreneurialism are so pervasive that tax breaks and consultancy spending are often not offered up for public debate. Those who protest such governmental moves are often condemned for putting their own parochial interests in front of the city's general well-being.

It is not difficult to see the problems of such urban governance. The city is seen to be coerced into certain decisions that are beyond the realm of debate and/or contestation. And, importantly, anyone who contests whether the dictates of entrepreneurialism should be ignored and/or subverted is viewed as illegitimate. Contemporary urban governance therefore contains within it a strange paradox. Its legitimacy rests on the claim that it is democratic. And yet its actions all too often rely on a technocratic rationale; that action must cohere to the dictates of external forces.

Entrepreneurial governance is therefore a powerful process of social coordination. It influences extend far beyond instruments of government. It operates within widespread understandings about what cities should do. You can find it in comments about 'what the city must do', or 'that wage cuts are inevitable in this economy' or 'too much debate is getting in the way of the redevelopment proposal'. It has become embedded in what many people understand as the public interest, often to the extent that democratic decision-making can be replaced by technocrats (Swyngedouw 2009).

Since the advent of entrepreneurialism in the late 1970s, the transformation of urban governance has therefore not simply occurred in city hall. It has also occurred in public understandings of what the city does, who runs it and what it operates for (i.e. in the 'distribution of the sensible'). This has given rise to a situation whereby the perceived ways in which a city can be organized are defined very narrowly. Put simply, most people believe cities have few choices but to adhere to certain logics. This inevitability – what some call the 'post-political' form of politics (Mouffe 2005) – can be evidenced in a variety of discussions that see certain forms of governance as inevitable: income inequality must be excused since redistribution is anti-business; tax breaks must be extended, since businesses will flee; higher city rankings must be aspired to, since not pandering to such metrics would present the city as undesirable and unproductive (Davidson and Iveson 2014).

When we constrain the governance of our cities in such ways, it is difficult to identify what makes them democratic. At one level, the city operates as a system of social coordination that makes them entirely facilitators of market processes (Harvey 1989a). This generates a technocratic set of processes and institutions that aim at enabling inter-city competition and growth. On the other hand, urban governance operates as a set of ideas: that urban governance is democratic and (largely) liberal. This elevates a set of ideas associated with democracy (e.g. participation, consultation and empowerment) that are often incongruous with the logics of entrepreneurial governance. How then do these two things co-exist? To answer this question, we must consider the ideological characteristics of urban governance.

The ideologies of urban governance

As we have seen, Rancière (1999) understands democracy as being founded on equality. Politics happen when moments of social change (re)produce equality; where an inequality is articulated and its remedy necessitates a change in that society which produced the inequality. Outside of these rare moments, society operates through a set of regulatory mechanisms that Rancière labels as 'police' and 'policing':

> The police is thus first an order of bodies that defines the allocation of ways of doing, ways of being, and ways of saying, and sees that those bodies are assigned by name to a particular place and task; it is an order of the visible and the sayable that sees that a particular activity is visible and another is not, that this speech is understood as discourse and another as noise ... Policing is not so much the 'disciplining' of bodies as a rule of governing their appearing, a configuration of *occupations* and the properties of the spaces where these occupations are distributed.
>
> (Rancière 1999: 29; emphasis in original; cited in Davidson and Iveson 2014: 5)

From this definition of 'the police' it is clear that Rancière sees society as having an overarching regulatory arrangement; what we might call governance. Policing is concerned with the ways we do things, how we talk about things, and what is deemed appropriate (i.e. what is sensible). It precedes active disciplining in that regulation is deeply embedded in the ways we think and engage with the world around us.

The ideas that shape our governance arrangements, like democracy, liberty and freedom, should therefore be considered part of the police. As such, ideas that counter these – autocratic, collectivist, monarchy – might be deemed insensible. But, as much recent urban scholarship has argued, these ideas are often not enacted. As Rancière claims, contemporary police orders contain a deep paradox, where their claims to legitimacy are countered by the argument that the state must take the only actions possible. It must act because of necessity, not because of democratic decision-making.

In order to understand how this paradox of urban governance operates, we can turn to another contemporary political thinker, Slavoj Žižek. Žižek uses the concept of ideology to examine the ways we think. Unlike some other Marxists, Žižek does not view ideology as a form of false consciousness – something that stops us seeing how things really are. Rather ideology is the way we structure our understanding of reality. It is the irremovable glasses we use to make sense of the world around us. As such, ideology can be said to take the form of fantasy, where we apply a set of ideas to interpret the world. Crucially, ideology is a social product. It can be thought of in the terms of language. Language is something we all share and we use it all the time to understand the world. We also know that language is unstable. Some words can change their meanings, leading to a cascading effect throughout our linguistic structures.

Ideology should be understood as 'those discursive forms that construct a horizon of all possible representation within a certain context, which establish the limits of what is "sayable" (Laclau 2006: 114). The parallels to Rancière's police are evident here. But I am interested in the ways that Žižek finds ideology to be an active shaper of reality. That is to say, ideology has its own internal workings. Žižek's extensive writings on ideology identify a huge array of ways that ideology operates. But here I want to focus on just one aspect: namely, the ways in which ideology enables ideas about governance to be upheld (e.g. democracy) even when they are clearly not translated into practice (i.e. post-democratic urban governance).

To start understanding this seeming paradox we must acknowledge that we hold ideology at a distance. Žižek develops Marx's idea of false consciousness to illustrate this. For Marx, workers had a false consciousness because they did not realize the exploitative nature of their employment. A significant contribution of Marx's thought was to reveal to the working classes the hidden exploitation within capitalist production. Žižek finds this analysis dated. He claims that today we know lots of exploitation, amongst many other problematic processes. And yet this does not result in social change. The comprehension of a social antagonism does not necessitate a remedying action, even if our ethics and principles might demand it. He therefore claims that contemporary ideology is characterized by the fact that 'even if we do not take things seriously, even if we keep an ironical distance, we are still doing them' (Žižek 1989: 33).

FIGURE 13.1   Vancouver's Olympic Village under construction in 2008.

We can apply this analysis directly to questions of urban governance. Take, for example, the continual bidding by cities for large-scale events such as the World Cup or Olympics. Time and time again, cities run into financial difficulties when preparing for the games. Cost overruns, unforeseen expenses and inefficiencies are all familiar news stories to those living in host cities. The most (in)famous examples of such speculative failures include Montreal's 1976 Olympics, where it took the provincial government 30 years to pay off the debts associated with Olympic developments. In other host cities, the difficulties associated with calculating the full economic impact of the games have often led to costs being incorporated into long-term budgeting. For example, Sydney's Olympic Park was constructed in the late 1990s to host the 2000 games. Yet efforts still continue today to make the site economically self-sufficient. This most symbolic of urban development agendas almost always involves a large initial investment with an unknowable return. This is hardly an investment deal that would be attractive at your local bank.

Across the numerous examples of loss-leading and debt-inducing events, one is therefore encouraged to ask the question: why do cities keep bidding for the mega-events? An obvious answer to this question is that cities do not have much of a choice. It is generally understood that the range of policy and development choices that cities can make is severely limited. So, bidding and hosting an event is simply all the city can do. It either develops in this manner, or does not develop at all. To some extent, this answer is correct. But this answer is restricted to the realm of government and governmentality. We need to look at questions of governance if we are to factor in our answer why there remains a political and popular desire to pursue development projects that offer little in terms of secure returns.

CASE STUDY 13.2 BIDDING AGAINST THE ODDS

The creation of spectacular urban events, such as festivals, sporting events and conventions, was recognized by Harvey (1989a) as a staple element of entrepreneurial urban governance. Cities, Harvey argued, were compelled to be speculative actors; they had to risk public monies to support and grow the public finances. This has meant that many cities have bid to host events such as the Olympics and FIFA World Cup. Although most bidding cities promise that these events will generate revenues, most – if not all – of these events have cost the public taxpayer significant amounts (see Flyvbjerg and Stewart 2012). Often times, this involves the construction of mega-event budgets where the city or state government provides assurances to private investors. Any cost over-runs (i.e. deviations from the budget) are therefore absorbed by the city or state government. Recent examples include the city of Vancouver (Canada) having to take ownership of the 2012 Winter Olympic Village – at the cost of over US $300m – due to the collapse of the project's real estate developer and hedge fund financing (see Figure 13.1).

Translated into Rancière's terms, we might therefore consider the complex collection of understandings, logics and legislative structures that make hosting an event the obvious option for city development part of 'the police'. Some types of economic and urban development that operate outside of these structures are literally policed, in the sense that it becomes contrary to societal rules. Mega-event bidding and development is simply the 'sensible' option. Other options, which might include the state funding small-scale cooperative enterprises or the state itself forming companies to produce things, are deemed 'insensible'. To stand up at a public meeting and suggest such alternatives might give you the impression that you are talking another language. Not only that, you might be accused of endangering the general well-being of the city by proposing unworkable development solutions.

Yet Rancière's understanding of 'the police' is not restricted to apparent governmental structures, since it extends to governance. The police justifies itself according to certain ethics and logics. In democratic societies government and governance is legitimized by the moral principles developed from democratic thought (i.e. equality). Government and governance should therefore reproduce the conditions of equality. This means that when an inequality is presented, the police order (i.e. society) should confront that clash in order that the inequality be corrected and society remain legitimately democratic: 'For politics to occur, there must be a meeting point between police logic and egalitarian logic' (Rancière 1999: 34). What this understanding of politics relies upon is a foundational enlightenment premise – that when a (social) contradiction becomes evident, this contradiction will be worked out and a new entity (i.e. society) emerges that transcends the contradictory state.

This dealing with contradictions can be illustrated simply. Imagine you eat something and it makes you sick – you are dealing with an antagonism: your food does not operate as nutrition. The logical response is therefore to stop eating the thing that makes you sick. This basic philosophical premise is foundational in our police orders. Watch the daily news and it is full of problems being reported, most of which are accompanied by an implicit or explicit message that something must be done (i.e. the contradiction dealt with). If we return to urban development as mega-event, we find our philosophical foundations operating in a strange manner. We produce these developments (i.e. eat this food), but they more often than not fail (i.e. make us sick). So why do we keep doing this? What is happening in our governance processes that continue these illogic activities?

For Žižek, the answer lies in how our ideology now works. Our police operates and is legitimated by premises (i.e. equality, liberty etc.) that are no longer taken seriously. Governance therefore tends to be deeply cynical. The idea that current modes of urban development run contrary to what we believe our governance structures should be doing is therefore already factored into our understandings. So, yes we know that hosting a mega-event will not produce the jobs, societal benefits and infrastructure that are promised, but we proceed as if we do not know this. Indeed, Žižek argues the agents of governance incorporate this societal cynicism into their actions:

> Cynicism is the answer of the ruling culture to this cynical subversion: it recognizes, it takes into account, the particular interest behind the ideological universality, the distance between the ideological mask and the reality, but it still finds reasons to retain the mask … This cynicism is therefore a kind of perverted "negation of the negation" of the official ideology: confronted with illegal enrichment, with robbery, the cynical reaction consists in saying that legal enrichment is a lot more effective and, moreover, protected by law.
>
> *(Žižek 1989: 29–30)*

Those broader societal processes that produce a system of governance are here found to have a strange logic, where the illogical outcome of actions (i.e. contemporary forms of urban 'development') does not result in a governance tension. Having contemporary urban development processes not work – for most people, and for those most in need – does not therefore seem to be an action-necessitating governmental (i.e. political) and/or governance (i.e. societal) problem.

Conclusion

Using Rancière's theories of the police and politics to conceptualize governance allows us to see how societal ideas shape the ways in which we approach, interpret and contest urban development. It also shows us that in democratic societies, there is a requirement that all socio-political actions re-inscribe the equality of peoples in order that government and governance remains legitimate. Where equality is not re-inscribed – such as when certain perspectives about what to develop are not permitted, or the particular outcomes generated by development are unjustifiable – then this type of government should be deemed illegitimate according to our governance ideology.

When we look at the paradigmatic modes of urban development and urban politics, it is clear that we often pursue urban changes that are contrary to our understandings of what they are meant to achieve. And yet, governments still bid for mega-events and most city residents continue to support them. For philosophers like Slavoj Žižek, this presents a critical problem: how do we solve societal and political problems when the manifestation of problems does not necessitate a corrective response? How is it that we cynically go on with types of urban governance that are divisive, unproductive and, at times, outright corrupt?

Over the past 40 years, urban governance has been critical in the production of social inequality, political indifference and the concentration of power. Entrepreneurial urban politics have benefited only a small section of city dwellers, particularly in those places where these politics have been pursued the most aggressively. Yet this mode of governance has become engrained in the ways that we think cities run. Both inside and outside of government buildings, we understand the parameters of urban governance to be narrowly defined. When the whole rationale of urban governance becomes undermined by its outcomes, we are yet to see a substantive countermovement; what Rancière might call 'politics'. As Žižek argues, this might be something to do with the ways that we all think about societal contradictions and antagonisms. That is to say, from a cynical viewpoint that has us carry on as though we do not know how illogical things really are.

The challenge for those studying urban governance today is therefore to (re) discover a type of politics that makes urban governance up for debate and discussion – one where our police order is open for questioning and reform. Whether this means returning to the ancient premises of democracy (à la Rancière) or coming up with a new theory of everything (à la Žižek) is unclear. Yet it seems critical urban scholarship has an important role to play in this project by exposing and interrogating those moments of disagreement, contestation and political imagination that have always been present in the city.

Key reading

Davidson, M. and Iveson, K. (2015) 'Recovering the politics of the city: from the 'post-political city' to a 'method of equality' for critical urban geography', *Progress in Human Geography* 39(5): 543–559.

This paper outlines a method for understanding contemporary urban governance using Rancière's theory of politics.

Harvey, D. (1989) 'From managerialism to entrepreneurialism: the transformation in urban governance in late capitalism', *Geografiska Annaler. Series B*, 71: 3–17.

A seminal paper that identified the main characteristics of neoliberal, entrepreneurial urban governance.

Rancière, J. (1999) *Disagreement: Politics and Philosophy* (translated by Rose, J.), Minneapolis: University of Minnesota Press.

The text in which Rancière explicitly sets out his theory of (democratic) politics. In recent years, this theory has become very influential with urban scholars interested in understanding the tensions within contemporary urban governance.

Ward, K. (2000) 'A critique in search of a corpus: re-visiting governance and re-interpreting urban politics', *Transactions of the Institute of British Geographers*, 25: 169–185.

A paper that carefully examines the precise nature of the shift from government to governance within cities during the 1980s.

Žižek, S. (1989) *The Sublime Object of Ideology*, London: Verso.

Žižek's first major English language text that sets out his Lacanian-inspired reading of Marxism and Marx's theory of ideology.

14

INFORMALITIES

Melanie Lombard and Paula Meth

Introduction

Urban informalities are a reality of all cities, despite their association with cities in the Global South. The concept commonly describes spatial, economic and political practices viewed as 'not-formal', such as slum housing and street trading. These processes and practices appear to occur outside of, or arise in contrast to, the formal mechanisms often entailed by state legislation and regulations. 'Informality', then, is a power-laden concept, used to criticize and condemn ways of living which are viewed by urban elites or urban authorities as something other than formal, or less than legal, and therefore inferior. In contrast, others have viewed informalities within cities as cause for celebration, an example of the agency and resilience of poor urban residents. Yet, as this chapter reveals, understanding informalities means more than simply describing practices in poor cities: it suggests revealing processes which are evident in most cities. Furthermore, the very debates over what is and what is not 'informal', and associated responses, may tell us more about wider city politics, city visions and the normativity of legal mechanisms used by city managers, than simply describing particular parts of cities which appear different to more formal, legal and wealthy areas.

This chapter predominantly focuses on literature and cases on the urban poor in the Global South following the historical association of informality with this group and region. However, we contend that there is much recent work to counter these assumptions, as the urban rich across the world operate through informal modalities too; and moreover, that poverty-related informal practices occur within cities in the Global North (see Chapter 8, this volume). The chapter first examines why urban informalities matter, before moving on to situate the concept of informality within wider urban theory. It illustrates these sections with two case studies – one from South Africa and one from Mexico – before concluding by highlighting the example of 'beds in sheds' in the UK context, as an important reminder of apparently informal practices in contexts far closer to home.

Why do urban informalities matter?

The concept 'informal' can be applied to urban housing and settlements, infrastructure (water, electricity, sewerage etc.), economic practices, and social processes (education, health etc.), as well as political practices. Generally it is classified into one of three key categories: spatial, economic, and political. While our focus is mainly on spatial informality as the form most closely associated with the city, arguably most (if not all) forms of informality are inherently urban, given the purchase of regulatory frameworks in urban areas. In reality, there is also a great deal of overlap between these three categories.

Spatial informalities are commonly associated with informal housing and settlements involving a combination of informal land acquisition, self-help construction, and incremental service provision. Land acquisition may be through a process of squatting or invasion of public or private vacant land. It may involve coordinated group action to 'invade' land, or may be the result of gradual occupation over time. Alternatively, settlers may acquire land through the process of informal subdivision, where land is bought and developed in a process that contravenes local planning or building regulations. Self-help housing may be built by household members through 'sweat equity', or financed by the family with help contracted in, but is usually incremental. Housing is often built from 'illegitimate' materials according to a country's building regulations. Similarly, service provision may involve a long process of petitioning local authorities for water, electricity and sanitation, as vacant land usually lacks provision, while simultaneously residents often have to self-provide services through illegal connections or onerous sourcing of clean water. Informal housing is the dominant mode of settlement across the developing world, with the United Nations Human Settlements Programme (UN-Habitat) in 2001 identifying 78.2% of the urban population in these regions as living in slums (UN-Habitat 2003: vi). Case study 14.1 illustrates some descriptive elements of spatial informality in South Africa, identifying significant challenges for the residents. However, as debated below, such a literal and descriptive interpretation of informality can be limited, as well as politically controversial.

Economic informalities within cities usually relate to what is labelled the 'urban informal economy', a broad term capturing both informal enterprises (such as street vending, taxi-driving etc.) as well as informal employment practices (which may occur in the formal economy) (Chen and Vanek 2013). The significance of the informal economy globally cannot be understated. In India alone, if informal employment within agricultural work is included, then informal employment accounts for 90% of the country's employment (ibid). Work experiences, levels of productivity and entrepreneurial success within the informal economy are highly varied. For many, conditions are dire, and rewards are limited, but owning an informal business can also be a lucrative, efficient practice. Much inequality in the informal economy is gendered, as it dominates women's options for employment globally, but they are often clustered in insecure, poorly paid sectors. Again it is important to remember that many of the poor working conditions attributed to the informal economy are also present within the so-called formal economy, which returns us to our earlier comment on the political risks of assuming particular definitions of urban informalities.

Finally, political informalities are often defined in contrast to formal politics. The latter refers to 'the operation and constitutional systems of government' (Painter and Jeffrey 2009: 7), whereas informal politics, being a far broader term, refers to political practices which are part of normal social interaction, including 'forming alliances, exercising power, getting other people to do things, developing influence and protecting and advancing particular goals and interests' (ibid). This distinction is clearly slippery and our chapter does not support a rigid binary interpretation, but recognition of informal politics is important, as it intimately shapes everyday urban life, along with the usual formal political structures, and often specifically relating to informal economy and shelter. In particular, clientelism is recognized as a form of informal politics that is particularly prevalent in the context of urban informal settlements (Auyero 1999).

Spatial, economic and political informalities seldom exist in isolation from each other; moreover, they regularly co-exist with what are commonly understood to be formal practices. A key message of this chapter, thus, is the interconnectedness of different informal modalities, as well as the intersections and co-constitutions of formal and informal processes. In fact, formal practices are often dependent upon informal ones to function (e.g. informal employment contracts within the formal sector and informal political negotiations within sectors of government). Despite this interconnectedness, for urban scholars, researching informalities has methodological implications, as regular approaches and methods (such as relying on published housing statistics) can entirely overlook informal practices. Informalities are often understood through ethnographic, context-specific work, but statistical data, though often lacking, is also of considerable importance. Chen and Skinner (2014) argue that continued efforts must be made to encourage the conceptualization, measurement and categorization of informal economies via national and international statistical communities. They remind us that 'data have power' and that it must be made available to those supporting informal workers (Chen and Skinner 2014: 231).

As already hinted at in these sections, merely *describing* practices as informal can risk overlooking the politics shaping and defining urban informalities. However, this chapter argues that such descriptions are still significant in a political sense, as they confer recognition on the inherent inequity in the poor quality of urban living environments and working practices, with the potential to yield policy responses to improve urban habitations and economic conditions. For millions of residents, living informally is disturbing and unsafe as Case study 14.1 illustrates. Thus recognizing the material realities and challenges of living and working informally has been a critical achievement of organizations such as UN-Habitat who have attempted to shape national and urban policy; although policy interventions have not always entailed positive outcomes for poor residents, particularly as more recent initiatives around the eradication of slums have resulted in severe dislocation and suffering. Informal enterprises are frequently the target of urban eradication policies conducted in the name of improving 'hygiene' and reducing exploitation, but such interventions can significantly disrupt individuals' livelihoods and thereby exacerbate urban poverty.

CASE STUDY 14.1 LIVING INFORMALLY IN CATO CREST, DURBAN

As argued below, much of the current debate around informalities focuses on the abstract construct and its political ramifications, risking a neglect of the actual conditions of informality. Thus, this first case study illustrates the lived experiences of informality in South Africa, revealing how this poses very real material challenges for residents' wellbeing. It also points to context-specific examples of both informal economic and political practices and reveals the inter-connections between all three 'forms' of informality.

The case study is of Cato Crest, an area in the city of Durban that was settled informally from the 1980s onwards. Houses are built from materials sourced locally: mud, timber, scrap metal, plastic, tyres and packaging (see Figure 14.1). Over the past decade the area has benefited from state investment in housing and infrastructure and is in the process of being gradually formalized. The settlement is now a mix of informal and formal housing. For residents living informally the material properties of their houses shape their everyday lives, usually for very negative reasons. Houses are small and poorly constructed, with children and parents often sharing single rooms for sleeping, eating and washing. The lack of privacy for adults and teenage children is a dominant concern (Meth 2013b). The houses lack safe electricity, water supply or sewerage facilities. Illegal electricity connections are common and electrocution (often of children) is reported and feared by residents. Residents rely on paraffin for cooking and treacherous fires have led to multiple tragic losses of life, income and home:

> A young lady came to my house ... crying that her house was on fire and her children were in the house. It was so sad ... Days went by, and this woman told me that she hears her child crying in her ears every day, asking her to take him as he is burning.
>
> *(Thandeka ♀)*

Vulnerability to fire is fuelled by the relative high density of the shacks. An additional concern is that the internal and external environments are degraded. Rats are a significant problem, as are cockroaches, snakes and lizards. Paths around the houses are muddy and houses themselves are not water-proof. Water seeps through roofs, walls and under doors. Areas of housing are surrounded by sections of undeveloped land which residents describe as 'bush'.

These materialities shape residents' physical and mental health concerns as well as their senses of identity as political subjects and citizens. Siboniso's quote illustrates this:

> Here in Cato Manor I arrive in 2000 … but good and bad things I learn from here. What I saw for the first time …[was] people living in the shacks… I thought [they] were mad or sick. It was take long time to me to be used [to] that place… After a long time I learn the life of that place. What I learn no one cares for you, every one do whatever [they] like to do.
> *(Siboniso ♂)*

More generally residents describe feeling neglected and abandoned, particularly by the state (Meth 2009). They believe they live like animals. Their living conditions shape an overall sense of a loss of humanity, exacerbated by the lack of health care, policing and employment in the settlement. Many residents are unemployed or work in the informal economy, selling food products, services or clothing. Some do this locally to be near children during the day, others commute to the city centre to capture bigger markets. Their home spaces shape their livelihood strategies. For example a lack of electrification prevents the purchase, preparation and storage of food items requiring refrigeration, and tiny homes preclude the storage of saleable items.

The settlement suffers from high levels of insecurity, which also undermines residents' wellbeing as it contributes to high levels of fear, anxiety and sleeplessness. Crime levels are high (including murder and rape) and are exacerbated by the materiality of the settlement, including the penetrability of housing, particularly through roofs, doors and walls (see Figure 14.1). Responses to crime vary widely and reflect the juxtaposition of formal and informal political structures and practices. Some residents turn to the police, while others turn to locally elected committees, who adopt formal structures with chairs and secretaries. Many turn to local politicians in the form of their ward councillor, whose political party dominates most structures in the settlement. But residents also make use of vigilante committees, often ad hoc in structure and illegal but not always operating against or without the knowledge of the police. Religion is also a key source of solace and both Christianity as well as traditional witch doctors provide residents with a sense of purpose and order.

FIGURE 14.1   Penetrable roofing in Cato Crest, South Africa (Source: Paula Meth).

What the debate below illustrates, however, is that the state – along with wealthy urban residents – also produces, practices and prospers from urban informality, and thus informalities are not simply a 'poverty' problem, associated with the poor and poor parts of the city. However, operations and functioning of structures at the urban, regional and national level ensure that the urban poor often bear the brunt of informal outcomes. Indeed, as Case study 14.1 illustrates, material realities of informalities are significant. Informality here is not simply a political construct, but rather an everyday lived reality. The chapter moves on now to examine the treatment of 'informality' within urban studies more broadly, teasing out how key debates have developed since early understandings.

Informality and urban theory

The identification of the 'informal sector' is often traced to Hart's (1973) distinction between formal and informal economic sectors, in his study of informal trading as a response to underemployment in Accra. This concept was adopted by the International Labour Organisation (ILO) in the 1970s to describe small-scale, unregulated economic activities carried out by people beyond the purview of the state. This suggested that the informal sector could be seen in opposition to the large-scale, regulated, 'modern', formal sector, as part of a dualistic framework.

Although controversial, this conceptualization has had an enduring effect on how informality is understood, which is explored further below.

Debates on spatial informality date back even further, to 1950s Latin America, where informal settlement was observed in cities as a consequence of rural–urban migration linked to industrialization and agricultural decline. Abrams' *Man's Struggle for Shelter in an Urbanizing World* (1964) was one of the first investigations into urban expansion and its consequences in Third World Cities. It was written in a context of increasing concern about rapid urbanization in the Global South which was characterized as an 'urban explosion', as the urban population in developing countries increased from 285 million in the 1950s to 1.2 billion by 1985 (Kasarda and Crenshaw 1991: 467). The scale and speed of this transition, combined with low levels of economic development in many countries, left governments unable or unwilling to respond to housing needs. Faced with this lack of government provision and an inability to buy land or housing on the formal market, many families turned to 'self-help housing' as a means of meeting their basic need for shelter. As outlined earlier, this involves acquiring land, constructing housing and resolving service provision through various informal means.

While such processes have been commonplace throughout the Global South for many decades, their characterization has often been crude, simplistic and politically significant. Descriptions of these settlements in accounts of the time see their populations as 'urban informals' (Abrams 1966), 'marginals' (Park in AlSayyad 2004: 9), or even 'human flotsam and jetsam' (Lloyd 1979: 209), who are '[l]iving almost like animals ... overwhelmed by animality' (Schulman 1966 in Mangin 1967). These negative portrayals of informal settlements were reinforced by Lewis's (1967: xiv) work on the 'culture of poverty', which describes the inner-city slums of Mexico City and Puerto Rico as exemplified by 'family disruption, violence, brutality, cheapness of life, lack of love, lack of education, [and] lack of medical facilities'. Yet such portrayals may, in some cases (see Case study 14.1), resonate with residents' own views of informality, raising important questions about who defines informality, how and why.

Importantly, such understandings of these settlements supported policy 'solutions' such as preventing migration (by law or by interventions aimed at increasing 'pull' factors in the provinces), eradicating existing settlements, and preventing new ones from forming through eviction and displacement to new housing or to places of origin (Mangin 1967). By the 1960s, policies of eviction, demolition and eradication were widespread, as local governments sought to discourage further settlement on the basis of the apparent drain on resources and 'culture of poverty' that prevailed in these places. In response, a debate arose as to whether squatter settlements (as they were termed) were a 'problem or a solution' (Mangin 1967). Working in Peru, Turner (1972) suggested that they allowed people the 'freedom to build' in circumstances where governments did not have the will, resources or flexibility to provide the right kind of shelter, drawing on a great potential resource in the desire, energy and initiative of families to house themselves. It was Turner who developed the notion of 'self-help housing', to describe the process whereby the owner–occupier constructs some or all of the accommodation, with or without (professional) help.

At the same time, based on her research in the *favelas* of Rio, Perlman was suggesting that the 'myth of marginality' was used for the social control of the poor, who were not marginal, but integrated into society 'on terms that often caused them to be economically exploited, politically repressed, socially stigmatized and culturally excluded' (Bayat 2000 cited by AlSayyad 2004:9). The myth, drawn from academic theories and local prejudice, characterized the urban poor as rural migrants, living in ruralistic shantytowns, manifesting symptoms of social disorganization, and political radicalization. Contrary to this popular view, Perlman found that *favela* dwellers were socially well-organized and cohesive, culturally optimistic with aspirations for their children's education and their housing, economically hard-working, and politically neither apathetic nor radical: 'In short, *they have the aspirations of the bourgeoisie, the perseverance of pioneers, and the values of patriots. What* they do *not* have is an opportunity to fulfil their aspirations' (Perlman 1976: 242–3, original emphasis).

In response to these arguments for the recognition of informal settlement, and residents' agency and capabilities, international agencies and national governments came to accept the idea that self-help housing could be a solution rather than a problem. As a result, sites-and-services and upgrading policies were implemented in many countries during the late 1960s and early 1970s. 'Sites and services' refers to the provision of basic infrastructure (latrines and community water sources) on dedicated plots of land for low-income households to develop via self-help processes; while 'upgrading' relates to in-situ improvements to existing infrastructure in a settlement, including housing and services such as roads, drainage and water supply. The widespread adoption of the self-help idea through policy in this way, particularly by international agencies such as the World Bank, meant that improving rather than replacing informal settlements became the priority for intervention (Davis 2006).

However, critics suggested that 'self-help releases government from its responsibility to provide adequate housing as a basic need for its low-income population' (Moser and Peake 1987: 5), and constituted the double exploitation of labour (at work and in housing construction), subsidizing wage costs (Ward 1982). 'Self-help' therefore arguably heralded a new era of the privatization of housing supply, championed by the World Bank. Despite these critiques, the idea was given a new lease of life by the suggestion of legalization as the solution to informality, based on the work of the Peruvian economist Hernando De Soto. In his well-known book *The Mystery of Capital* (2000), De Soto suggests that provision of legal titles is the solution to informality, and indeed the means to unlocking the potential for wealth creation by the poor. He suggests that the 'dead capital' of $9.3 trillion dollars held by the poor in informal assets can be 'brought to life' through the legalization of property titles, allowing it to be exchanged or used for credit, hence reinvigorating the economy. Despite the widespread application of legalization programmes in cities and countries of the Global South, critics of this approach argue that De Soto's claim to have the 'solution' to informality has not been substantiated, due to his oversimplification of the causes of poverty and informality (Fernandes 2002).

To make sense of these debates, Rakowski's (1994) suggestion of dividing them into two broad tendencies, designated as 'structuralist' and 'legalist', is helpful. While structuralists view informal settlements as the result of capitalism's uneven development (e.g. Mangin 1967; Perlman 1976; Davis 2006), legalists see informality as an alternative, rational and even 'heroic' economic survival strategy (e.g. Turner 1972; De Soto 2000). These 'heroic' notions of entrepreneurial informal settlement residents, conveyed by concepts of self-help and legalization, can be juxtaposed with accounts of informality as a manifestation of urban crisis, the inevitable result of urbanization under capitalism.

However, whether portraying urban informal settlements as crisis or heroism, such framings tend to view formality and informality as fundamentally separate (Roy 2005; see also Rodgers et al 2012; Angotti 2013). In fact, the dualistic nature of the formal/informal concept has been a recurrent theme in debates, attacked for its crude and simplistic suggestion that the two sectors are separate rather than interacting (Bromley 1978). Accounts portraying informality as the opposite of formality tend to negate the reciprocal relationship that often exists between 'informal' and 'formal' sectors. In reality, this relationship is often so messy and tangled as to make the two supposed opposites anything but clearly delineated. Others suggest that the category is too broad to be meaningful, as the 'informal sector' has been applied to a wide range of contexts, activities and people who do not share specific characteristics (Moser 1994).

Recent debates have also been concerned with the 'return of the slum' into the language of international organizations and government policies as well as academic debates (e.g. UN-Habitat 2003; Davis 2006), with theorists arguing that this offers an over-simplified account of these complex places (Gilbert 2007: 698; see also Varley 2013; Huchzermeyer 2011). An often-cited example of this is Davis's *Planet of Slums* (2006), which sees informal settlements as 'a fully franchised solution to the problem of warehousing this century's surplus humanity', in a world where exclusion occurs at local, national and global levels (Davis 2006: 200–1). Critics suggest that the discursive resurgence of 'slums', which stereotypes residents and looks to environmental improvements to address poverty, may also provide local authorities with the justification they need for eviction and eradication policies (Gilbert 2007; Meth 2013a), something that appears to be on the rise in many regions (Fernandes 2011). For example, in the case of South Africa's Slums Act 2007, the UN's 'Cities Without Slums' campaign was used to justify the demolition of slum settlements and the removal and displacement of their (mostly black) populations (Huchzermeyer 2007).

While these debates may seem to reprise similar arguments to those which accompanied the earliest conceptualizations of informality, postcolonial urban theorists have attempted to advance the concept of urban informality by reversing its normative inference and using it to foreground the agency of the populations it is all too-often used to marginalize, a tendency identified by Roy (2011b) as 'subaltern urbanism'. Concepts such as Bayat's (2004) 'quiet encroachment', Benjamin's (2014) 'occupancy urbanism', Simone's (2004) 'fluidity', and Yiftachel's (2009) 'grey spaces' have been instrumental in reframing conceptions of urban informality; and as part of a wider 'postcolonial urban studies' (Robinson 2006), they offer a response to the global dominance of western theory. In particular, Roy's conceptualization of urban informality as 'an organising logic, a system of norms that governs the process of urban transformation itself' rejects informality as a separate sector and sees it as 'a series of transactions that connect different economies and spaces to one another' (Roy 2005: 148). In this reading, urban informality extends beyond the urban poor to encompass the actions of different sectors including middle- and high-income urban residents, the state, and business interests. However, only some forms of informality are criminalized – normally those used by the poor – leading to an 'uneven geography of spatial value' (Roy 2011b: 233). Similarly, McFarlane's (2012) reconceptualization of informality as practice opens up new ways of understanding how informal and formal actions relate to each other. His study of the causes of and responses to the 2005 floods in Mumbai, and the discourses around this, reveal the extra-legal practices undertaken by the state and developers, which arguably caused the disaster. Informality here is used as a form of urban critique: 'a basis for rethinking not just informality, but planning itself, in cities across the global south' (McFarlane 2012: 106).

Such conceptions are helpful in thinking through urban informality's potential to reveal power relations. However, mindful of the somewhat abstract nature of certain debates, this chapter provocatively paraphrases Mabin's (2014: 28) critique of 'southern theory' to ask whether some current theories of informality are 'more opaque than helpful'. Mabin (ibid: 31) suggests 'southern theory' falls short of its claims empirically, as it fails to offer 'profound and substantial research on what is going on' in urban contexts; and theoretically, as it does not offer the promised innovation. Although often inspiring and stimulating, there is, perhaps, a danger within such theoretical fields of over-abstraction, and the subsequent eclipsing of the empirical phenomena which they purport to illuminate: of only seeing 'slum as theory', as Rao (2006) puts it. These provocations support our earlier argument in this chapter, about the need for the concept to maintain a focus on the material challenges of informality, and its associated politics, given that it describes the everyday lives of millions of people. This is highlighted in Case study 14.2, with examples from Mexico on the construction of informality in legal terms, showing how state-driven categories both respond to and reproduce patterns of urban informal settlement.

CASE STUDY 14.2 INFORMALITY AND LEGALITY IN CINCO DE NOVIEMBRE, QUERÉTARO, MEXICO

As discussed above, insecure land tenure is a key characteristic of many informal settlements, deriving from lack of rights to land. This is a primary cause of such settlements' categorization as 'informal', highlighting the central role of legal categories in framing informality, illustrated by a second case study from Mexico. Mexico has one of the longest-running programmes of land tenure legalization in the world, implemented as a response to the informal growth patterns that have shaped cities, largely through subdivision on peripheral agricultural *ejidal* land. An *ejido* comprises land owned communally by farmers under Mexican law dating from the agrarian reforms of the 1920s and 1930s, which could not be legally sold until reforms in the 1990s; *ejidos* continue to be the main source of land for informal urban development. It has been argued that the process of land tenure legalization has effectively sanctioned the informal *ejidal* land market and hence the ongoing use of informal development (Azuela and Duhau 1998), as discussed below.

Cinco de Noviembre is an informal settlement on the edge of the metropolitan zone of the medium-sized city of Querétaro, Mexico. The neighbourhood was founded in 1995, when a local *ejidatario* illegally sold a plot of land to leaders of a social movement seeking housing for its members. They divided it into 400 housing lots, of which the original owner retained 10, including a plot for him to occupy. All of the remaining plots were sold, but habitation density of the neighbourhood remains low: around only 60% of plots are occupied, mainly by extended households with low incomes. The neighbourhood was connected to the formal electricity network in 2004, nine years after it was established, but has no formal water supply or sewerage connection. Currently, water is supplied by tankers paid for by the residents, while formal services are being negotiated with the municipality by the neighbourhood committee. Many other basic services are still lacking, including rubbish collection, street paving, and telephones.

The inadequate services and under-occupation in Cinco de Noviembre are linked to the neighbourhood's legal status. Due to residents' lack of legal titles, some services are impossible to obtain: as the neighbourhood is not legally recognized by the municipality, residents are unable to formally petition the local authorities for services such as water. In response to this situation, residents submitted a request for legalized land tenure to the state government, which initiated proceedings four years ago. The residents anticipate that legalized land tenure, in the form of legal titles, will allow them to obtain those services that the neighbourhood currently lacks through recognition from the municipality.

However, the legalization process in Cinco de Noviembre stalled several years ago. Since land reforms in the 1990s, the process requires the consent of the *ejidatario*, who still formally owns the land. Despite being initially supportive of the development, the *ejidatario* had a change of heart five years ago when local land values increased dramatically due to sales of nearby land to formal developers involved in building middle-income housing. Regretting his decision to sell for a relatively low price on the informal land market, the *ejidatario* has tried to buy residents out, threatened them with eviction, and taken the case to a tribunal (which threw it out). He is now manipulating the regularization process in order to negotiate more money for the land, withholding his consent as leverage to request additional payments from the residents.

For the municipality, the unregularized and hence rural status of the neighbourhood justifies withholding services, despite residents' demands. The lack of land tenure thus underpins the 'legal limbo' of the neighbourhood, compounded by its physical situation on the 'peri-urban' periphery, which highlights the neighbourhood's peripheral position, politically and administratively as well as spatially (see also Lombard 2014). Despite their de facto possession of the land, residents are legally seen as 'trespassers' (Azuela 1987), leading to their subordinate situation and lack of recourse for problems relating to service provision. This case study thus shows how informality is shaped by the legal categories of the state (ibid: 524). The illegal land sales on which informal settlements are based have generated the process of land tenure legalization and the vast bureaucracy underlying it, which define informality and ultimately social practices, based on 'participants' perceptions of the legal situation' (Varley 2013: 17). In Mexico, informal development is 'institutionalized' by the state (ibid); and the ensuing legal categories, defined by the state, reflect and reproduce power relations inherent in local and national structures.

FIGURE 14.2   Rudimentary services, self-built housing and empty plots in Cinco de Noviembre (Source: Melanie Lombard).

Conclusion

Urban informalities are both evident in, and help to explain, urban processes in cities around the world. This chapter argues that 'informality' is more than an urban condition; the concept allows us to understand and critique complex processes and politics at the urban scale. This can be seen particularly in the case study from Mexico, which detailed the dynamic co-production of informality through formal legal mechanisms and informal practices of spatialized power. However, this chapter also claims that current debates on informality risk obscuring the lived realities of informalities (often most acutely suffered by the urban poor) because they tend to focus too much on abstract concepts and the intellectual project of undoing binaries. Using the case study from South Africa, the material everyday realities of informality, including the interconnections between spatial, economic and political practices, were highlighted. Both of these descriptive accounts bear witness to the daily grind of living informally, whether in terms of a lack of basic services and security that the residents of Cato Crest suffer, or legal-spatial insecurities of residents in Cinco de Noviembre.

Drawing on such empirical cases, the lens of informality adopted in this chapter suggests that urban scholars must go beyond simplifying interpretations and responses (especially those that conflate informality with poverty) to recognize the significance of process. To illustrate this point, recent attention to manifestations of urban informality in Europe and North America suggests that the phenomenon can no longer (if it was ever) be considered a 'Southern' issue. For example, in London the phenomenon of 'beds in sheds' appears as a response to high housing demand in areas where affordable shelter is scarce. In February 2012, the BBC News website published a video report highlighting the shocking extent and conditions of 'London's modern day slums' (BBC 2012). Increasing media attention on this issue has highlighted the illegal construction and/or conversion of sheds, garages and outbuildings for residential use in suburban back gardens, with estimates of up to 10,000 such dwellings in existence across Britain (Neiyyar 2013), mainly in London and the surrounding areas. Often contravening planning and building regulations, conditions in these dwellings are generally understood to be substandard, with many lacking adequate sanitation (Gentleman 2012). Meanwhile, in Greece, new types of informal shelter are being observed, such as the newly built areas of informal housing in Thessaloniki, which lack basic services (Karagianni 2013). In Spain, squatting as a response to housing need has been increasing in Barcelona (Pradel 2014). Such studies suggest that in the context of austerity urbanism, informal shelter is becoming more prevalent in European cities. Along with the need for an expanded understanding of urban informality, this highlights the need to retain at the forefront of debates the concept's critical and political edge, which within cities demands a focus on those most affected by injustice.

Key reading

Chen, M. and Skinner, C. (2014) 'The urban informal economy', in Parnell, S. and Oldfield, S. (Eds) *The Routledge Handbook on Cities of the Global South*, London: Routledge.

Chen and Skinner are actively involved in both researching urban informality as well as representing and advocating for such livelihoods. Their chapter is informed by their extensive hands-on experience as well as informed analysis, and offers an excellent overview and starting point for the reader.

Roy, A. (2005) 'Urban informality: toward an epistemology of planning', *Journal of the American Planning Association*, 71(2): 147–158.

Roy's work is at the forefront of critical debates that have advanced the concept of urban informality, reversing its normative inference and using it to foreground the agency of the populations it is all too often used to marginalize. This paper articulates one of her important contributions to the field, questioning narrow definitions of informality and drawing attention to the ways in which planning, as a discipline and profession, is implicated in its production.

Turner, J. (1972) 'Housing as a verb', in Turner, J. and Fichter, R. (Eds) *Freedom to Build: Dweller Control of the Housing Process*, New York: Collier-Macmillan.

Turner's research on urban informal settlements in Peru led to his development of the enormously influential concept of self-help, which remains widely used in debates today.

UN-Habitat (2003). *The Challenge of the Slums: Global Report on Human Settlements*, London: Earthscan.

This extensive report is a very useful starting point for a reader offering a contextualization and description of slums as well as an assessment of their developmental dimensions (social, spatial and economic) and policy interventions, using global case studies to illustrate its arguments.

15

LEARNING

Colin McFarlane

Introduction

How might learning operate as conceptual vehicle for understanding cities, how they change, how they are lived, and the ways in which they are contested? How do we make sense of what Christine Hentschel (2014: 360) has called 'an urbanism in motion', an urbanism where 'what we come to term 'knowledge', 'infrastructure' and 'resources' are never simply 'there', but must be translated, distributed, coordinated, perceived and inhabited'? To begin to respond to these questions, and to appreciate how 'learning' can or cannot help to make sense of cities, we need first to consider what we mean by learning.

While the history of urbanism is predicated upon the question of *how we might come to know* the city, the question of learning itself has remained black-boxed. This is likely to be due in part to a sense that 'learning' is an incidental process – that the creation and transformation of knowledge and perception is a background story to the central drivers of both urban change and the urban condition. It is also likely to be partly a result of a lingering sense that learning, if not quite *a*political, is somewhat removed from the formation of political struggle and the practice of urban contention.

Learning is, as anthropologist Tim Ingold (2000: 155) has described it, a kind of 'wayfinding'. This is not the clichéd populist notion of learning as 'journey', but learning as a process in which people 'feel their way' *through* a world that is itself in motion, continually coming into being through the combined action of human and nonhuman agencies' (Ingold 2000: 155). Whether you are a resident, a visitor, a policymaker, or an activist, the changing nature of the city – of how it is experienced over time and space – and of the particular contexts, agendas and interests of different individuals and groups, inevitably alters knowledge and perception of it. Here, knowing is an uncertain, embodied and often conflictual process that emerges inescapably through engagement with the world around us, as Ian Borden *et al* (2001: 9) put it in relation to the city: 'Knowing the city is ultimately a project of becoming, of unfolding events and struggles in time as well as in space' (see Chapter 19, this volume). If, as Richard Sennett (2008: 289) has argued, 'people need to practice their relations with one another, learn the skills of anticipation and revision in order to improve these relations', in cities that practice often takes the form of conflict and struggle, and occurs in contexts of radical inequality.

Learning is an expansive process of coming to know not just what urbanism *is*, but what the stakes of urban life are for different people. As a form of 'wayfinding' and struggle, the form and politics of learning cannot simply be restricted to the domain of specialist and expertise knowledge and their effects, as important as these are. We need alongside this to constantly ask who we – critical urban researchers – learn from and with, i.e. we need to attend to where critical urban knowledge comes from and how it is learnt. The spatialities of learning matter a great deal here. Learning is translocal in its geographies, and involves an ongoing labour in forging and developing connections between different sources, routes and actors. The shifting relations of knowledge, space and power have been key questions at stake in recent influential accounts of the translocal nature of urban policy and planning, including – and this is to mention just a few – a series of edited collections such as Roy and Ong's (2011) *Worlding Cities*, McCann and Ward's (2011) *Mobile Urbanism*, Harris and Moore's (2013) collection on planning histories and circulating knowledge, Bunnell *et al*'s (2012) collection on global urban frontiers, Cochrane and Ward's (2012) collection on researching the geographies of urban policy mobility, and McFarlane and Robinson's (2012) collection on comparative urbanism (and see Robinson 2006).

In this chapter, I examine this expansive domain of learning as a set of processes and spatialities. I consider how learning has been conceived in urban and cognate debates, develop a conception of learning as produced through distributed assemblages that become political in different kinds of ways, briefly explore examples of urban learning in practice, and outline a critical urbanism of learning.

Conceptualizing learning

Debates about the role of learning have a long history, particularly in relation to economic development, whether Schumpeter's (1934) *Theory of Economic Development*, Hayek's (1945) paper 'The use of knowledge in society', Machlup's (1962) *Production and Distribution of Knowledge*, or contemporary debates on the 'knowledge economy' or the World Bank-led 'knowledge for development' initiatives (e.g. King and McGrath 2004; Johnson and Wilson 2009). There is also, of course, a similarly complex history of debates on the nature, production, meaning and value of learning, with disparate influences from theories of epistemology (Greco and Sosa 1999), science studies (Callon *et al* 2009), organizational studies (Amin and Cohendet 2004), cognitive anthropology (Ingold 2000), and critical education and pedagogy (Freire 1970; see Case study 15.1 for a brief overview of some of the debates around learning and urban change).

CASE STUDY 15.1 PERSPECTIVES ON LEARNING

If urban learning has been a neglected topic, it would be wrong, of course, to suggest that it has been entirely ignored in accounts of urban change. Learning has been discussed in debates on urban economies and, increasingly, in debates around travelling urban models or policies (e.g. McCann and Ward 2011; Temenos and McCann 2012; Peck and Theodore 2010; Roy and Ong 2011). There has been a great deal of debate in economic geography, for example, about 'learning regions', 'regional innovation', 'institutional thickness', innovative 'buzz', skills development, the possibilities of knowledge mobility, and on the role of 'clusters', 'quarters', 'creative economies', and tacit knowledge (e.g. Amin and Cohendet 2004; Cumbers and McKinnon 2004, 2006; Florida 2005; Gertler and Wolfe 2002; Glaeser 1999; McKinnon 2008; Scott 2006; Storper and Scott 2009). This work has critically engaged with, for example, the efforts of states and supranational bodies to identify and develop specialist clusters within cities and regions, often taking the form of research and development and venture capital initiatives 'which attempt to inculcate a culture of innovation and learning' and seek to 'build and reinforce a sense of cluster identity amongst constituent firms and organisations' (Cumbers and McKinnon 2008: 959). There is a great deal of urban policy debate around, for instance, city cluster learning such as Agenda 21, or learning network formations from UN-Habitat to Infocity.

As states and supranational institutions have increasingly focussed on learning as central to competitive advantage within global economies, a key debate in relation to urban clusters has been around how to create linkages and networks through clustering in ways that facilitate learning through exchange and interaction. But as important as these debates have been, they have tended – not surprisingly given their economic focus – to restrict urban learning to questions of economic innovation, urban and regional competitiveness, and organizational learning, and have offered less in terms of critical engagement with power inequalities and exclusion, or on how learning operates as a coping mechanism or as a tactic of resistance.

The constitutive role of learning in processes of urban change and politics has been identified in debates on urban policy transfer, from Anthony Sutcliffe's (1981) *Towards the Planned City*, Ian Masser and Richard Williams' collection (1986) *Learning from Other Countries: The Cross-National Dimension in Urban Policy-Making*, and Anthony King's (e.g. 2004) surveys of colonial urbanism, to Joe Nasr and Mercedes Volait's (2003) collection *Urbanism: Imported or Exported?* and McCann and Ward's (2011) collection *Mobile Urbanism*. Literature on 'urban policy mobilities' (e.g. McCann 2007; Peck and Theodore 2010; Harris and Moore 2013) is one important example here. This disparate work has considered, for instance, how certain cities learn from particular policy discourses, such as discourses of 'knowledgecities', 'creative cities', or neoliberal, revanchist and punitive ideologies of urban development (e.g. Florida 2005; Hollands 2008; McCann and Ward 2009; Peck and Theodore 2010; Peck 2005, 2006; Ward 2006). If travelling urbanism is a far from new phenomenon, urbanism is, none the less, increasingly assembled through a variety of sites, people, objects and processes: politicians, policy professionals, consultants, activists, publications and reports, the media, websites, blogs, contacts, conferences, peer exchanges, and so on. But despite this surge in critical literature on travelling urban knowledge and policy learning, there has been relatively little attempt to consider how learning itself might be conceptualized.

If these debates are varied and distinct, all of them contain one central claim or assumption about learning: that learning is a process of potential transformation. Learning, even where it is explicitly described as uncertain – as in, for instance, strands of organizational theory that emphasize creativity and invention – refers to a process involving particular constituencies and discursive constructions, entails a range of inclusions and exclusions of people and epistemologies, and produces a means of going on through a set of guidelines, tactics or opportunities. As a process and outcome, learning is actively involved in changing or bringing into being particular assemblages of people-sources-knowledges. It is more than just a set of mundane practical questions, but is central to political strategies that seek to consolidate, challenge, alter and name new worlds.

What then is learning? And how should it be differentiated from the notion of knowledge? A useful starting point is to see knowledge as the sense that people make of information, i.e. information that is anchored in practices, beliefs, and discourses (Nonaka *et al* 2000). This is not to subscribe to the traditional understanding of knowledge as something that people 'possess'. Rather, knowledge is located in space and time and situated in particular contexts; is mediated through language, technology, collaboration and control; and is constructed, provisional, and constantly developing (Amin and Cohendet 2004: 30). Most importantly, if knowledge is the sense that people make of information, that 'sense' is a practice that is distributed through relations between people, objects and environment, and is not simply the property of individuals or groups alone (ibid). Learning emerges here as the process of making, contesting and reproducing knowledge – it is a name for the specific processes, practices and interactions through which knowledge is created, contested, and transformed, and for how perception emerges and changes.

A conventional conception of learning would imply a cognitive formal process of training or skill acquisition as a linear addition of knowledge, such as in learning a musical instrument or a scientific technique. In practice, however, learning emerges as a distributed assemblage of people, materials and space that is often neither formal nor simply individual. For example, people come to learn the local urban transport system, or where to buy the best fresh fruit in urban markets, or which parts of the city are safe or dangerous at particular times of day and night, not through formal training but through gradually developing a sense of how things work and change – we are back here to 'getting a feel' for how things are and how we might orientate ourselves to them. Rather than being confined to the individual, learning as a process is distributed through relations between people-materials-environment. For example, urban transport is learnt through multiple space-times and encounters, from developing an understanding of timetables and routines or through sharing stories and experiences of the nature and rhythm of transport networks, to the embodied experience of riding on the bus or train in terms of the quality of materials or the numbers and behaviour of fellow passengers, and the contingencies of everyday interactions that may leave a lasting impression. Learning here can be incremental or sudden, and is as much about developing a perception through engagement in the city as it is about creating knowledge (see Case study 15.2 for an example of learning in relation to a social movement based in informal settlements).

CASE STUDY 15.2 LEARNING INFORMAL SETTLEMENTS

Slum/Shack Dwellers International (SDI) is a collection of non-governmental (NGO) and community-based organizations (CBOs) working with urban poverty, particularly housing and sanitation, which operates throughout Asia, Africa and Latin America. It is a translocal experiment in building a new form of urban sociality; a learning movement based around a structure of what its leaders call 'horizontal exchanges' involving small groups of the urban poor travelling between neighbourhoods to learn from one another. The movement espouses a range of tactics that its leaders describe as indispensable to a development process driven by the urban poor. These include daily savings schemes, exhibitions of model house and toilet blocks, the enumeration of poor people's settlements, training programmes of exchanges, and a variety of other tactics. In seeking grassroots participation and horizontal exchange, SDI seeks to place urban learning at the centre of social and political relations. Its insistence upon learning from and between the urban poor emerges from the context of a failure – deliberate or otherwise – of the state to ensure collective provision of urban infrastructure, services and housing.

SDI groups attempt to deal with the crisis of social reproduction in many cities of the 'Global South', from Mumbai (where much of the work of the network started) to Cape Town, Phnom Penh and Karachi. This crisis has been sharpened by the restructuring of the relationship between capital and the state in many cities in the Global South, which has encouraged the privatization of infrastructure such as water or electricity supply, a general escalation of real estate prices, a growing trend towards gentrification, the collapse of various national and local welfare provisions, and increased forms of political repression in the form of the demolition of 'slum' settlements to make way for land redevelopment. SDI's tactics are translocal in that they are place-based and emerge from particular local histories, but take shape through exterior interactions rather than being place delimited. They are not new in themselves, although they are inventive ways of thinking and doing and configuring and reconfiguring relations between actors. In this relational translocalism, the object of struggle remains the locality (for example, the local municipal corporation), and this is informed in part by the struggles of other groups across the network. SDI leaders argue that urban development strategies that are designed for the poor, but not *with* the poor, tend to fail: 'SDI believes that the monopoly over information and knowledge exercised by officials, technocrats and professionals needs to be broken and poor people themselves need to gain control over knowledge in order to deal more effectively with their situation' (Patel *et al* 2001: 51–2). Its leaders articulate an entrepreneurial form of collectivist politics that emphasizes the capacities and skills of the urban poor.

In SDI, learning is first and foremost a sociomaterial practice of working in groups (McFarlane 2011). A key example of this, and one central organizing strategy of SDI, is the construction and exhibition of full-size model houses. Stories about how to construct model houses circulate SDI through an organized system of exchanges in which groups of the urban poor from different cities share ideas and experiences. In exchanges, visiting groups often join in with constructions and exhibitions as they are happening.

Learning of, and in, cities, occurs not simply through formal, linear and cognitive processes, but in experiential immersion in urban space-times. It involves not simply the absorption of codified information – plans, policies, maps, data-sets, and so on – but a myriad set of incremental and haptic processes that are relational and which stretch beyond the individual subject to the social and material instantiations of urbanism. Learning occurs through a distrusted assemblage of people, things, practices, rhythms and spaces, and is part of the architecture through which we come to perceive urbanism, whether we are policy makers, researchers, residents or activists (see Chapter 4, this volume). Everyone learns cities, although the ways in which people learn, the resources with which they can learn, the routes through which that learning takes places, and the political framings and objectives that organize that learning, are profoundly diverse and often deeply contested.

In *Learning the City: Knowledge and Translocal Assemblage* (2011), I argued that learning emerges through the interaction of three processes: translation, coordination and dwelling. Translation refers to the distribution and comparison of knowledges, ideas and resources across multiple space-times, from activists sharing ideas about how to protest against the state to policy makers seeking to learn from different cities. Coordination refers to the building of functional systems that anchors these multiple translations, and may include discourses or maps or policy frameworks or traffic lights systems, that attempt to deal with complexities. And dwelling refers to how learning is lived, and how over time we attune or educate ourselves through learning assemblages in different ways of seeing and inhabiting cities. These three processes shift and entangle in learning assemblages, through slow processes of bricolage, sudden improvisations that may, none the less, draw upon historical repertoires through which people get a feel for how things work and might get done, or in tactical learning through which activists might produce ideas or approaches that disturb, resist, or initiate alternatives ways of seeing the urban condition (see Chapter 3, this volume).

A critical urbanism of learning

Attending to how people differently learn cities can open out elements for a critical urbanism. Three moves are important here, and here I am taking inspiration from Peter Marcuse's (2009) discussion of rights to the city. First, *evaluating* urban knowledges that are presented to us as inevitable by entrenched elites and dominant ideologies of the city; second, *democratizing* learning processes by examining who is and who isn't included in them; and third, proposing alternative sets of urban knowledges, practices and imaginaries oriented to more socially and ecologically just forms of urbanism. The emphasis on learning assemblages here, on doing and making across multifaceted groups, resources and imaginaries, is about working towards a more inclusive urban commons (see Chapter 6, this volume).

Such a critical formulation of learning also depends on closely following the lessons learned from postcolonial critique. Ananya Roy (2013: 134) has argued that a focus on learning can shift attention from the formation of expertise and authority to the tacit and tactical imaginations engendered by urban inhabitation learning as a 'theory of postcoloniality' (as an assemblage that both rehearses and ruptures colonial mimesis). This also means thinking critically about how we as researchers learn the city. Learning always exists in a dialectic of learning and *un*learning, and here a postcolonial approach to learning is vital if we are to think critically about unlearning inherited and stubbornly persistent Euro-American modes of thought and knowledge production. In part, this is also about learning new ways of thinking about our existing forms of conceptualization. An important intervention here is Swati Chattopadhyay's (2012) book, *Unlearning the City: Infrastructure in a New Optical Field*.

Chattopadhyay (2012: ix) argues that the central problem with theorizing cities today is a 'paucity of vocabulary', but that this does not mean we necessarily need new concepts, but instead to rethink – or 'defamiliarize' – ourselves with existing terms. A key term here, she argues, is infrastructure. Part of her task is to consider infrastructure as a visual and material cultural politics, and this cultural politics is also a subaltern politics in that it exists beyond representation. She takes infrastructure from its usual reference points – sunk systems of water, waste, power, transit, and increasingly digitally enhanced developments and gated areas, and beyond a problematic of the state and market – to spaces at odds with dominant connotations of global urbanism, such as street art and festivals, or practices of play such as cricket (see Chapter 14, this volume). The attempt here is to place infrastructure in 'a new optical field' (ibid xvi): 'To enable a different vision of the city, we must unlearn our habits of thinking about urban materiality, particularly the material constitution of physical infrastructure.'

For example, Chattopadhyay aims to rethink the street as a space of *habitation*, of ephemeral performance and experience, rather than mere usage. Cricket becomes a key example, and not just as a pastime but as a public space of political performances. Cricket, as a street vernacular, is also a meeting space between friends, neighbours, and spectators watching from adjoining buildings, as well as a relay of bodily affects, arguments, and casual conversations. Here, the infrastructure of the street is temporarily changed: the infrastructure, through activity, is instantiated in a particular way that is not quite everyday nor quite exceptional: 'It is created out of a series of conjunctures, of bodies and objects, movements and views, noise and warmth, walls and roads, events and memories' (ibid 119). Infrastructure here shifts from roads as a primary urban form of transit to streets as a variegated set of experiences, a 'texture of the conjunctural' (ibid 120).

Another example Chattophadhyay highlights is subsistence painters and artists that work across urban India employed in vehicular art on trucks or busses – typically elaborate, colourful, and endowed with political, thoughtful and/or humorous messages. While these artists have their own distinctive signatures and flights of imagination, they draw on a rich repertoire of existing cultural resources and motifs, from religious symbols to political debates and everyday sayings. The improvisational quality of the urban artwork emerges through artists borrowing both from a craft tradition (which is rural and urban) as well as a literary tradition of poetry, political sloganeering, and street talk customs, representing a particular conjuncture of dwelling in India's rapidly changing urban modernity. Here, craft traditions that are learnt through embodied immersion in everyday practice and skill are translated as they encounter different contemporary moments and instances (political debates of the day, moments in popular culture, etc.). This inseparable mixture of habits of craft and literature with popular images and slogans is obviously a means for making a living, but it is constituted by forms of knowing that are at once spiritual, popular, traditional, fantastical, and modern. Here again, infrastructure needs to be rethought: mobility not just as a functional domain of moving people, goods, and capital, but as a site of marginal play and resistance that is contingently interpreted (see Chapter 19, this volume).

In Chattopadhyay's reconceptualization of infrastructure in relation to contemporary urbanism in India, subaltern practices exist on the 'edges of visibility', beyond definition and representation and in excess of authority, but can become 'popular' and visible to state and capital as they become agents of social change. For Chattopadhyay, in the spaces and switches between subaltern and popular lie a reconceptualization of infrastructure – as vital infrastructures of urban change – and a challenge to how urban theory might unlearn the city and develop new formulations of existing processes and formations.

Chattopadhyay, in raising the important question of subalternity as that which is beyond representation, poses a problem for thinking about urban learning, albeit a problem that she herself does not examine in the book. What kind of political strategies might be learnt when the scope for political agency is historically shattered? Where life is a fundamental struggle for urban inhabitation – struggles for water, sanitation, food, shelter, basic health – then the question is not just how learning takes place, but also one of the sorts of learning that are possible and how that informs the sorts of politics that seem affordable. For Christine Hentschel (2014: 361), this is a vital question: 'One wonders then, sometimes, about all the moments and constellations when learning seems impossible, too painful or simply not worth trying'. What is the lesson learnt from the street boy in prison, the dwellers' association destroyed, the skateboarder paralyzed after miscalculating the length of a wall? How can a theory of learning encompass being stuck, immobile, untranslatable, or the drive to just keep on doing the same? A critical urbanism of learning, then, needs to probe not just at 'truth claims' made by elites that position themselves as having the answers to urban problems, whether in the form of creative or smart or entrepreneurial cities, not just at including as many voices and perspectives as possible in the production of alternative urban visions. There is also a key challenge in understanding the limits of learning, in relation both to unlearning existing knowledges and to a politics of subaltern recognition with all the attendant challenges of representation, authority and positionality that come with that.

Conclusion

Cities are distinctive sociospatial formations constituted by a diverse set of specific forms, processes and experiences that change how people know, perceive and inhabit urban space. We can think here of their density, changeability, sometimes overwhelming complexity and illegibility; their role as spaces of tradition, security, and history, or of ease, manipulability, and sociality; the possibilities they generate for equality, opportunity, struggle, conflict, exploitation, and hardship; their spatial sprawl and connection, and their co-locating of multiple sites and experiences, from informal settlements, where most people arriving in contemporary cities will live, to transport or economic corridors, city centres, industrial and service areas, shopping centres, abandoned spaces, parks, subcultures, and often radically varied architecture; their changing rhythms from those of commuters and tourists to schools and nightlife; and their propagation of new politics, lifestyles, imaginaries, and technologies. It is through encountering elements of these myriad forms of urban change through the particular lives, contexts, and agendas of policymakers, activists or residents, that learning the city takes place.

To live in cities is to come up regularly against the unknown, as well as to the limit points of how different groups can and cannot learn as a result of ever deepening urban inequalities. However, urbanites approach the unknown not with a blank slate but with what they already know, and in the context of their own lives, resources, plans, hopes, fears, etc. As Steve Pile (2001: 263) has put it: 'Knowingness and unknowingness are constitutive of the city: each clads buildings in layers of visibility and invisibility, familiarity and surprise'. The critical purchase of conceptualizations of urban learning lies not in a straightforward call to know more of cities, but to expose, evaluate and democratize the politics of knowing cities by placing learning explicitly at the heart of urban debate. Learning is critical to how urbanism is produced, lived, and contested, and to how we might produce more socially just urbanism.

The question of learning – how it operates, what its politics and geographies are, how it might be levered into a more socially and ecologically just urbanism – cuts across some of the key preoccupations of contemporary urban studies, and in at least three ways. First, the translocal nature of learning and some of the implications of that translocal learning for policy or activism has become a key set of concerns. Second, the everyday forms of learning through which people attempt to cope with or advance their opportunities in the city have become a focus of a variety of work. And third, formally organised efforts to foster learning between different urban constituencies, whether in particular kinds of forums or city partnerships, is an ongoing focus of critical urban debate.

I want to close the chapter by identifying two implications of the conception of urban learning assemblages for how we think and research the city. First, the conception of urban learning assemblage demands attention to how learning features in the efforts of different groups to assemble the city. For instance, current debates about the 'creative' or 'smart' city value particular exclusive groups, spaces and forms of urban development, particularly around well-educated elites living in premium residential spaces and working in high-end service economies, including in science, technology, research, media and finance. Through attending to the historical processes of assembling knowledge, the concept of urban learning assemblage can serve to expose which groups and ideologies have the greater capacity to render urbanism in particular ways over others through, in part, privileging particular forms of urban learning. At stake here is the critical relationship between the actual and the possible city, between the city that has been produced and the city that might have been, or that might otherwise arise.

Second, and following this, one opening or direction for thinking about cities through the concept of urban learning is to do with the idea of *potential*. In the face of exclusive claims by elite groups to know what is best for the city – those who would advocate, through ideologies of neoliberalism, exclusionary forms of creative, smart or intelligent cities, or gentrified and gated urbanism – and in contexts of often extreme poverty, violence, inequality, and exploitation, we need to hold on to the potential of other, more socially just urbanisms. Cities are places of unexpected encounters, progressive ideas, forms of knowledge and activism, and can generate not only inventive ways of perceiving and acting in urban space, but also new forms of urban learning and possibility.

As David Harvey (2008: 33) argues, for example, in urban social movements there is the potential to 'reshape the city in a different image from that put forward by the developers'. By forcing different knowledges and perspectives in public spaces, through protest, or in negotiation with state actors, and by embarking on and promoting alternative forms or experiments in urban learning, it is in these movements that much of the drama of urban politics will be played out. These movements are not necessarily aimed at the state, but often inhabit a wider politics of urban learning that aims to raise awareness and understanding of issues as diverse as urban asylum and racism to international disaster and crises and threats to public spaces. In these movements and struggles, argues Harvey (2008: 40), the diverse project of 'right to the city' both as a working slogan and political ideal, is continually posed anew through a focus 'on the question of who commands the necessary connection between urbanization and surplus production and use': 'The democratization of that right, and the construction of a broad social movement to enforce its will is imperative if the dispossessed are to take back the control which they have for so long been denied, and if they are to institute new modes of urbanization.' This concerted and sustained campaign of rights to the city is a fraught and contested project of learning the forms, contours and imaginaries through which a better city might be assembled.

Key reading

McCann, E. and Ward, K. (eds.) (2011) *Mobile Urbanism: Cities and Policymaking in a Global Age*, Minneapolis: University of Minnesota Press.

Peck, J. and Theodore, N. (2010) 'Mobilizing policy: models, methods and mutations'. *Geoforum* 41:2, 169–174.
    Policy mobility and urban imaginaries are key to the politics and form of urban learning processes, and this book/paper is an excellent guide to how policymakers take on, translate or resist travelling ideas, visions and models.

Roy, A. and Ong, A. (2011) (Eds) *Worlding Cities: Asian Experiments and the Art of Being Global*, Oxford: Wiley Blackwell.

Robinson, J. (2006) *Ordinary Cities: Between Modernity and Development,* London: Routledge.
    Not explicitly concerned with learning, but important to considering the political, ethical and methodological challenges of learning from and across cities often 'off the map' of mainstream urban theory.

16

MATERIALITIES

Alan Latham

Introduction

This chapter asks: what kinds of materials are cities made of? And what are the qualities of those materials? This might seem obvious. They are made of things like houses, apartment buildings, roads, factories, offices, cars and trucks, ports, airports, railway stations and mass transit systems, and constructed of bricks, concrete, steel, glass, wood and stone and so on. Right? Cities *are* made of materials like this. But if we look around most urban environments we will also see all sorts of other things. Trees, grasses, shrubs, animals of all different kinds, from domestic cats and dogs, to horses, chickens, and cattle, and wild animals like possums, foxes, rats, mice, raccoons, boars, coyotes – even monkeys and elephants depending which city you find yourself in. You would, if you looked closely enough, and in the right places, see insects and grubs galore. And if we looked even closer we would also start to notice that a whole range of other things are congregating in and circulating through urban environments. Things like energy – in a whole range of forms and for a whole range of entities – as oil, gas, petrol, as electricity, as food, as lighting. Or water: running through pipes and drains, through streams and rivers, stored in reservoirs, filling ponds and swimming pools, festering in puddles, seeping into foundations and walls, falling onto roofs, flooding along roads. Or waste of all different kinds: from buildings as they are torn down and rebuilt, from factories and offices, from the smoke stacks of power plants, from homes and restaurants, from the tail pipes of trucks, automobiles, and buses. But cities also function through a whole series of apparently immaterial elements. Things like culture, social norms and conventions, law, atmospheres even. Or even stranger elements like code, or information. In much of the social sciences it is these immaterial elements which have been viewed as the proper domain through which cities should be understood. It is not that social scientists think the other elements are not important. Rather they think that it is not important to them; after all social science is about the *social*, isn't it?

The following chapter presents a brief outline for studying the materialities of urban life. One aim of what follows is to set out a number of different framings of the material and materiality within contemporary urban studies. But the chapter's main focus is more direct. It is to explain the importance of thinking carefully about the materials and materialities through which urban environments and more broadly cities are constituted. This is not to discount the force of the immaterial elements of cities. Rather, it is to make a particular argument about how we should think about the social. Through thinking through the materialities of urban environments the following pages attempt to demonstrate how the material, and materiality, is always shot through with the immaterial. And following from that, the chapter argues that in an important sense the material and materiality do not just hold the social together – in a very real sense they *are* the social.

Six propositions for studying urban materialities

What follows are six interlinked propositions for thinking about the materialities of cities and urban environments. These should not be treated as definite statements. Rather think of each proposition as a suggestion to help think with the material thickness of urban life.

Proposition one: materialities are not just buildings and infrastructure

Put another way materiality is not simply the material stuff that grounds the social world. The notion that the material world is in some way central to how to go about understanding cities and urban environments is of course second nature to a great deal of urban theory (see Merrifield 2002; Tonkis 2005; Parker 2015). Architectural and planning theory, for example, are acutely aware that their thinking is addressed to the potential construction and reformation of real concrete places (see Chapter 3, this volume). And they are aware that it addresses, in some sense at least, working with construction materials, concrete, steel, glass, wood, and so on. Equally, economically oriented urbanists are aware that when they describe land markets, or industrial clusters, or local labour markets, they are referring to abstractions grounded in actual existing sites: sites woven together with buildings, roads, railways and all sorts of other physical things. Indeed, within much of critical urban studies the fact of this embedded capital is central to how we should understand cities.

Perhaps the classic statement of this avowedly materialist position is provided in the work of David Harvey (1973; 1989c; 2000). For Harvey contemporary cities are the product of the dominant mode of economic organization within society. Or put more bluntly, cities are the places where capitalism finds its ground. In this framing, the material world – the world of buildings, roads and other physical infrastructures – is not just the background for the action of social, economic and political life. It is the very stuff through which these elements are produced and reproduced. As we create our cities, our cities create us. To put things in a more philosophical key we are both the subject and objects of our environments. This is to restate, extend, and urbanize the argument developed by Karl Marx in *Capital* and in *Grundrisse*. Referring to the relationship between production and consumption Marx wrote: 'production not only creates an object for the subject but also a subject for the object' (Marx 1973: 93). Now what is arresting about this argument, both as configured by Marx and as later developed by Harvey is that it makes clear that buildings are more than just buildings, factories and offices more than just factories and offices, infrastructure is more than just infrastructure. The relations between people, objects, and spaces that they realize is the medium through which capitalism makes and reproduces itself.

The difficulty with such a framing of the material and materiality is that it both goes too far and it does not go far enough. It goes too far in stating that the ultimate source of all social power is rooted in the dynamic power asymmetries of capitalism. One has to ask why must we locate all of the power of the social and its materiality in the dynamics of the mode of production? Why not also take religion, or science, or the force of tradition or nationalism, or even biology and evolution and other similar entities seriously? But for the concerns of this chapter the bigger problem is that such Marxian framings do not go far enough in acknowledging and accounting for the force – or agency if you wish – of materials and their materialities. Materials and their materialities are more than just objects worked on by human agency. They also, as philosophers from Spinoza (1992), to Whitehead (1920), and Stengers (2010) have argued, generate their own forms of agency and action; agencies often both simultaneously interwoven with and independent from human intent.

Proposition two: cities are constituted through a diverse range of different materialities woven together in variously continuous and discontinuous ways

So, following the science historian Andrew Pickering (2001), we need ways of thinking about materialities that go beyond the social sciences' established notions of subject and object. This could be reframed as a need to think about the range of materialities that hold urban environments together. Of course we can return to the materiality of stones and building material, and point to the different ways buildings are constructed and cared for, the ways they age and transform (see Edensor 2012). There is a long-standing tradition of landscape research that focuses on just this kind of difference. Lars Lerup (2001, 56-7) drawing on the landscape historian J. B. Jackson (1953) talks about how the American house transformed as it moved west across the continent. Losing its ground and weight, it discarded basements for concrete-piles, brick and stone for chipboard and cinderblocks. In cities like Houston the vernacular house comes to float on the landscape's surface. This is a landscape where 'the entire foundation of the ground level ecology is soft, rhythmic and unstable, held together by the roots of the canopy of trees, creating the absurd impression of a city suspended from the treetops from which its cars, riders and roads gently swing'.

What is interesting about Lerup's description of Houston is how it pushes past clear definitions of the built and un-built, the human made and the natural. This landscape is a novel mixing of building, nature, dwelling, technology, and more. A landscape 'dominated by motion and time and event', where each element 'hide[s] an essential vulnerability: trees die, cars and markets crash, and the air slowly kills' (Lerup 2001: 58). This description of Houston is compelling in how it threads together a plurality of different material and materialities through which the city functions. This is an environment that includes both that which is built and in some sense planned by people, but which also incorporates the worlds and agencies of nature, and the cybernetic force of transportation systems and markets (forces that in all sorts of ways exceed the will of any single individual or group). It is a rendering that places into question what exactly is inside and what outside the materiality of the urban environment. In Lerup's view urbanists don't just have to think about the human built environment and people's goings-on within it. Urbanists need to take into account all sorts of non-human elements and agencies that course through and within cities; elements whose materialities urbanists have rarely bothered to question. Entities like temperature, or weather, or atmosphere. Entities like air, which in Houston 'functions much like ... skin, an immense enveloping organ, to be constantly attended to, chilled, channeled, and cleaned. Pools of cool air dot the plane, much like oases in deserts' (Lerup 2001: 58).

Proposition three: materialities within cities possess all sorts of often surprising agencies

To say that material in all its plurality plays into the life of cities is to say that cities are also animated by an equally plural range of agencies. Here we loop back to the terrain set out in point one. There we approached the notion that just as we make our cities, our cities make us. One way of thinking about that intertwining is to consider the way urban environments and cities function as enormous sites where materials are assembled, processed, distributed and consumed. This work of transformation involves pulling the natural world into the ambit of human production and reproduction, generating a complex kind of second or urbanized nature. Just as cities are built through nature, cities transform nature. Crucially, however, this second nature is a nature dominated by and made subservient to human power (Smith 1984; Gandy 2003; Swyndeouw 2006; Heynen *et al* 2006). This is a useful heuristic. But again it does not give enough space for the plurality of agencies that make up this 'second nature'.

One way to think more expansively about the agency of material is to think beyond ideas of nature and the natural. This is to connect with the work of writers like Michel Serres (1982), Sarah Whatmore (2002), Bruno Latour (2005), Jane Bennett (2010) and others who have explicitly worked at inviting the busy-ness of the non-human into the world of the social. These thinkers push the notion of agency beyond the idea of human intentionality, to focus on the fact that non-human materials *affect* us (as individuals, as groups, as societies) in all sorts of ways

– obvious and not, consciously and not. And this fact of affecting represents a kind of agency.

This can be understood in a number of ways. It can be understood in the ways that humans are constantly picking up and swapping properties with non-human (Olsen 2010). It can be understood in the ways that as we come to know better how to feed and nourish ourselves, as we come to understand the complex microscopic worlds of microbes and biology, and as we have created evermore complex medical collectives of experts, hospitals, and medicine, we have as an urbanized species become physically larger (taller and heavier), longer-lived, and less fecund (Floud *et al* 2011). This is not just to point to the politics of biopower highlighted by Michel Foucault (1979) and others (Cisney and Morar 2015). It is also to point to a very different experience of being human in a world where our bodies and lives are formed and manifested in such environments (Shove 2003; Sloterdijk 2004). This agency can also be seen in the way epidemics such as the Black Death that swept through Europe in the fourteenth century transformed social, political and economic history. Travelling along trade routes and pooling in the cities that acted as hubs for a growing inter-urban system of commerce, the microbial disease vector *Yersinia pestis* piggybacked on fleas, which piggybacked on rats, themselves piggybacking on merchant boats and carts, which themselves sprung from the aggregating power of cities (McNeill 1976).

In a gentler way, the agency of the non-human can be seen in the ways that the presence of trees, scrubs, and all sorts of plants draw ecologies around them. Not only do they help clean and purify the air of cities, there is also good evidence that they help contain and smooth our moods (Van den Berg *et al* 2007). They also provide habitats for a vast array of wildlife that populate urban environments. This includes species like foxes, rats, mice, raccoons, cockroaches, and mites, that live off the waste of human society. It also includes a perhaps more surprising populations of animals such as wild boars, coyotes, peregrine falcons and deer, to name just a few species who to many people's surprise manage to find productive niches within urban environments (Lorimer 2015). Indeed, rather than the urban representing some kind of tamed second nature, this suggests a wilder, more turbulent and surprising ecology. A world of 'edgelands', feral landscapes, and unofficial countrysides (Farley and Symmons Roberts 2011; Mabey 1973; Monbiot 2014).

Proposition four: urban materialities enfold a diverse range of temporal and spatial scales

Along with cities the previous paragraph talked of the microbiological, along with the bodies of parasites, the hulls of ships and trading routes. It talked of the relation between food, public health, medical knowledge, and human bodies. And of the ecologies inhabited by animals like peregrine falcons, foxes and coyotes, the world of edgelands, feral landscapes, and unofficial countrysides. These are just a few of the materialities circulating and aggregating in urban environments. None the less, to think of all the diverse assemblages of entities outlined in the previous paragraph is to confront entities that confound taken for granted notions of scale.

Urban studies, of course, has been much interested in scale. Writers like Neil Smith (1992b), Neil Brenner (1999), and Erik Swyngedouw (2004), have highlighted the importance of scale and its social and political construction to contemporary urban dynamics. And, indeed, largely through the theories of these authors, concepts of scale have come to infuse and animate much of urban studies. What is striking about many of the assemblages of materiality listed above is how they confound established notions of scale. They do so partly through the way they extend into 'scales' not really accounted for in conventional renderings. They include the molecular and the microscopic, along with ecological ranges that stretch across and beyond boundaries of cities, regions and nations. Indeed, this mixing up of scales, temporal and spatial, suggests a more topological diagramming of the materiality of cities (Amin and Thrift 2002; Marston *et al* 2005; Latham 2011). This is a way of thinking that places into question notions of what is big or small, enduring and ephemeral.

We could use any of the above cases as examples. But it might be more interesting to turn to the example of energy and mobility. Here we can talk about the ways that the emergence of urban environments suffused with the energy of fossil fuels has created environments that flicker, hum and thump with machinic intensity; propelling light, heat, air, metal and rubber, data, and much more through urban environments. How the intimacy and immediacy of much of this energy – wrapping up, cocooning, carrying vulnerable and tiring human bodies – is sourced through the burning and recycling of solar energy accumulated over millennia, and transformed and stored over geological timescales. Thinking further we encounter the strange energetic entanglements of the increasingly sedentary lifestyles of many who live in cities, with their weird enfoldings of the animate and inanimate, the biological and the inorganic, of speed and stasis, of the very small and the very large (Ng and Popkin 2012; Roberts 2010). How the 'natural' scale of the human body becomes a kind of extended mobile hybrid: the automobile, the house, and so forth. How inhabiting urban environments that demand less and less of the individual muscles of human bodies pulls us away from a genetic inheritance founded in all sorts of ways on dynamic corporeal movement (Lieberman 2013; Latham 2015).

CASE STUDY 16.1 CITIES AND THE MATERIALITIES OF THERMAL COMFORT

Materiality brings into view a range of relations that are easily overlooked, or not considered properly urban. Consider the question of temperature and thermal comfort. Urban knowledge-based economies are to a very large extent based around an architecture of large office buildings. Think of the skyline of global cities like New York, Hong Kong, Dubai, or Shanghai. Think also of the landscapes of suburban business parks and edge cities. The functionality of these economic sites and the buildings that populate them is dependent on intricate systems of air-conditioning that regulates their temperature and humidity independently of the immediate weather outside. Not only does the heating and cooling of these buildings consume vast amounts of energy, the ubiquity of such systems of temperature control within the urban environment seems to be leading to the emergence of a global temperature norm. Whereas previously people in different countries expressed wide variations in the range of temperatures they found comfortable, research has shown that such preferences are beginning to converge around a temperature of 22 degrees Celsius. In many places urban dwellers seem to be becoming less tolerant of, and less skilled in dealing with, natural variations in weather and climate. This in turn is pulling people to spend more and more time inside safely cocooned from such variation. Wealthy, advanced economies, are not just urban societies, they are also increasingly societies of the indoors (Hitchings 2011; Nicol *et al* 2012). So, even as global change pushes us to consume less energy, contemporary patterns of urban life and the practices that they enfold pull us towards futures dependent on complex machinic infrastructures of atmospheric control.

Proposition five: cities are constantly involved in generating new organisations of materialities

It should be clear from the previous four propositions that cities involve not only manifold entanglements of human and non-human agency, but that they are constantly involved in generating novel nexuses of such entanglements. It is hardly surprising then that living in cities involves dealing with – and organizing – these entanglements and the agencies that emerge through them in all sorts of ways. Thinking of this points attention to the various networks of infrastructures that have developed to support contemporary urban life. We can think of infrastructures of sanitation and waste disposal. Water provision. Electricity networks and systems of power supply. And we can think about how those infrastructures incorporate the very small and intimate, along with the very large and extended. They involve things like regimes of personal hygiene and bodily care (Shove 2003), along with the microscopic like viruses, bacteria, mineral trace elements and so on. At the same time they involve technological systems that spread over hundreds if not thousands of kilometers (Hughes 1993). This also points us towards thinking about the ways these infrastructures enfold and are animated by all sorts of kinds of information.

To think about information, yet again, requires thinking past the idea of materiality as simply the material (cf. Proposition 1). Following the suggestion of Pickering (2001), it requires understanding urban materiality as a dynamic process of sense-making. Infrastructure functions as infrastructure – as systems that support social life

– through the interweaving of material and technological elements with all sorts of social and political institutions that maintain and sustain them (Graham and Thrift 2007). Networks like electricity grids, or sewerage systems do not simply run themselves. They themselves rely on a vast entanglement of further technologies and conventions: postcodes, design standards, billing and metering systems, to name just a few examples. Matthew Gandy (2005) refers to this infrastructural skeleton as a kind of 'cyborgization'. But this structuring of the city's fabric can be understood as something more lively than that. Think of the ways that the spread of computer code through the fabric of cities – within infrastructures, within personal devices, within the objects like automobiles and house we inhabit – is involved in generating urban spaces that are coming to take on an increasingly sentient quality (Thrift 2014).

Proposition six: thinking about the materialities of cities does not just mean returning to the concrete

Throughout the previous five propositions – and indeed in the introduction of this chapter – there has been a continual invocation of material plenitude within urban environments. That there is so much *stuff* that makes up urban worlds has been repeated over and over. And it would be tempting – logical even – to take from this that thinking seriously about the material and materiality implies a valorization of the concrete-ness of cities. That, however, is not the argument I wish to make. Yes, there is a sense that attending to the materialities of our urban environments demands attending to a certain kind of concrete-ness. At the same time, paying attention to the concrete does not imply a retreat from abstraction in urban theory. Rather, it pushes us to consider how abstraction might be used to allow us draw out the force of the material and materiality (Latham and McCormack 2004: 707; McCormack 2012a).

What does that mean? Well we can think of at least four conceptual vehicles that might help make sense of the force of material and materiality in urban theory. First, materiality is *emergent*. The materiality of the urban is emergent in that it is product of a mesh of heterogeneous, small-scale, self-organizing, processes (Johnson 2001; Latham 2003b). This is not to say that cities are places where anything can happen, or anything goes. Of course cities are also planned and designed. Rather it is to repeat that cities are about the manifold relations between a whole range of human and non-human forms of life, a complex of relations that lacks any underpinning structural logic. Second, cities' emergent materiality can be understood as *machinic* (Amin and Thrift 2002). This notion of the machine is foreshadowed in the urban critic Lewis Mumford's (1934) description of even ancient and medieval European cities as enormous transversal machines. Developing this intuition, Felix Guattari (1995) talks of how the machinic should be understood not as the outcome of, but as the prerequisite, of technology. It is the presence and force of certain social, linguistic, corporeal, and cognitive machines that allows specific materialities to emerge in cities. This conceptualization of the machinic is powerful because it moves us away from thinking about technologies and bodies as having an already predefined and configured materiality. Instead, it provides a vehicle for understanding this relation in terms of its 'emergence between corporeal and machinic materialities, relations that may display systematic tendencies without being structural' (Latham and McCormack 2004: 707).

Thirdly, the emergent, machinic, materiality of cities can be understood *diagrammatically*. That is to say, cities and the spaces that make them up can be understood through diagrams some of which map out imminent relations of power, whilst others act as a generative device for imagining possible 'territories of practice' (Somol in Latham and McCormack 2004: 708). Such diagrams can also be thought of in much more ordinary or everyday key. Think of the forces drawn together through the lines of a football field, of the curbs of a sidewalk, or a children's playground. Lastly, the materiality of cities can be thought of as *expressive*. Not expressive in the sense of an individual subject. Nor just in terms of a festival or event. It is the expressiveness in terms of the emerging relations and event-ness of bodies and materials. To quote Brian Massumi (2002: xxi):

> Expression is not in a language-using mind, or in a speaking subject *vis-à-vis* its objects. Nor is it rooted in an individual body. It is not even in a particular institution, because it is precisely the institutional system that is in flux. Expression is aboard in the world. … It is non-local, scattered across a myriad of struggles over what manner of life-defining nets will capture and contain that potential in reproducible articulations, or actual functions.

Conclusion

So, cities are a complex amalgam of different materialities. They are made up of the 'natural' as much as the built, the small as well as the large, the non-human as much as the human, the improvised along with the designed. In the midst of this complexity the concept of urban materialities is productive precisely because it pushes urbanists to think with and through these diverse assemblages of agency that populate and animate urban environments. But the usefulness of materiality goes beyond simply that of a sensitizing heuristic; the concept of materialities is not just a device that nudges us towards attending to the heterogeneity of urban life. Thinking about urban materialities forces us to think carefully about other domains of the urban. Domains like the political. The politics of cities and urban environments are intertwined with the diverse agencies of the material in all sorts of ways. From questions of how people move around cities, to who deals with their waste, to how access to the multiple infrastructures of power, information, and communication that weave through a city's fabric is managed and controlled, materialities of all kinds animate urban politics. What is more, the emergent agencies that populate urban worlds demand that urban researchers recognize that the political is not predefined or configured, but something that is discovered, acted into. Something full of surprises. And this is something that is equally true for the domains of the social or the economic.

Key reading

Amin, A. (2008) 'Collective culture and urban public space', *City*, 12(1), 5–24.

A fabulous account of what a post-humanist, materially oriented urban studies might look like.

Lorimer, J. (2015) 'Spaces for wildlife: alternative topologies for life in novel ecosystems', Chapter 8 in *Wildlife and the Anthropocene: Conservation After Nature*, Minneapolis: University of Minnesota Press.

Explores the novel ecologies that cities open up for wildlife of all different sorts.

Molotch, H. (2014). 'Below the subway: taking care day in and day out', Chapter 3 in *Against Security: How We Go Wrong at Airports, Subways, and Other Sites of Ambiguous Danger*. Princeton University Press.

An engaging study of the New York subway.

Simone, A. (2004) 'People as infrastructure: intersecting fragments in Johannesburg', *Public Culture*, 16(3), 407–429.

Describes the materialities of life in Johannesburg and explains why attending to this everyday materiality matters.

17

MOBILITIES

Ola Söderström

Introduction

We are standing on the pavement, with Lake Hoan Kiem behind us and the road in front of us. It's already dark in Hanoi, Vietnam, and I am looking at the continuous flow of motorbikes – a real emblem of the city – wondering how I will make it across to reach the other side of the road. My friend, who has been living in the city for years, offers me the obvious solution: 'Just keep walking at the same pace; don't stop and the bikers will avoid you.' From that evening onwards I know how to cross the road, and align my pedestrian mobility with the mobility of motorbikes and other vehicles in Hanoi.

As this chapter shows, these are but two of the numerous mobilities that one encounters when living in and researching the city. The anecdote above also shows that mobilities are constitutive aspects of cities, that they are diverse and connected and that they imply the use of place-specific skills. These are some of the issues developed in this chapter, in which I argue that mobility in urban theory has been approached in three different ways: as the mobility of people, things and information in cities; as a constitutive element of cities, both making and shaping them; and as a characteristic that represents a city in the form of models or policies that travel from place to place, acting as exemplars to follow (or not).

This chapter shows that there is no linear narrative running through these three approaches. Although it had been a core element of early urban theory from the nineteenth to the turn of the twentieth century, mobility ceased to be seen as a unified concept and central element of urban theory after World War II, and instead came to be approached primarily as a set of different movements within cities. More recently, the idea that mobility – or, rather, interconnected and interdependent mobilit*ies* – is the very stuff of cities has re-emerged in the context of a 'mobility turn' in the social sciences. This chapter focuses therefore on the

FIGURE 17.1 Negotiating mobilities in the centre of Hanoi (Source: Ola Söderström).

theoretical continuities and discontinuities between early and recent theorizing and concludes with a brief case study of Hanoi, Vietnam, to show how these different threads of thought around the mobility in and of cities can be woven together.

The pulse of cities

The Latin word *mobilis* is the root of the word 'mobility', signifying the ability of different entities to move or be moved (Oxford English Dictionary). The word became a concept at an early stage in urban theory, playing a central role in the work of German theorists such as Georg Simmel and Sigfried Kracauer in the late nineteenth and early twentieth century. For Simmel, who theorized the city in the context of the rapid expansion of German industrial urbanization in the second half of the nineteenth century, the modern metropolis was the crucible of a new form of social life. Mobility, embodied by figures such as 'the stranger' on which he wrote a seminal text (Simmel 1908), was the main source of the new mentality and form of social interaction he saw emerging in the large German cities (see Chapter 10, this volume). Berlin in particular, where Simmel lived and taught, was, as Walter Ruttmann's stunning 1927 film (*Berlin: Die Sinfonie der Großstadt*) shows, a city of hectic, criss-crossing movements of people, underground trains, cars and trams. However, mobility in the art and social theory of the period was not reducible to human movement but considered as a total phenomenon and a major feature of industrial capitalism. In such a holistic view of the city, the rapid circulation of money was connected to the flickering landscape of urban lights, and the fast pace of life and hurried states of mind of urban dwellers. For Sigfried Kracauer, writing in the late 1920s and early 1930s, mobility (*Beweglichkeit*) was the flip side of the melancholia that he associated, like Simmel, to the urban way of life. The spectacle of the city compensated for the melancholia generated by an uprooted and lonely urban life, but it could at the same time, he argued, generate further feelings of melancholia if the city dweller became aware of its vacuity (Agard 2008).

The strong relations between the 'German school' and the Chicago School of urban sociology are well known. For instance, Robert Park followed Simmel's teaching in Berlin in the 1890s, while Kracauer was acquainted with various Chicago sociologists. There is thus a continuity and development of Simmel's and Kracauer's theorization of mobility in the work of many authors of the Chicago School. For the latter, mobility was, as it had been for their predecessors, a feature that defined the very essence of cities. Park (1950, quoted by Lindner 1996: 113) thus characterized North American cities as 'a complex of atoms in motion' and, like Simmel, distinguished two dimensions of urban mobility: geographical movement and the multiplicity of sensorial stimulations (Park *et al* 1925). The German school and the Chicago School also shared a fascination for figures embodying the specificity of urban societies of their times, such as the American hobo (homeless migratory worker, see Case study 17.1).

CASE STUDY 17.1 THE HOBO: A FIGURE OF THE US INDUSTRIAL CITY

Author of a path-breaking 1923 study, *The Hobo: The Sociology of the Homeless Man*, Nels Anderson (1889–1986) left home in his teens and spent the early years of his working life as a hobo worker himself (Anderson 1961). He then studied sociology at the University of Chicago where he wrote, under the apparently very loose supervision of Robert Park and Ernest Burgess, a report on Chicago's hobos for the city's Council of Social Agencies. This report was handed in as his master's thesis and eventually became a classic of urban studies. In *The Hobo*, Anderson ethnographically described Chicago's 'Hobohemia', a then waning world of young, generally single and often educated men moving from place to place in search of temporary jobs along the railway lines. With an amazingly vivid journalistic style of writing, Anderson brought flesh to the more abstract interest of the German school in mobile men as figures of the industrial metropolis. He also showed that these mobile men, and mobility in general, were functional to the development of the USA and its frontier job market, rather than one of its accidental pathologies (see also Cresswell 2010).

Mobility was present as a topic of empirical enquiry and/or theorized by all members of the Chicago School. However, its theorization was most developed in the work of Ernest Burgess on the growth of cities (Burgess 1925). Burgess saw mobility as a synthetic indicator of the social and physical dimensions of urban growth, to be distinguished from routine movement in space, as it 'implies a change of movement in response to a new stimulus or situation' (Burgess 1925: 58). Thus for example, daily movements to work might keep a situation constant, but mobility described the emergence of new movements triggered by two main factors: 'the state of mutability of the person [...] and the number or kind of contacts or stimulations in his environment' (ibid: 59). Translated into present terms, this means that mobility is related to potential mobility ('mutability' in Burgess's terms) and connectivity (phone calls for instance). In sum, mobility is the 'pulse of community' ... 'indicative of all the changes that take place' (ibid: 59).

Both the German school and the Chicago School had a negative vision of mobility, relating it to isolation, mental health problems, confusion and destabilization. This moral subtext is related to the fact that these urban theories were situated within a more general critique of capitalist modernization. For some contemporary scholars, this theoretical strand is thus imbued by a 'metaphysics of sedentarity' (Cresswell 2006; Jensen 2009). This normative bias should not obfuscate the fact that these scholars provided a rich, sensorial and cultural analysis of mobility as well as a substantial socio-economic interpretation, containing a wealth of theoretical insights, many of which have resurfaced only recently, as we will see below. This approach to mobility as a 'total social fact' is an important legacy of this early theorizing, which in many ways got lost in the decades after World War II.

In the years of post war reconstruction, the countries of the Global North witnessed a boom in the use of the automobile and the development of transport infrastructures. These massive material developments had equally massive management and simulation needs in order to measure, organize and forecast the mobility of people and goods. New metaphors – where, for instance, bodies in movement became analogous to atoms or molecules – were used to reduce the complexity of mobility to aid the search for general 'movement laws' (Abler *et al* 1971). Previously seen as a total, interconnected social and spatial fact, mobility was broken down into a series of distinct movements of people, vehicles and information. The emergence in the 1940s of a specific traffic science and the development of transport geography from the 1960s onwards became the main disciplinary vehicles of this conception of mobility, where its 'complex characteristics [...] and its contexts are simply taken as variables to be swapped into an equation' (Adey 2009: 45).

In other words, the comprehensive theorization for which urban mobility was known at the turn of the century progressively withered in urban studies when after World War II 'a range of structural or static theories took over' (Urry 2007: 26). While this does not imply a disappearance of mobility as a theme of empirical inquiry or as an element in the theorization of change (see, for instance, Castells

(1976b) and Bunge (1971), who developed strong critiques of the social impact of the automobile), it does mean that the concept of mobility in the second half of the twentieth century had lost its former centrality and explanatory status in urban theory. The phenomenon of urban mobility came to be considered as a series of different movements in cities governed by quasi-natural laws rather than a socio-spatial phenomenon that constituted the very stuff of cities.

The city as constituted by related mobilities

The critique of the social and environmental costs of the automobile, as well as the 1973 oil shock, led in the mid-1970s to an interest in the reasons behind modal choices, i.e. the choices made by individuals and society with regard to modes of transportation. This was a first step towards a reconciliation of spatial and social logics (Gallez and Kaufmann 2009) in urban mobility theory while these two aspects of mobility had tended to be analysed separately during the three preceding decades.

Since the 1980s, the profound transformations of society related to contemporary neo-liberal globalization have prompted the development of new approaches in social theory at large, founded on the idea that present times are epitomized by increasing mobility. In urban theory, relational perspectives in particular have come to the fore (Jacobs 2012). This relational view, articulated around flows, movements and connections rather than territories, scales and boundedness (Amin and Thrift 2002; Massey 1991), envisages the city as a constantly mutable and reordered phenomenon. While modernist planning defined cities in terms of discrete dimensions, such as the four urban functions of inhabiting, working, recreation and circulation, contemporary theory defines them in terms of processes of becoming. In other words, there is a trend from ontological approaches to spaces and cities to ontogenetic ones (Adey 2006: 78).

The recently established interdisciplinary field of mobility studies suggests that these flows, movements and connections are 'materially reconstructing "the social as society" into "the social as mobility"' (Urry 2000: 2). Scholars in the field therefore announce and welcome a mobility turn and the formation of a 'mobilities paradigm' within the social sciences (Sheller and Urry 2006). The plural – mobilit*ies* – is crucial here because what is at stake is not the focus on a specific form of mobility but the study of a vast array of interdependent mobilities.

The success of mobility studies in various domains of the social sciences (and particularly in sociology and geography) has boosted a variety of research projects on urban mobility. There has been a significant increase in research on mobility *places,* such as airports and planes (Adey 2008a and b), automobiles (Laurier *et al* 2008; Urry 2006), subways (Butcher 2011) and motorways (Merriman 2007), *practices* such as driving (Thrift 2004a), 'passengering' (Bissell *et al* 2011), cycling (Spinney 2011) and walking (Edensor 2010) and *methods* that are specific *to the study of mobilities* (Büscher *et al* 2010; Fincham *et al* 2010) such as mobile video ethnographies. What mobility studies have brought to studies of urban mobility lies less in the subject matter than in the approach, or in how research topics are 'respecified' (Laurier *et al* 2008: 2). A study of automobility offers a good example of this change in perspective. Analysing car traffic is nothing new, but rather than looking at cars from the exterior, as indicators of density of flow between two points in space, mobility scholars are interested in what happens in the car – how a group of passengers interact, for instance, during a car ride (Laurier *et al* 2008: 2). In other words, under the influence of approaches such as ethnomethodology or non-representational theory (see Chapters 2 and 21, this volume), the embodied, interactive and emotional experience of mobility has tended to replace molecular or atomist visions of the phenomenon.

We can thus think of the transformations in the theorization of mobility since the early twentieth century as a spiralling movement in which present perspectives reconnect with earlier ones while at the same time moving forward by bringing new ways of thinking and doing research. Recent work on urban mobility shares a number of ideas and intuitions with the German school and the Chicago School. First, the breadth of the concept: mobility is a concept distinct from movement (Cresswell 2006) that encompasses different types of movement and their interrelations; second, the importance of mobility for urban analysis, as it is constitutive of urbanism; and third, a shared methodological attention to the sensorial, emotional and psychological aspects of urban life.

The mobility turn has developed and formalized these ideas. Burgess's concern for potential mobilities is included in the recently developed concept of *motility* – the capacity to be mobile – which includes access to the means of mobility, the skills to use them, and their appropriation (Kaufmann 2002). In the same way, the idea that mobility is a specific form of 'spatial capital' (Lévy 2003) important in the social stratification of contemporary cities can be interpreted as a formalization of the importance of the concept in early twentieth-century urban theory. There is also a methodological continuity with an accent on qualitative, often ethnographic research protocols, although present work uses a much wider spectrum of methods, from video ethnography (Laurier 2010; Spinney 2011) to travel diaries (Latham 2003a) and mobile ethnographies (D'Andrea 2009). The point at which recent theories depart more clearly from earlier work is in the normative evaluation of urban mobility. Although it still haunts certain analyses, the moral dimension of early theorizing is much less present today, where mobility scholars seek to avoid both its *a priori* condemnation and celebration.

Finally, the extent to which mobility studies have a critical purchase has been a matter of debate. The conceptualization of mobility has been criticized for being too broad (Adey 2006), too descriptive (Faist 2013), or paying too little attention to the social production of mobilization and immobilization (Franquesa 2011). Neo-Marxist political economists, for instance, consider the idea of a generalized mobility theory with scepticism and locate the core of critical urban research – particularly in an age in which finance plays a crucial role in the fate of cities (Moreno 2014) – within the analysis of the dialectical relation between the mobility of capital and its fixation in urban real estate or infrastructures (Harvey 2006).

But if we admit that mobility studies do not constitute a unified theory, and move away from general theoretical positioning to empirical research, we realise that urban mobility studies are critical in at least three areas: when looking at the co-dependence of mobilities and immobilities, for instance, between highly mobile Singaporean businessmen and women and their domestic workers (Yeoh and Huang 2010; also see Case study 17.2); when focusing on clashes and power games between different mobility cultures, such as cyclists and car drivers (Jensen 2007); and when considering the shaping of conduct through technologies of mobility, such as new road infrastructures (Merriman 2005; Söderström 2013). The concept of mobility has also gained prominence recently in other domains of urban studies, particularly in planning theory and urban political geography.

FIGURE 17.2   Foreign maid agency in central Singapore (Source: Alice Chen Huiyu).

CASE STUDY 17.2 POWER AND CONTROLLED (IM)MOBILITIES IN SINGAPOREAN HOUSEHOLDS

In Asian cities with rapidly growing economies, many Asian women migrate to work as maids in middle-class or elite households in other countries. In Singapore, domestic workers are mainly from the Philippines and Indonesia and form a labour force of over 200,000 women. Several aspects of domestic work in Singapore and in other South-east Asian cities, from distance-mothering to uses of public space, have recently been studied by urban geographers. In one of these studies, Yeoh and Huang (2010) investigate power relations between women belonging to different classes in Singapore, coming together in so-called 'global households' (i.e. households reliant on global movements of people). They show that in these households the mobility of employers is premised on the controlled (im)mobility of their domestic workers. The presence and absence of maids, and their movements outside the home, are strictly regulated. In Yeoh and Huang's study, mobility is a relational phenomenon, particularly with regard to class (and sometimes ethnic) relations. Far from a celebration of universal fluidity and boundlessness, it brings attention to how the mobilities, and also the immobilities, of individuals are interdependent in contemporary cities.

Cities on the move

Since the early 2000s, a series of publications studying the mobility of public policies in geographical space have developed new understandings of policymaking in an age of fast-paced globalization (Peck and Theodore 2010; McCann and Ward 2011; McFarlane 2011). In discussing policy *mobility* instead of policy *transfer* – a term generally used in political science – geographers have highlighted the selective, power-laden, spatially complex and often unpredictable process through which policies travel: aspects that are less present in research within political science (McCann 2011). Pursuing and deepening the debate recently, however, other authors have questioned the concept of mobility, pointing out the ways in which it is actually too limited to fully grasp the variegated ways in which a policy in one place can be influenced by a policy somewhere else (Allen and Cochrane 2010; Robinson 2011; also see Case study 17.3). It has been suggested, therefore, that researchers need to use a repertoire of different conceptual descriptors and methodologies in order to capture the material, immaterial, imaginative and sometimes elusive ways in which policies cross geographical spaces (Roy and Ong 2011; Söderström and Geertman 2013).

CASE STUDY 17.3 MOBILITY AND BEYOND IN GLOBAL URBAN POLICYMAKING

In a contemporary context of multifarious connections between places through all sorts of technologies, from air travel to ICT, the ways in which urban policies in different cities influence each other is necessarily a complex phenomenon. Pondering over what it is that puts urban policies in different cities in relationship with each other, Jenny Robinson (2013) criticizes the predominant materialist vision of 'relationality' in contemporary urban studies. She claims that policies 'move' across space in 'all sorts of ways – through forgotten conversations at meetings, long-distant reading of publications and reports, unpredictable friendships, and collegial networks' (Robinson 2013: 9). The methodological consequence of this perspective is, she argues, to examine how policies are 'arrived at' rather than following policy connections between places. This implies a spatial imagination that is topological rather than topographical, i.e. where Euclidean distance is less important than the means through which people, ideas and things are either brought into proximity or moved apart (Allen 2008). Applying this to the making of Johannesburg's 2006 Growth and Development Strategy, she shows that policymakers are at pains to identify the origins of their ideas because they have drawn on a very broad set of meetings, visits, readings, talks, etc. Robinson's study thus highlights the limits of the concept of mobility in urban policy analysis. It also draws our attention to the limits of the concept in urban studies in general: cities cannot be reduced to people, ideas and things on the move, and nor should their study focus merely on the use of mobile urban methods.

Hanoi: mobile city

In this final substantive section, I will lead you back to the streets of Hanoi to glimpse how the different conceptions of mobility described in this chapter so far can be woven together to account for the city's recent trajectory of change. The pulse of Hanoi has changed in spectacular ways during the past thirty years. The dense and hectic swirl of motorbikes circling Lake Hoan Kiem that I learned to navigate has progressively given way to the reign of automobiles. What was in the 1980s a city of cyclists later became one of bikers, and is now dominated by the car. These changes in the modes of transportation, so striking for occasional visitors to the city, are expressions of other, less visible changes in wage distribution, driving skills, traffic regulations, social stratification and senses of collective identity. These changes find their origin in important political decisions that have been made since the reform (*Doi Moi*) in 1986: the adoption of a new constitution in 1992, the removal of the US embargo in 1994 and the end of the obligation for foreign companies to partner with national companies in 2006. In brief, Hanoi is rapidly globalizing, but in ways that are specific to the city's regional situation and political regime.

One way of understanding the city's specific trajectory of globalization is to examine how different transnational flows and relations have generated urban change (Söderström 2014), or, in other words, how the city is constituted by mobilities. First of all, through the various phases of economic liberalization, the city has become an attractive place for the fixation of new capital flows, and in particular Foreign Direct Investments (+800% between 1993 and 2010). In the absence of a significant industrial sector, these FDIs have targeted the construction sector, radically transforming Hanoi's morphology, in particular its spatial extension and the shape of its skyline. Originating more and more frequently in other South Asian countries, these flows of capital have materialized, for instance, in the form of new roads or elite edge cities. Second, Hanoi has become a place of either settlement or transit for increasing flows of people, due to the liberalization of intra-national migrations in 1997, the return of Vietnamese of the diaspora (*Viet Kieu*) and the explosion of tourism (+300% in Vietnam between 1998 and 2008).

These new patterns of human mobility have also had dramatic consequences on the city's built environment, leading most notably to a hotel and restaurant boom in the historic centre. Third, although the state still holds a firm grip on the media, information flows have greatly impacted modes of consumption and routine practices in the city. Through the gradual hybridization of the ubiquitous Starbucks, KFC and H&M urban landscape with more traditional forms of public life, clothing, housing, food, shopping and leisure activities are being slowly but surely, and very profoundly, transformed. There is, in consequence, a stark 'contrast between the ascetic, carceral Hanoi of the 1980s and the sensuous, lively Hanoi of the present' (Thomas 2002: 1616).

Fourth and finally, the increasing connection of Hanoi to global flows is manifest in the introduction of new planning principles and architectural types: Korean and US planners have recently introduced the idea of a green belt, while other urbanists have worked with Vietnamese architects and planners in order to preserve Hanoi's public spaces (Söderström and Geertman 2013). The travelling architectural types of the shopping mall or the gated community, to name but two examples, were introduced by French and Indonesian industrial and architectural firms in the mid-1990s (see Chapter 3, this volume).

These are not free-floating global flows, of course. As recent mobility scholars have argued, they must be articulated within systems of meaning and power

FIGURE 17.3 A street in the gated community of Ciputra, Hanoi International City (Source: Ola Söderström).

(Cresswell 2010). Understood in this light, the city appears to be negotiating a geopolitical shift, with increasing economic and policy relations with Korea, Indonesia and Japan supplanting formerly predominant relations with Western and Eastern Europe. Speculative developments have, however, been facilitated by high levels of corruption in the state apparatus, and have led to the destruction of important parts of the city's environmental and architectural assets. Hanoi has also witnessed the development of social inequalities, with an upward social mobility for a political and economic elite and a downward social mobility for sectors of the population with limited resources, such as rice farmers on the fringes of the city, who often find themselves victims of land grabs (Labbé and Boudreau 2011). At street level, the search for profit, the transition from extended family households to nuclear ones and the consumption frenzy that has seized Hanoi since the mid-1990s all point towards the formation of neo-liberal subjectivities.

This analysis produces a rather grim picture of Hanoi's recent trajectory. However, it is important to remember that urban analysis from a mobility perspective leads us to consider cities as being constantly reordered. Hanoi as a mobile city, despite an autocratic regime, is not moving as a monolith in one direction. As well as a place of the negative changes noted above, it is also a place of positive change, where people can distance themselves from the new routines of the consumption society, where associations involved in the defence of the city's assets are becoming more numerous and more vocal, and where planners are resisting imported blueprints for the city's future.

Conclusion

By looking at the ways in which mobility has been theorized and studied in urban research from the early twentieth to the early twenty-first century, we have seen the mutability of its importance for and relations to the urban phenomenon. At times it has been a central element of urban theory (in seeing cities as mobilities), and at others one of its numerous dimensions (seeing mobility as a set of movements within cities). We have seen how, following a spiralling movement, recent theorization of mobility have revisited and developed theses formulated a century ago. We have also made our way through various empirical contexts: from the European and US industrial metropolises (Berlin and Chicago), to the emerging centres of the postcolony (Singapore, Johannesburg and Hanoi). My analysis indicates that mobility is not only theorized differently in different historical contexts, but also in different geographical contexts. In other words, the theorizing of mobility is situated: the specificity of cities brings to the fore salient (and sometimes new) relations between mobilities and cities, such as the paradoxes of the 'global household' in Singapore. Because urban studies are historically and spatially situated, the theorizing of mobility is bound to evolve, respecifying its objects of study in interesting ways or elegantly reinventing the wheel in the years to come.

Key reading

Adey, P. (2009) *Mobility*, London: Routledge.

Cresswell, T. (2010) 'Towards a politics of mobility', *Environment and Planning. D: Society and Space*, 28 (1): 17–31.

Above are two recent and influential texts that introduce contemporary mobility studies and propose an analytical framework.

Park, R. E., Burgess, E. W. and McKenzie, R. D. (1925) *The City*, Chicago: University of Chicago Press.

Simmel, G. (1908) *Soziologie: Untersuchungen über die Formen der Vergesellschaftung*, Leipzig: Duncker and Humblot.

Above are two foundational texts that develop a holistic view of the relations between city and mobility.

18

NEOLIBERALISM

Ugo Rossi

Introduction

Neoliberalism is a political philosophy and an 'art of government' that has become hegemonic within capitalist countries after the economic crisis of the mid-1970s and the consequent dismantling of economic and institutional relationships based on the Fordist-Keynesian mode of production and regulation. In the subsequent decade, the fall of the Soviet Union opened the way for the triumph of neoliberal capitalism in the alleged absence of viable alternatives. Neoliberal ideas, therefore, gained ground at a time of economic and geopolitical turmoil and change between the 1970s and the 1980s. The following two decades, then, have seen the expansion of the neoliberal project in both qualitative and quantitative terms: on the one hand, it has deeply restructured existing economic and institutional relations in capitalist societies; on the other hand, it has achieved an increasingly global reach at a time of advanced globalization.

Cities have played a central role in the ascendancy and the geographical extension of neoliberalism. The aim of this chapter is to show how cities have been crucial to the realization of the neoliberal project. The chapter is divided in two main sections: the first section is dedicated to reconstructing the historical trajectory of neoliberalism; the second section presents the two-sided relationship between cities and neoliberalism, looking at both the way in which neoliberalism has incorporated long-term features of the urban phenomenon in capitalist societies, and at the way the urban phenomenon has been reshaped by the advent of neoliberalism itself.

The historical trajectory of neoliberalism

The hegemony achieved by neoliberalism in contemporary societies over the last three decades has coincided with the triumph of the idea of 'freedom', understood as an aversion towards any limitation imposed by the state to markets and individual enterprise. This principle was already present in the founders of classical political economy, most famously expressed by the *laissez-faire laissez-passer* motto attributed to the physiocrats contrasting market protectionism in eighteenth-century France. Between the First and the Second World War, free-market ideas started being revisited by scholars and intellectuals disappointed by the failures of classical laissez-fare liberalism, symbolized by the financial crash of 1929. The new neoliberal ideas demanded a substantial change in the theorization of the relationship between state and the market and thus of the idea of freedom. As Foucault (2008) argued in his famous lecture on neoliberal governmentality at the Collège de France, while classic liberalism had called on government to respect and monitor market freedom, in the neoliberal approach the market started being viewed as the organizing and regulatory principle underlying the state itself (Lemke 2001). This means that the logic of the market was bound to permeate all aspects of social life according to the emerging neoliberal wisdom.

The starting point in the trajectory of neoliberalism can be identified in 1938, when a heterogeneous group of scholars and entrepreneurs sharing the belief in free market, along with a hostility towards collectivism and socialism, took part in the Walter Lippmann Colloquium in Paris. The event was dedicated to the identification of a line of action aimed at relaunching the liberal cause in both intellectual and political terms. Neoliberalism's rise began at a time in which Keynesianism with its emphasis on state intervention and an expansionary fiscal policy became predominant within Western governments in response to the Great Depression, most notably in the United States with the New Deal embraced by President Franklin D. Roosevelt in the 1930s. However, due to the start of the Second World War the intellectual and political agenda resulting from the Lippmann Colloquium could not be implemented.

Shortly after the end of the Second World War, a decisive contribution to the resurgence of the still-embryonic neoliberal project came from the Mont Pelerin Society, founded in 1947 in Switzerland by a group of intellectuals comprising prominent economists and philosophers such as Friedrich von Hayek (the first president of the Society), William Röpke, Milton Friedman, Karl Popper and Ludwig von Moses. Both the Lippmann Colloquium and the Mont Pelerin Society played a decisive role in the ascent of post-war neoliberalism, bringing together scholars from different national and intellectual backgrounds (Audier 2012). In the following decades, the neoliberal project was reinvigorated by the growing popularity of the Chicago School of Economics within and outside academia. In 1962, in particular, the publication of Milton Friedman's *Capitalism and Freedom* attracted the attention of a large readership going beyond the narrow circles of post-war neoliberalism. In 1976, Friedman received the Nobel Prize, gaining international acclaim as one of the most influential economists of the twentieth century. This popularity led him to become a key adviser to the Reagan Administration in the 1980s, while his supposed contribution to economic reforms in Chile during Pinochet's military dictatorship remains more controversial.

The adoption of ideas derived from neoliberal thinkers within the policy sphere followed in the wake of neoliberalism's becoming a 'common sense' system of values in Gramscian terms after the decline of Keynesianism. To analyse neoliberalism's trajectory from 'common sense' to a coherent governmental rationality and practice, it is worth starting from Jamie Peck and Adam Tickell's identification of two phases of neoliberalism, a characterization that is now widely accepted within the general literature dealing with this topic, especially in human geography: the so-called roll-back and roll-out stages of neoliberalism over the 1980s and the 1990s (Peck and Tickell 2002). The former is seen as the destructive moment in the ascendancy of neoliberalism within the globalizing world. In this phase, neoliberalism dealt mainly with the economic sphere, imposing the privatization of state-owned enterprises and the restructuring of the labour market according to the flexibility imperative. The latter phase, on the other hand, exemplifies the creative moment of neoliberalism touching on a larger set of societal realms such as social and penal policy, the sense and practice of urban citizenship, the cultural sphere, in the attempt to entrepreneurialize the governance of economy and society as a whole (see Chapters 11 and 13, this volume).

The rollback stage of neoliberalism is customarily associated with the Reagan and Thatcher governments in the USA and the UK during the 1980s. In explicit contrast to demand-side policies drawing inspiration from Keynes, the Reagan Administration established a new pattern of supply-side policies, commonly known as Reaganomics. While Keynesianism supported measures aimed at stimulating the aggregate demand with a multiplier effect on the general economic activity, even if this entailed relatively high rates of inflation, Reagan's neoliberal economic policies looked at inflation as a serious problem threatening economic stability and social prosperity, caused by uncontrolled deficit spending. The 'war on inflation' thus became one of the major concerns of the Reagan Administration and prominent justification for budget cutbacks hitting welfare state institutions such as transportation, healthcare and education (McCormack 2012b). In order to revitalize the stagnating economy, Reaganomics tried to stimulate economic growth acting on the supply side, rather than on the demand side as recommended by Keynesian economists. The supply-side policy consisted essentially of two lines of action: on the side of businesses, and on the side of the workforce. The former benefitted from generous tax breaks and incentives, while the latter experienced a process of flexibilization and re-orientation towards either low-skilled jobs or knowledge-intensive professions, leading to new phenomena of socio-spatial polarization in contemporary globalizing cities. Moreover, the high interest rates policy adopted by the US Federal Reserve in the early 1980s in order to reduce the money supply and to get rid of inflation, as recommended by the monetarist economists, had consequences at an international level, aggravating the debt of developing countries, particularly in Latin America, and opening the way for a decade of neoliberal reforms under the rubric of the so-called 'structural adjustment programmes' (Peck 2001).

The 1980s are, therefore, the destructive stage in the ascent of neoliberalism. The 1990s, on the other hand, are the decade in which neoliberalism has devised the constructive part of its governmental project, aimed at rebuilding the capitalist state on entrepreneurial bases and at re-shaping in innovative ways the relationships between business interests and the public sector: the roll-out phase of neoliberalisation, to use the terminology proposed by Peck and Tickell. While conservative leaders such as Ronald Reagan and Margaret Thatcher in the 1980s embraced the ideal of the purest free-market society, in the 1990s the centre-left parties governing the US, the UK as well as key countries in Continental Europe such as Germany and Italy, sought to reconcile the enhancement of social cohesion with the stimulation of economic competitiveness and to turn the public sector into a more accountable and entrepreneurial organization, rather than dismantling it in a straightforward manner.

Peck and Tickell's characterization of the neoliberal era dates back to the early 2000s, taking into account the two previous decades of the neoliberal ascent. Since then, two further stages can be observed: the first lasted until the late 2000s, being characterized by the toughening of police action and the adoption of exceptional security measures (such as the Patriot Act signed by President George W. Bush in 2001) towards ethnic minorities and the new 'dangerous classes'; the second has derived from the financial crash of 2007–08 and the ensuing global recession, taking the form of a *déjà vu* of the austerity policies experienced during the 'destructive' phase of the neoliberal age in the 1980s. We can define this phase in terms of 'late neoliberalism', due to the central role played by 'negative' forces such as war and economic crisis on the evolutionary pathway of the neoliberal regime.

Let's start from the first stage in the late neoliberal trajectory. In the 2000s, in reaction to the September 11 attacks and the subsequent wave of global terrorism, the return of conservatives to political power in the major Western countries, most notably the Bush Jr. administration (USA, 2001–2008), but also those of Merkel in Germany (2005–present), Berlusconi in Italy (2001–2006 and 2008–2011), Sarkozy in France (2007–2012) and finally of Cameron in Great Britain (2010–2016), led to the accentuation of the penal wing of the neoliberal state. At this time, both the military conducting the global 'war on terrorism' in the Middle East and elsewhere, easily degenerating into brutal human rights' violations (as disclosed by the scandals in the detention camps of Guantanamo and Abu Ghraib), and the police repressing the popular riots in the French *banlieues* monopolized the attention of critical geographers and urbanists (Gregory 2004; Dikeç 2006). The Islamophobia generated by the association of global terrorism with Islam along with the refusal of an allegedly laxist approach to the integration of ethno-cultural minorities inspired by multicultural thinking led British Prime Minister David Cameron, in 2011, to argue for a 'muscular liberalism'. This idea was intended to reverse the process of recognition between the majority and the ethnic minorities: minorities must adapt to the moral and social values of the majority, rather than the other way around, as in progressive multicultural approaches (Kundnani 2014).

However, these debates rapidly lost international resonance for essentially two reasons: first, the emphasis laid by the Obama Administration on the need for a renewed dialogue between West and Islam; second, the advent of the global economic crisis and its devastating effects on European and North American societies, which redirected the attention towards economic and social issues. In this context, despite its failures, neoliberalism has given proof of resilience not only in economic terms but also in broader cultural and political terms, dealing with its crisis of legitimacy by presenting itself as a governmental rationality ensuring fiscal discipline and economic stability, particularly in the European context. Ironically, the new era of austerity has taken shape shortly after Western national governments had resorted to the proverbial 'socialization of losses', using public money to bail out suffering banks and credit institutions in the aftermath of the 2007–08 financial crisis (Blyth 2013).

CASE STUDY 18.1 GEOGRAPHIES OF THE GLOBAL ECONOMIC CRISIS: FROM CITIES TO NATION-STATES

A distinctive feature of the recent (2008–2009) global economic crisis and the related tide of austerity measures lies in its geographical ubiquity. The crisis started in the United States but then rapidly migrated to Europe, hitting with particular intensity such diverse national political economies as the United Kingdom, Iceland, the Baltic States and, finally, the countries of Southern Europe, historically the most fragile economies of the European Union. The coinage of the so-called PIIGS (Portugal, Ireland, Italy, Greece, Spain) offers immediate evidence of the geographical ubiquity of the crisis, as this (infamous) label brings together a majority of Southern European countries along with a country in the northern periphery of Western Europe: Ireland. However, more general empirical evidence shows that the effects of the crisis reverberate well beyond the confines of Southern Europe or the PIIGS circle. In 2013, about five years after the start of the economic crisis, the core economies of the Eurozone that were more persistently and seriously affected by a lack of economic growth in terms of GDP were those of Italy and The Netherlands, two countries that are clearly different in terms of socio-economic structure and sectoral specialization.

The wide range of countries that have been hit by the economic crisis has led commentators and analysts to offer schematized understandings of the various ways in which the crisis has manifested itself within highly differentiated capitalist national economies and societies, both in structural and policy terms. Not only the wider public and mainstream analysts, but also economists and other social scientists concerned with the study of capitalist economies from an explicitly heterodox viewpoint, such as critical political economists, have commonly interpreted the crisis from the point of view of national political economies (Streeck and Schäfer 2013). This is at one and the same time surprising and unsurprising: it is no surprise as, even in a context of globalization and the rescaling of societal governance, state-centred views have kept dominating common understandings of capitalist economies, as the influential literature dealing with the varieties of contemporary capitalism shows (for a critical review see Peck and Theodore 2007); even so, it is surprising, as since its inception the 'urban roots' of the crisis (Harvey 2012) have been clear not only to scholars professionally engaged with this scale of enquiry (such as human geographers and other socio-spatial scholars) but also to the common wisdom, given the centrality of the mortgage crisis in the financial crash of 2007–2008. The following 'sovereign debt crisis' has overshadowed the urban rootedness of the global crisis, leading the nation-state to regain prominence in the eyes of commentators and analysts.

CASE STUDY 18.2 GEOGRAPHIES OF THE GLOBAL ECONOMIC CRISIS: SOUTHERN EUROPE

In his acclaimed book on austerity, international political economist Mark Blyth has offered a clear schematization of the nationally differentiated forms in which the crisis has hit the PIIGS countries, leading to the current 'age of austerity' (Blyth 2013). His book shows how the five PIIGS countries (Portugal, Ireland, Italy, Greece, Spain) are far from representing a homogenous pattern of economic trajectory, from an era of expansion and allegedly irresponsible public spending to one of economic collapse and austerity, as the conventional (neoliberal) wisdom about the sovereign debt crisis maintains. Rather, these countries should be divided into three typologies: the first is represented by Greece, an exceptional case of a problem country long affected by a perverse combination of low productivity growth and irresponsible expansionary policies, which led to unsustainable debts and deficits; second, there are Ireland and Spain, whose economic pathways reflect more closely an ideal-type of neoliberal mode of economic development epitomized by the USA and the UK, founded on the following, interrelated characteristics: expanding financialization, rising private debt of consumers and households and an unprecedented property bubble; third, the 'slow-motion growth' countries, namely Portugal and Italy, which have been affected by sclerotic growth for many years already before the global crisis; their historically export-led manufacturing economies have been damaged by the adoption of the euro currency, because devaluations of the exchange rate that used to compensate long-term deficits and to reanimate national economies in response to economic slowdowns (as happened in 1992 in Italy) have no longer been possible, thus forcing governments to resort to international bond markets that led to increasing public indebtedness.

Blyth's interpretation is supported by strong evidence, with official data clearly showing Italy's high government debt (around 120% of the GDP), while Spain for instance had a more moderate public debt at the advent of the crisis (60%) that thereafter has risen to the 97% of the total output as a consequence of the recession, according to the latest estimates. Even so, if we zoom in our perspective on a lower scale than that of the nation-state, the process of variation looks more complex and nuanced. From this point of view, cities and the ways in which they have been dealing with the crisis and the subsequent wave of austerity measures offer a valuable vantage point to understand the complexities and intricacies of contemporary neoliberal economies. In urban scholarship, even those critical investigations on the urban manifestation of contemporary austerity that have thus not embraced a state-centred view have mostly focused on specific national contexts, while scholarship embracing a cross-national, comparative perspective, attempting to question or at least to complicate the conventional state-centred understanding of the global economic crisis, is more limited. The notion of variegated neoliberalisation (Brenner *et al* 2010) helps us uncover the specificities of place in a context of heightened circulation of urban development models and policy imperatives of both austerity and growth.

Cities as laboratories for neoliberalism

The previous section of this chapter has showed how the trajectory of neoliberalism has evolved since its appearance around the late 1930s. Neoliberalism started its trajectory as an elitist intellectual movement, it constructed its hegemony as a 'common sense' system of values, and finally it has consolidated itself as an art of government coping with an increasing number of social issues.

Cities are of special significance for the understanding of neoliberalism. In this section, we show how cities and neoliberalism are linked by a relationship of mutual learning and reinforcement. On the one hand, neoliberalism has learnt from the urban phenomenon and particularly from the way in which this has taken shape within capitalist societies; on the other hand, the urban phenomenon has learnt from neoliberalism and particularly from its tendency to commodify every aspect of social life. For illustrative purposes, we will deconstruct two commonly used definitions associating cities and neoliberalism: urban neoliberalism and neoliberal urbanism. Customarily, in the scholarly literature these two terms are used interchangeably. Here we differentiate between the two terms, associating the notion of 'urban neoliberalism' with what we define as the urbanisation of neoliberalism, while the notion of 'neoliberal urbanism' will be referred to as the neoliberalisation of the urban experience.

Urban neoliberalism, or the urbanisation of neoliberalism

In the first instance, it can be argued that there are distinguishing and long-term features of the urban experience within advanced capitalist countries that have become key aspects of the neoliberal regime of societal governance. In general terms, while during the Keynesian 'golden age of capitalism' national policies focussed on issues of wealth redistribution and social welfare, cities were already committed to the imperative of growth, especially US cities, as repeatedly underlined by specialists of post-war urban politics in the United States (Judd and Swanstrom 2014). Subsequently, the growth imperative has been incorporated into a more coherent policy and conceptual and ideological framework, under the banner of the entrepreneurialization of urban governance, analysed for the first time by David Harvey in the late 1980s (Harvey 1989a), which we now recognize as a distinctive trait of 'urban neoliberalism'. More specifically, two aspects historically associated with social life in capitalist cities have played a crucial role within contemporary neoliberal societies: consumption and housing.

Since the rise of capitalism, the practice of consumption has characterized the functioning of urban societies. As unrivalled concentrations of consumers and sellers, cities and larger metropolitan areas have become crucial to the growth of consumption. With the advent of mass production in the twentieth century, increasingly standardized forms of consumption gained ground in Western societies and particularly in their cities and metropolitan regions. The twentieth century saw department stores, whose early examples were in large industrial cities such as Paris, New York and Chicago, proliferating at the expense of small shops and traditional open-air markets (Sennett 1978). The society of mass consumerism largely characterized the central stages of the so-called 'golden age of capitalism' in the three decades following the end of the Second World War (see Chapter 9, this volume).

After the economic and geopolitical turbulences of the 1970s, the decline of Fordism and the related system of mass production paved the way for the advent of a neo-capitalist development pattern – commonly known as post-Fordism – based on lean production on the one hand (particularly represented by Toyota's just-in-time technique) and on diversified consumption, on the other hand. With the globalization of the world economy, the decline of Fordism in the Western economies was accompanied by the relocation of production to low-income countries in the Global South and the emergence of newly industrialized economies, particularly in the rampant East Asia. In the West, the predominance of the manufacturing sector was replaced by the growing importance of the service-oriented economy, formed essentially by two sectors: the producer service firms (law, accountancy, banks, etc.) and the consumption-oriented activities (retail trade, restaurants, leisure venues). In this context, there was a shift of Western cities being organized as spaces of production in the Fordist and Keynesian era to cities as spaces of consumption under the post-Keyenesian and post-Fordist era. Since the late 1980s, cities started also to attract a massive influx of international and domestic tourists, even some cities that previously had no such tradition, such as Bilbao in Spain, whose new tourism wave was thanks largely to the new Guggenheim museum of contemporary art.

The adoption of a neoliberal rationality of economic and societal governance has created, therefore, conditions for the triumph of consumerism as a mode of being and living reflecting increasingly pervasive individualization processes. In this context, the wider financialization of capitalist societies and of consumption itself through the diffusion of credit cards and other forms of consumer credit has given a decisive contribution to the penetration of the culture of neoliberalism into individual consciousness and behaviours. While in the Fordist era and the so-called golden age of urban-industrial capitalism, mass consumerism was oriented towards the acquisition of durable goods, such as the family car, home appliances and furnishings, the post-Fordist and neoliberal era witness increasingly individualized and diversified forms of consumption. Neoliberalism has given a twofold contribution to the intensification of individualized consumerism: first, it has allowed globalization to function as an hegemonic force within the world economy, opening markets and economies to foreign direct investment and the import of external commodities from abroad, even in previously state-planned economies such as China and India; secondly, it has triggered processes of commodification involving an increasing number of societal domains, which were under public control at the time of welfare capitalism, such as healthcare, transportation, education and professional training (Comaroff and Comaroff 2000). Under the neoliberal regime, commodification and the related phenomenon of consumerism have affected, therefore, not only conventional consumption goods but also public services and, increasingly, any aspect of our everyday life: put it simply, we are now customers not only of shops and department stores, but also of hospitals and schools, as well as of online social networks that have radically transformed our everyday life. Large cities and metropolitan areas have been key sites of this multifaceted societal commodification.

Housing in capitalist cities is another societal realm in which commodification precedes the advent of neoliberalism, giving rise to a process of mutual reinforcement, by which neoliberalism has seconded and deepened a phenomenon that is intimately associated with capitalist societies. Historically, the housing sector has functioned as a contra-cyclical regulator of economic growth, as testified by the fact that Western economies – the US economy above all – have become dependent on high and constantly increasing house prices, particularly in central locations for business and political reasons. Moreover, the segmentation of the housing market is instrumental to the reproduction of capitalism's social divisions. After the Second World War, at the peak of the Keynesian age, the politics of social welfare mitigated the commodification of housing through the provision of public and subsidized housing, particularly in the working-class areas of industrial cities and towns. However, the advent of neoliberalism has revived capitalism's long-standing attitude towards the commodification of housing. With the precipitous reduction in the supply of public housing and the end or relaxation of rent regulation in many Western countries, access to housing has increasingly taken the form of homeownership granted by bank loans implicating growing household indebtedness. The expansion of mortgage markets and the consequent financialization of housing have then become distinguishing features of ascendant neoliberal capitalism and its crisis in the late 2000s.

Also, social discrimination is a long-standing phenomenon in the housing market, which neoliberal regimes have normalized within a market logic. In the

United States, African Americans have long been discriminated in their access to the housing credit sector as banks denied them mortgages, particularly in so-called 'red-lined', undeserving neighbourhoods (Aalbers 2016). Even so, the Federal Housing Administration created in 1934 by the US Congress never seriously attempted to put an end to this state of affairs through an explicit anti-discrimination policy. The subsequent 'subprime' mortgages adopted since the 1990s, whose collapse was behind the financial crash of 2007–08, have formalized this system of government-sponsored segregation, as these lending schemes have been prevalent in neighbourhoods with high concentrations of low-income racial minorities.

Neoliberal urbanism, or the neoliberalization of the urban experience

The previous section of this chapter has shown how long-term distinctive features of capitalist cities, such as the imperative of growth, the relentless expansion of consumption and the commodification of the housing sector have become key aspects of the neoliberal era, acquiring renewed significance as engines of capitalist restructuring and development in post-Fordist times. In this section, to illustrate the notion of neoliberal urbanism, we will take a look into the reverse process: namely, how neoliberalism has subsumed within its rationality societal phenomena that previously existed partially or fully outside the capitalist circuit of valorization in contemporary cities: creativity and culture are instructive examples in this respect.

Creativity has a historically ambivalent relationship with capitalism, oscillating between incorporation on the one hand (from conventional technological innovation processes to today's post-Fordist knowledge-intensive industries) and autonomy or even opposition (creativity as a transgression and practice of contestation) on the other hand. Neoliberal urban regimes have attempted to neutralize this ambivalence, mobilizing powerful mechanisms of seduction and co-optation of alternative subjectivities, for instance through the institutionalization of autonomous creative spaces and the spectacularization of social and cultural diversity. Critical urban scholarship has showed how even an intrinsically radically phenomenon such as the squatter movement in European cities has not been immune to processes of normalization and incorporation into the capitalist logic of what French situationist Guy Debord famously defined as 'the society of the spectacle' in the late 1960s.

Creativity's process of appropriation into capitalism's cultural circuit has particularly intensified over the last two decades within the context of advanced liberal societies. An important contribution to this process has come from the publication of *The Rise of the Creative Class* in 2002, a best-selling book authored by Richard Florida, an economic geographer and urban planner at that time based in the deindustrializing city of Pittsburgh, Pennsylvania, in the middle of the US 'Rust Belt' (Florida 2002). In this book, Florida provided not only a new socio-cultural explanation but also a powerful narrative for what had been already identified by economic development scholars as the key role of technology and human capital as drivers of regional economic growth. In brief, Florida argues that the members of the creative class – notably, workers and professionals who make use of creativity in their jobs: from architects and engineers to software developers, designers and other emerging figures in the information society – tend to favour urban environments characterized by the tolerance of socio-cultural diversity related to the presence of ethnic and sexual minorities as well as artistic communities. Moreover, according to Florida, creative class members share values and attitudes based on ideas of meritocracy and individuality along with the willingness to be part of collaborative working environments.

While left-leaning commentators have warned against the risks associated with the uncritical adoption of Florida's theory and the related public discourse, such as the legitimization of gentrification dynamics in urban settings attracting creative class members, since its publication *The Rise of the Creative Class* has been enthusiastically welcomed by local politicians and other politico-economic elites first in the USA, from which its empirical evidence was derived, and subsequently in a rapidly increasing number of 'wannabe' creative cities across the world. As a governmental technology enabling the cross-national mobility of urban development patterns, neoliberalism has allowed creativity to become a global policy narrative mobilized by policymakers and administrators in order to create consensus around newly proposed or already existing urban regeneration initiatives across the globe. Moreover, the individualistic and meritocratic ethics underlined by Richard Florida is behind the widespread careerism among creative-class members, which weakens their social ties, undermining a sense of togetherness and the willingness to engage in spontaneous forms of cooperation. In 2013, San Francisco – a city witnessing skyrocketing house prices in recent years – has seen poorer members of the creative class protesting against the unsustainable costs of living caused by the presence of the affluent creative professionals employed in the Silicon Valley and residing in San Francisco. In recent studies, Richard Florida and his fellow researchers recognize the fact that in economically dynamic cities such as Washington, Boston and New York, the creative class's colonization of downtown districts forces service and working-class residents to move outside the central city into the least desirable parts of town, leading to novel forms of socio-spatial segregation and inequality (Florida *et al* 2014).

While the urbanization of creativity is likely to become an increasingly contested issue in contemporary capitalist cities, the mobilization of culture for urban regeneration purposes still exerts a powerful influence on social consciousness, as cultural festivals and events are portrayed as opportunities for enjoying the vibrancy of city life for both residents and external visitors. With the entrepreneurialization of urban governance in the 1980s, culture has been incorporated into a logic and circuit of valorization, as at that time cities started using their heritage and cultural endowments to attract public and private investments in a global context marked by heightened interurban competition (Rossi and Vanolo 2012). Competition over titles attributed to cities as a whole (for example, the European Capital of Culture), or to especially valuable portions of their built environment (historical centres, monuments, etc.) such as the UNESCO's world heritage nomination have led – according to critics – to an intensified economic colonization of the cultural realm. Italian writer Marco D'Eramo has recently written about what he calls the 'UNESCOcide' of contemporary cities:

> It is devastating to witness the death throes of so many cities. Splendid, opulent, hectic, for centuries, sometimes millennia, they had survived the vicissitudes of history: war, pestilence, earthquakes. But now, one after another, they are withering, emptying, becoming reduced to theatrical backdrops against which a bloodless pantomime is staged [...]. UNESCO's 'World Heritage' listing is the kiss of death. Once the label is affixed, the city's life is snuffed out; it is ready for taxidermy.
>
> *(D'Eramo 2014: 47)*

D'Eramo's association of the notion of urbicide with the tourist exploitation of contemporary cities may sound exaggerated to many readers. However, it conveys the pervasiveness of the culturalization of the urban environment, which is another way in which neoliberalism pursues its project not only of making everything profitable in economic terms, but also of annihilating the publicness of contemporary cities through myriad processes of enclosure brought about by the commodification of urban spaces.

Conclusion

After having outlined the trajectory of neoliberalism over the post-war decades, this chapter has looked at the two-sided relationship of neoliberalism with the urban phenomenon. In particular, it has been argued that there is a process of mutual reinforcement between cities and neoliberalism. This means that neoliberalism has subsumed within its logics and functioning long-term features of social life within capitalist cities, such as the commodification of housing and the relentless expansion of consumption. The Keynesian social compact during the so-called golden age of capitalism temporarily mitigated these processes of commodification, which neoliberalism has subsequently brought to new life. At the same time, neoliberalism is not only drawing from pre-existing characteristic features of capitalist urbanization, but has also expanded the economic base of contemporary cities, by commodifying societal phenomena that previously existed partially or fully outside the capitalist logics of valorization, such as culture and creativity.

This ambivalent relationship of mutual reinforcement explains the key contribution of critical urban studies to the understanding of contemporary neoliberalism. It is likely that ongoing processes of so-called 'planetary urbanization' (Brenner and Schmid 2014), with the geographical extension of urbanization processes and related neoliberalization dynamics spreading across the globalized world, will not only intensify the intimate relationship between cities and neoliberalism but will also bring to light novel aspects and dimensions in which this relationship takes shape. Moreover, the ambivalence is not limited to the described relationship between cities and neoliberalism. Indeed, cities are sources of crisis and austerity but also of promised resurgence for contemporary capitalism, as the latest global economic crisis has demonstrated: a crisis with clear urban roots, owing to the unprecedented financialization of housing markets that has particularly affected indebted households living in large cities and metropolitan areas. On the one hand, during the crisis, cities have been key sites for the implementation of austerity measures adopted in compliance with the fiscal consolidation targets imposed by international financial institutions and national governments to municipalities and regional administrations. On the other hand, cities are persistently seen as key spaces of resurgence for capitalist economies, as urban environments concentrate the innovative forces of contemporary societies (Rossi 2015). The current or future, depending on geographical contexts, phase of post-recession transition will show us whether neoliberalism will still take the lead in the governance and regulation of the constitutively dialectical development of contemporary capitalist cities.

Key reading

Brenner N. and Theodore N. (Eds.) (2003) *Spaces of Neoliberalism: Urban Restructuring in North America and Western Europe*, Oxford: Blackwell.
    This book collects articles originally published in *Antipode: A Radical Journal of Geography* which have had profound influence on the ways in which critical geographers and urban scholars have understood the urbanization of neoliberalism in subsequent years.
Ferguson, J. (2006) *Global Shadows: Africa in the Neoliberal Order*, Durham, NC: Duke University Press.
    In this book, anthropologist James Ferguson reflects on the marginalization of Africa in a context of neoliberal globalization, particularly investigating the contradictions of democratization programmes through the lenses of the theory of governmentality.
Foucault, M. (2008) *The Birth of Biopolitics: Lectures at the College de France, 1978–79*, Basingstoke: Palgrave MacMillan.
    In these lectures, Michel Foucault analyses the ascent of neoliberal ideas in Western countries, in both the German version reasserting the role of the state as a guarantor for fiscal stability and market efficiency, and in the American version, laying emphasis on the idea of freedom and the preeminence of the market logic.
Harvey, D. (2005) *A Brief History of Neoliberalism*, Oxford and New York: Oxford University Press.
    In this book, eminent Marxist geographer David Harvey reconstructs the historical trajectory that has led neoliberalism to become the hegemonic ideology and economic practice from the mid-1970s onwards.
Peck, J. (2010) *Constructions of Neoliberal Reason*, Oxford and New York: Oxford University Press.
    In this book, geographer Jamie Peck offers a genealogical analysis of the rise of neoliberalism and its politico-economic and intellectual project, particularly exploring the institutional conditions that have allowed the neoliberal logic to expand across the world.

19

PLAY

Quentin Stevens

Introduction

Scholars use the term 'play' to convey and explore disparate aspects of urban experience and to underpin varied arguments about social life (Sutton-Smith 1997). Three broad perspectives on play emphasize different social needs it can fulfil: its practical value as developmental learning; its cultural significance as luxurious, wasteful activity consciously separated from immediate practical needs; and its dialectical role as a lived critique within urban social life. The playfulness of any given social practice can be characterized by particular actions, aims and situations of its application. While play is often considered in opposition to work, seriousness and order, it also resists the dualism of Western thinking. It slips between categories and contexts, and studying it presents both conceptual and practical challenges. No definitions of play easily tie it down, but all are useful in thinking differently (indeed, playfully) about how people act and why they gather together in cities. In contemporary urban theory, play is often a rubric for the multifarious aspects of urban life that are not easily reducible to the rationalities and conventions of productive work, commodified leisure and social conventions. The theory and practice of play explore the experiential, social and psychological potentials and constraints that urban concentration and heterogeneity afford, through unplanned exposure to unfamiliar people, experiences and actions. This thinking finds application across a variety of disciplines and contexts.

The meaning and value of play

One scientific, rather insipid stream of thinking consigns play to a distinctive set of places, times and activities separate from the serious, normal, adult world. Play is seen as a practical activity that trains children for real life, through the development of physical, social and psychological skills. The play of adults, such as sports and hobbies, serves to dissipate or recuperate their physical and mental energies. That such activities are fun is largely dismissed as an epiphenomenon (Sutton-Smith 1997; Spariosu 1989).

But most interest in play comes from theorizations which, more true to the spirit of play itself, turn this scientific notion on its head. In this other view, play permeates every human action and interaction. Culture as a whole is something played (Huizinga 1949), our experience of the world remains playful throughout our lives, and in every performance people know that rules and limits can be tested. Rather than a distinct set of activities, many definitions of play identify particular dispositions and contexts that frame its physical, economic, social, and psychological purposes and impacts. For Huizinga, play is voluntary, superfluous and disinterested. It does not serve immediate practical needs. For Caillois (1961: 5): 'Play is an occasion of pure waste: waste of time, energy, ingenuity, skill, and often money.' Play is thus usefully employed as a theory for examining the urban activities that to most people seem irrational, pointless, or dangerous. The absence of wider instrumental purpose makes playful experiences ends in themselves. Play is the essence of what makes people's actions meaningful rather than merely utilitarian; it allows the exploration of identity. Huizinga also suggests play is set apart from ordinary life in both time and space, and creates it own order through its own rules. Rather than being completely separate, Lefebvre (1991) sees play as engaging dialectically within the order of everyday life: refracting, resisting, critiquing and transforming it. Play is not just children preparing for adult life. It is a low-risk means for all people to experiment with social practices, roles and rules. Some play follows strict rules, through which people explore the consequences of rules or test themselves against them. Often play involves changing or making up rules as you go along, to prolong the game, or to make it fairer, more challenging or more interesting. There are also forms of play that have no rules.

Huizinga (1949) devotes a whole chapter to the etymology of the word 'play'. It emerged from both Old English *plegan,* to exercise or frolic, and *plegen*, which shares with *pledge* and *plight* the sense of action that exposes one to risk. The noun 'play' also refers to a structured context within which something has free movement. These meanings are retained in contemporary thinking about people's play in cities. Play explores the freedoms the city provides. But play's forms often highlight recognition of urban society's limits and rules, by engaging with them and carefully testing them. What separates the intensities and freedom of play from real risks of injury and loss is that all participants retain control over the limits of their exposure.

Anthropological and sociological accounts

Contemporary thinking about play in the city has been informed by detailed anthropological and sociological studies of social role-playing. Anthropologists Geertz (1972) and Turner (1982) examined the temporary transgression and inversion of social rules and roles in the liminality of transformational rituals. Bakhtin (1984) and Huizinga (1949) similarly focussed on the special temporary social atmosphere of the medieval carnival, a scene of abundance and diversity that engendered freedom from constraints, encounters with the unfamiliar, inversion of social conventions, intense sensory experiences, and the exploration of new social possibilities. These non-urban, historical studies provide a contrast to the impersonality, rationality and vapidity of modern Western city life.

Sociologists have also studied similar characteristics in strangers' interactions in modern Western societies. The playfulness of urban encounters with difference has a key role in the thinking of Simmel (1950), Benjamin (Benjamin 2006; Gilloch 1996), Caillois (1961), Goffman (1972, 1980, 1982), and Sennett (1971, 1978, 1994). Play, even play with rules, provides means for individuals to step outside their usual social role, interact with those who are different, challenge conventions, and thereby explore being themselves. Play's intense, expressive experiences thus have a major role in shaping social identity. Simmel's seminal essay on 'pure' sociability emphasizes that it 'exists for its own sake', standing outside the instrumental pursuit of material gain (Simmel 1950). In common with the liminality of rituals in traditional societies, such playful engagements with difference and irrationality have a dark edge. Something that is play can all too easily turn serious and dangerous, especially in cities, where social differences are great, there are many unknowns, much is at stake, and situations change very quickly. What is play for one person may not be so for others. Geertz (1972) writes of 'deep play': addictive forms where individuals feel a need to play rather than freely wanting to, or forget they are playing and ignore risks and limits. Schechner (1993) writes of 'dark play' that involves deceiving or exploiting those who are not playing.

FIGURE 19.1   'Reclaim the Streets' party with music and jugglers, Melbourne (Source: Quentin Stevens).

Sennett (1978: 264) suggests that play-acting among adult strangers is a particularly sophisticated way of acting and engaging with others, constituting 'the essence of civility'. He identifies play as a fundamental aspect of the cosmopolitan public life that emerged in Europe in the late seventeenth century, when encounters among strangers became common (see Chapter 10, this volume). Following Huizinga (1949), Sennett suggest that this social play is a luxury that relies on the abundance and freedom that the wealth of cities allows – a wealth that is itself facilitated by frequent exchange among strangers. The 'nonfunctional socializing' of play (Sennett 1978: 118) has its own rules. Through such rules people enact and experience utopian conditions of social equality and fairness that they cannot experience in 'real life' (Caillois 1961). These rules require that players step away from their own prior advantages and pursuit of immediate survival needs and personal desires. Participation in play has to be more important than winning. Sennett (1978: 118) argues: 'the more (people) interact outside the strictures of necessity, the more they become actors'. This masking 'disguises the conventional self and liberates the true personality' (Caillois 1961: 21). Through expressive play, those who participate learn that other people are capable of a rich repertoire of social behaviour. The diverse performances that constitute social play provide ways for people to create and communicate social meaning and value.

Sennett argues that children's play-acting is merely a way of learning how adults socialize through play. It is a 'self-distanced activity' that acculturates children to society. Play does so by teaching children that social behaviour follows conventions, but also that these conventions can be tested and changed, as they learn 'to think of expression in the situation as a matter of the remaking and the perfecting of those rules to give greater pleasure and promote greater sociability with others' (Sennett 1978: 315). But even for infants, play is 'something more complex than sure gratification': there is a risk because the activity may or may not give pleasure (Sennett 1978: 318). Sennett notes that in 'reality', frustrations and constraints lead to apathy and withdrawal, whereas in play they lead to increased involvement. Play is a rare, nuanced realm of social experience where one can simultaneously experience frustration, risk, attentiveness and pleasure.

From anthropological and sociological perspectives, whether a particular individual's action is or is not play is largely determined by the wider social context and by the immediate responses of other people. The complexity of urban life means that these contexts for play are always changing and unpredictable. The crowd is one motif often used to characterize the playful possibilities of city life, in terms of its closeness, anonymity and consequent stimulation. Beyond the practical exchanges that occur within any given play event, players and 'audience members' – who are also playing roles – constantly signal to each other about the nature of their play: about quality of performance, fairness, inversions of conventions, changing rules, and missed cues. This higher level of social communication also has its own rules and playful transgressions (Bateson 1956).

The dialectics of urban play

Benjamin (1999, 2006) was an early critic of the predictability and alienation of modern industrial urban living and urban space, and an early theorist of the potential re-enchantment of urban experience through encounter with strangers, with unfamiliar spaces and sensations, and through imaginative, exploratory play (Gilloch 1996). Benjamin drew attention to the labyrinthine nature of cities, their rich range of characters, and their unexpected juxtapositions of old, new and forgotten images, within which individuals could discover new meanings and relationships (Savage 1995). In the 1960s, Lefebvre and the Situationist International similarly critiqued the increasingly instrumental post-war urban planning approaches that restricted the gathering of people, the mixing of difference, and possibilities for sensory experience and for action (Lefebvre 1991, 1996; McDonough 2002; Andreotti and Costa 1996). These approaches included demolishing historic quarters and sites; reducing urban density and eliminating central public places of intense social difference; zoning activities within cities and within individual buildings; developing large-scale transportation infrastructure at the expense of a connected pedestrian network; eliminating nature; stifling all strong sensory stimuli in the interests of health and safety; and regulating behaviour, both directly through war and policing and indirectly by promoting the nuclear family unit, commuting, consumerism, and centrally administered cultural policy. The Situationists saw passivity and boredom as the chief threats within technocratic urban society.

Contesting these conditions, Lefebvre (1996) proclaimed people's right to assemble freely in city centres, to be different, and to play. He argued that urbanism was not a consequence of industrialization, but was a separate process pursuing continuous social development through the discovery of new social needs and the fulfilment of human potential. Lefebvre regarded play, work and daily life not as separate, but as having a productive coexistence, where tensions, contradictions and conflicts between them generated new possible social and spatial forms in a constant dialectical development. For Lefebvre, play was thus among the necessary functions of urban space.

The Situationists proposed a series of activities and spatial forms that would both critique existing conventions of urbanism and playfully explore new possibilities. These included the *détournement* (repurposing) of existing urban spaces and objects to strip away their existing narrative power and to discover new potential within them. *Dérive* (drifting), like Benjamin's (2002) *flanerie*, involved playfully wandering through the urban milieu to rediscover its forgotten fragments and encounter new experiences. *New Babylon* was a utopian architectural proposal, a high-tech, open megastructure elevated above the existing city where people could freely wander, socialize, and reconfigure the environment to explore new sensory and social ambiences. These initiatives all sought to bring art, urban space and individual experience closer together, and to encourage people to constantly play to explore and transform space and life (McDonough 2002).

These thinkers presented play as a dialectical, lived critique of the technocratic apparatus of post-war urban planning, which provided a means to study, criticize and transform it. Like earlier sociologists, they also saw the characteristic spontaneity, non-instrumentalism and exuberance of play as optimizing people's engagement with the potentials offered by the distinctive density and diversity of urbanity so as to explore human potential and discover new social needs.

The character of play

Analyses of the various forms that play takes help to tease out what play can offer as a way of thinking about people's actions in public spaces, in social life, and in cities generally. Huizinga (1949), in using the Latin term *ludus*, emphasizes the structured nature of many kinds of play. Caillois (1961) contrasts *ludus* (play institutionalized as a game) against *paidia* (play that is unreflective, wilful, spontaneous and exuberant), and suggests that social practices of play follow a general progression from *paidia* toward *ludus*. He argues that rules of play are not ultimately constraints on action. Rather, new experiences gained through *paidia* stimulate the desire to 'invent and abide by rules' which 'discipline and enrich' it (Caillois 1961: 28-29). *Ludus* helps people utilize and develop patience, skills, knowledge and enjoyment, by voluntarily subordinating their behaviour to arbitrary rules and routines. Caillois' spectrum emphasizes that play's escape from instrumentality and compunction can lead in two directions: either resistance to rules, or observance of different, and in many cases more-constricting rules, in the self-discipline of games, hobbies, and the arts.

Caillois also defines four basic forms of playful activity: competition, chance, simulation and vertigo – his terms are *âgon*, *alea*, *mimicry*, and *ilinx* (see Case study 19.1). This typology identifies different ways that practices of play escape the behavioural constraints of work and social reproduction. Each of the forms of activity emphasizes different reasons that people are drawn into play, and particular opportunities that urban conditions offer to play in these ways. Caillois's four categories of play illustrate different ways that life is lived more intensely. Competition and simulation enable increased personal control over the body and over communicated meaning. Chance and vertigo involve escape from behavioural and experiential controls. In each case, play is an emancipation from the routines, constraints and preconceptions of everyday social existence. All four forms of play express social ideals that are difficult to achieve in everyday life. Each also suggests forms of heightened bodily and mental engagement with the urban milieu. Public places can allow people to fully 'be themselves' and transcend the roles that have been defined for them by work and domestic life. The four forms of play act out and amplify the potentials for freedom, equality, individual control and transformation that are latent in the diversity, intensity and relatively unstructured sociability of cities. Because urban settings gather and multiply the diversity of social life, public play in cities often combines several of these forms.

CASE STUDY 19.1 FOUR FORMS OF PLAY

Caillois (1961) identifies four fundamental forms of play: competition, chance, simulation and vertigo. Each illustrates a different way people transcend their everyday experience of the city and experience the urban milieu more intensely.

Competition in cities is stimulated by their aggregation of people with diverse abilities and ambitions, and the presence of audiences who judge and reward them. Urban settings also provide a range of physical challenges to explore. Competitive play encourages people to test and develop their knowledge and skills. In playful forms of competition, the goal is not defeating others, but mastery of the self. In diametrical contrast, playing with chance involves a complete escape from any kind of effort, experience, or skill. People willingly abandon themselves to unpredictable, uncontrollable forces. In doing so, they escape certainty and the burden of responsibility. The density, diversity, and dynamism of people, activities and spaces in the urban public realm create conditions of chance that engender playful activity: people wander freely through labyrinthine spaces and have unpredictable encounters with strangers and with new sensations and events (Savage 1995; Andreotti and Costa 1996; Stevens 2007).

Simulation involves pretending different characters or situations. This allows escape from one's conventional place in the world and the development and testing out of new identities, meanings and realities. Simulation often involves audiences, and the city is a key stage for seeing and being seen. Unlike competition and chance, simulation is not bounded by precise rules. It gives each individual opportunities to invent new rules and roles, which transcend predetermined, often instrumental social relations and conventions. As Sennett (1971) noted, these playful interactions can carry over into other settings and more structured relationships. This 'make believe' – invention, substitution and transformation – is at the heart of what cities are.

Vertigo emphasizes the active, bodily, sensory forms of playful engagement with our surroundings. People 'abandon themselves', escaping from normal bodily experience and self-control and being transported to new, more intense forms of experience. Playfully falling, sliding, jumping, climbing, dancing, spinning or moving quickly can generate exciting, intoxicating physical sensations of instability and distorted perceptions. There is also psychological or 'moral' vertigo, through disorderly acts such as breaking objects, making loud noises, fighting, acting lewdly or being close to strangers. Through such crude, uninhibited behaviour, people disrupt social propriety, seeking temporary respite from the pressures and responsibilities of the social order (Caillois 1961). Both kinds of vertigo overcome alienation, by bringing direct bodily engagement with the material world and a sense of agency within the social world. They allow for the arousal, exposure and satiation of normally forbidden desires.

FIGURE 19.2   Simulative play: posing as a display mannequin in a shopping arcade, Melbourne (Source: Quentin Stevens).

Analysing play

Studying practices of play requires close, detailed observation and description of what people do in particular places and times. But play is fluid, multifaceted and dialectical. Despite scholars' identification of certain activities and general forms that constitute play, it can be difficult to distinguish from the wider study of everyday life or of productive work. Forms of competition, chance, simulation and vertigo are often instrumentalized to serve other people's production and consumption needs. Play is most useful in an adverbial sense, as a theorization and analysis of how and why any action is undertaken playfully. The degree of play in any situation or activity can vary, often dynamically; things can suddenly turn serious. Play also recognizes that individual human actions actively constitute and reshape their contexts. Thus it is important to consider the nature and role of play events within wider social processes that unfold over time. As a rhetorical device, the concept 'play' places attention on the mood of people's experiences and actions (Sutton-Smith 1997), and on the social, sensory and spatial contexts surrounding them which links play to the study of atmosphere, affect, and human motivation (see Chapter 2, this volume).

The importance of mood highlights that play can also involve psychological states that do not readily lend themselves to external empirical observation. People's playful experiences in and of cities can include imaginative fantasy and competition, emotional engagement with risks, drug-induced vertigo, and sharing such experiences with others through speech, image and text. These experiences are not intrinsically spatial, nor necessarily linked to cities, but need to be considered in light of how they shape and are shaped by urban social and built contexts.

As a category of behaviour, play calls attention to the multiple aspects of action, perception, meaning and setting that are not expected and which may go unnoticed. As Caillois's (1961) four types of play indicate, such activities often tend toward extremes of intensity and effort or, at the opposite limit, release. An examination of play thus provides some calibration of the overall scope of city life and its risks. Play is a form of testing boundaries, both in social terms and in the physical world of the street (Stevens 2007). Thus observing play can be a very useful way of learning about what the physical and social constraints of the public realm do and do not achieve. Signs, regulations, controls, and barriers in public spaces tell us a lot about what people actually want to do in those settings. Such constraints also dialectically inspire people to action. While play often follows rules, it also often involves ongoing transgression and reformulation of them to achieve desired effects.

Recent research in architecture, planning, geography, sociology and performance studies has reported empirically on the rich diversity of informal, unplanned playful activities that occur in urban public spaces, focusing on the diversity of actors and their actions, the power relations that these uses illuminate and challenge, and how people's playful actions transform the meanings, histories, functions and forms of the urban settings that they occupy (Stevens 2007; Franck and Stevens 2007). These playful activities include guerrilla gardening (Reynolds 2008), street art, yarn-bombing (Moore and Prain 2009), dancing, street theatre, vandalism, and a range of forms of mobility through the city such as skateboarding (Borden 2001), cycling (Spinney 2010), and parkour (Lamb 2014). Geography has moved beyond a focus on children's play to examine how cities are lived and shaped through adult play, which has distinctive modes and motivations (Woodyer 2012).

CASE STUDY 19.2 PLAYFULLY CRITIQUING URBAN SPECTACLE

As urban managers compete to make their cities more liveable, vibrant and appealing, critical research focuses on how public play continues to provide a lived critique of the 'festivalized' city based on programmed consumption and sponsored creativity. Studies of people's playful responses to urban leisure, tourism, themed environments and the experience economy illustrate their resistance to manipulation, and their carnivalesque efforts to subvert and rework received spaces, roles and codes of behaviour. Within the planned liminality of vacation packages, the night-time economy (Hughes 1999), and

FIGURE 19.3 The dialectics of urban play: promenading tourists encounter skaters and graffiti underneath a concert hall on London's South Bank (Source: Quentin Stevens).

temporary seasonal activity spaces such as skating rinks (Bell 2009) and artificial beaches (Stevens 2010), researchers continue to observe playful deviance, the dialectical contestation of limits, and self-discovery.

London's South Bank illustrates dialectical, productive tensions that often arise between different ideas and forms of play in cities. First developed as a theatre precinct when such leisure activities were barred from the medieval city, and redeveloped with additional cultural institutions for the 1951 Festival of Britain, this now-gentrified waterfront also contains numerous galleries, bars, restaurants, and designer shopping. Both scheduled and informal street performers entertain people promenading through. But other kinds of playful activities also occur here: young adults doing parkour, skating, and holding impromptu dance parties on the sandy strip down on the river's edge. These are not the playful activities that the setting was consciously designed for. They involve more risk, contestation and sensory immersion. They illustrate a friction between the two sources of the word 'play' – 'to exercise, or frolic', and to put something at risk. These acts are often intended as performances to audiences of both peers and strangers. Such cases show that users are not constrained by urban spectacle. From villagers betting at a Balinese cock fight (Geertz 1972) to teenagers in the West Edmonton Mall (Shields 1991), individuals engage playfully, creatively and critically with the world of artifice, reinventing themselves, and learning new ways to bend society's rules.

Conclusion

Current research on play and the city has extended in quite varied directions and had a range of applications, showing that the boundary-testing nature of play itself inspires dialectical responses to conventional ways of thinking about a range of urban processes and problems. The concept of play makes an important and productive contribution to urban theory because it sits in a tension with everything we think of as proper and practical about city life. Play causes us to think critically about what activities, feelings and values the city does not support. It questions the idea that the city is or should be an efficient machine, or a market, or an ecology, or a political system. Those who refer to play are generally counterposing it to behaviour that is 'normal' – everyday, conventional, expected, calculated, efficient, constant. Play behaviour is in some way unusual, special and different. Those special qualities reflect a range of value judgements about urban society and urban space. By critiquing the predictability, instrumentality, efficiency and alienation of modern urban living and urban space, play emphasizes the variety, flux, and accordant unpredictability of urban experience. It forces us to recognize that people's behaviours and ambitions are not always pragmatic. While thinking about play illuminates the importance of social codes, rules and roles, it also recognizes their transgression and the dynamics of their communication and transformation.

Play provides a renewed focus on lived social experience and individual agency, countering abstract structuralist and representational readings of urbanism and simplistic ideas about urban functions, goals and identities. It draws attention to the rich, intense, multi-sensory embodied perceptions and opportunities for action that cities offer. It encourages examination and description of how activities are undertaken, and the social and physical contexts that frame them. Play connects to a wealth of earlier thinking from anthropology, sociology, psychology and philosophy about people's actions, values, meanings and social relations that can be used to better understand the complexities of contemporary city life. Theories and acts of play both critique the anomie and constraints of modern cities, and present an alternative view that affirms cities' freedom and their vitality as crucibles of exploration, creativity and human development. Play highlights the luxury of urban public spaces that are relatively open to users and uses, and the possibilities of social engagement with strangers for its own sake. In these respects, play has a broadly utopian and progressive mien. Its spontaneity and creativity brings a distinctive focus on new, recovered and expanded social needs and capacities.

Research and practice oriented toward play can be critiqued in terms of their theoretical presumptions and the methodological questions they raise. Caillois's (1961) discussions of cheats, spoilsports and artificially induced liminality through machines and through drugs point to underlying assumptions about the authenticity and normativity of play. A focus on observable, intense bodily play performances can tend to prioritize the activities of young men from dominant cultural groups, disregarding the different play forms and needs of others. What is play for some may cause annoyance, threat or injury to others. The concept of play opens up an

CASE STUDY 19.3 CITY GAMES

Critical thinking about play and the city is influential in contemporary computer game design (Walz and Deterding 2014; von Borries *et al* 2007). Increasing computing power has supported the design of expansive virtual urban play environments in games such as *Grand Theft Auto* and *Mirror's Edge*, and *SimCity* where players create and manage an urban landscape. Increasing internet speeds have allowed the development of massively multiplayer online role-playing games (MMORPGs) and virtual worlds, such as *The Secret World* and *Second Life*, where people can interact remotely with other real players. Such platforms have complex settings and gameplay that are conducive to exploration, player interaction, surprise, and user customization. The urban backdrops of many such games emphasize complex role relations, social transgression, and interactivity among players, audiences and their environment. Academic research is accordingly beginning to examine the playful potential of built environments, both real and propositional, analysing them with reference to specific theories or manifestos of play and city life. Some practice-based game research considers what role the 'real' public and real urban settings can have in computer-based games and play. These potentials are rapidly expanding due to mobile information and communication technologies, engendering for example geocaching games and flash mobs. Mobile phone applications are extending pervasive and location-based games into the real world, for example *Pac Manhattan* and *Ingress*. Research continues to explore new ways that people might 'play the city'.

affirmative, constructive view of the uncertainties of urban life, of the chances, risks, transgressions and constant change that attract people to cities. Accounts of play in public spaces often reify individual action, opposition and danger in the name of freedom, and imply that such uses are thus highly democratic. But such play may happen at the expense of other collaborative social uses of public spaces (Parkinson 2012). If play permeates all social activity and has a fundamental role in defining culture, more research is needed into the role of play as a category or mood of action within social life, and how social negotiation occurs over the purposes, limits, transgressions and recuperations of play in the urban realm.

Key reading

Andreotti, L. and Costa, X. (Eds) (1996) *Theory of the Dérive and Other Situationist Writings on the City*, Barcelona: Museu d'Art Contemporani de Barcelona ACTAR.
Comprehensive anthology of the Situationist International's original critiques and propositions about urbanism.

Caillois, R. (1961) *Man, Play and Games*, New York: Free Press of Glencoe.

Sociological study of the different forms that play takes, their social aims and effects, and play's role in social development.

Huizinga, J. (1949) *Homo Ludens: A Study of the Play Element in Culture*, London: Routledge and Kegan Paul.

Anthropological inquiry into play both as a specific form of cultural activity and as a way of understanding culture.

Lefebvre, H. (1991) *Critique of Everyday Life*, Volume 1, 2nd edition, London: Verso.

Presents play and leisure as a lived, exploratory critique through which people can identify unmet needs and challenge the anomie and alienation of modern urban life.

Sutton-Smith, B. (1997) *The Ambiguity of Play*, Cambridge, MA: Harvard University Press.

The most extensive analytical review of the broader cultural discourses that underlie various theories of play.

20

POLITICS

Andrew E. G. Jonas

Introduction

What political processes take place in cities? At first glance, this seems like a straightforward question. We know that most cities are run by locally elected representatives (mayors, city councillors, etc.). Sometimes we are encouraged to vote in a local referendum on which services are delivered, how taxes are raised, or where local jurisdictional boundaries are drawn around our community, city or metropolitan area (see Case study 20.1 and Figure 20.1). We also know that resident urban voters can elect national politicians, who in turn pass laws that influence how their cities develop economically and socially. But even if most of us have little to do with the everyday business of running a city, and perhaps even less to do with local or national elections, we are frequently exposed to media coverage of urban politics, which tends to focus on dramatic events like the riots which swept parts of the UK capital, London, in August 2011, or the rioting in the City of Ferguson, Missouri, USA, which occurred in August 2014 (see Chapter 7, this volume). These events might be limited to a particular city, or an area within it; yet they suggest that urban protests can also influence attitudes and policies in the wider society. Moreover, urban-based struggles for democracy, such as the Arab Spring uprisings in Tunis (Tunisia) and Cairo (Egypt) in 2011, or the ensuing Occupy movements in New York and London, suggest that there is something essentially global about urban politics as much as it is often about local issues.

The purpose of this chapter is to show how critical urban theorists in North America and Europe have thought about the local and global dimensions of urban politics. It will demonstrate that the analysis of urban politics has been central to the development of critical urban theory. Nevertheless, it has proven surprisingly difficult to pin down the nature of urban politics, both in terms of its substance and also its territorial scope. Debates about the substance of urban politics have ranged

CASE STUDY 20.1 REDRAWING URBAN POLITICAL BOUNDARIES

Although critical urban theorists often take for granted the boundaries of urban political jurisdictions, such boundaries are frequently contested and can undergo a reorganization. A case in point: in September 2014 voters living in suburban villages and parishes in parts of the East Riding of Yorkshire, England, and which are contiguous to the City of Hull, were asked whether they wished to remain in the East Riding or, instead, become part of an expanded City of Hull (see Figure 20.1). It transpired that a clear majority of voters in the suburbs were opposed to the proposed boundary change. The result of the referendum was that 96.5% voted 'NO' and 3.5% voted 'YES' to the proposed boundary change, with a turnout of 75.27%.

The boundary issue in Hull reflected ongoing tensions between urban development and maintaining local levels of public service provision. Earlier in 2014, Hull City Council had resolved to establish a commission of inquiry to examine the case for extending the boundaries of Hull to include the adjacent villages and parishes, which tend to have higher per head property values and household incomes than is the case in Hull. The Council hoped that the enlarged city would be able to compete more effectively to attract investment and redistribute its tax revenue base over a wider population, resulting in improved services for Hull residents. The East Riding of Yorkshire Council's response was to hold a referendum in the suburbs in order to gauge public opinion on the boundary question. Whilst a 'YES' vote would not have resulted in a boundary change it, none the less, was anticipated that there was significant public opposition to the proposed boundary change. In the event, this indeed was the case.

FIGURE 20.1 Boundary referendum poster in a village in the East Riding of Yorkshire (Source: Andrew Jonas).

from it being either about struggles around the collective consumption of public goods and services in the city or, alternatively, related to urban economic development. In regard to the former, critical attention focusses on the role of the neo-liberal state in cutting back on public services and welfare in the city, and how city authorities and urban communities have responded to austerity measures imposed upon them. In the latter case, critical scholars have pointed to the rise of the New Urban Politics in which urban growth coalitions – growth-oriented partnerships between the local state and locally dependent firms and private investors – promote the economic development of urban localities. In both cases, the analysis of political processes operating inside cities offers a powerful critical lens onto wider spatial restructurings of the state and the capitalist economy (see Chapter 13, this volume).

At the same time, however, critical scholars have started to question whether we should restrict the analysis of urban politics to what takes place inside the jurisdictional boundaries of the city. For example, Rodgers *et al* (2014) invite you to consider not what urban politics are about but *where* you might begin to look for them. In this chapter, I wish to take up this invitation and try to locate – or, better still, *re*locate – urban politics. The chapter makes the argument that the boundaries of urban political analysis are being stretched, conceptually and empirically. Today's critical urban scholars examine larger urban territories, such as city-regions, and political processes that extend well beyond local (i.e. urban) jurisdictional boundaries. This does not mean that they have completely discarded a territorial frame of reference in preference for a relational approach (i.e. an approach which emphasizes social relations extending far beyond a given territorial jurisdiction). Instead, critical urban political analysis involves thinking about how local political and economic interests come together within a city, or examining the workings of global processes throughout several cities as well as larger territorial units such as city-regions (see Chapter 8, this volume). In conclusion, there are a variety of ways in which you might want to take up the invitation to look beyond the jurisdictional limits of the city in order to understand the changing nature of political processes occurring both within and between urban areas.

Locating urban politics: from collective consumption to urban development

Studies of urban politics undertaken in the 1950s and 1960s often produced very detailed and, by-and-large, descriptive case studies and comparisons of conflict around issues such as housing, education, segregation, services, and planning in the city. This was a period of growing prosperity in Western Europe and North America; a time when cities were very much at the forefront of efforts by national governments to improve standards of living by delivering a wide range of urban services, such as public housing and education, and undertaking extensive programmes of slum clearance and urban renewal.

In the USA, urban renewal proved especially controversial because it often resulted in the displacement of thousands of low-income residents (mainly the urban poor and ethnic or racial minorities) from the inner city. In addition, it led to the creation of bureaucratic structures in local government, which were not directly accountable to urban citizens. Consequently, urban scholars at the time became quite preoccupied with the question of 'who governs?' (Dahl 1961), the answer to which in the urban context varied from reformist political elites with ties to key local business sectors or, alternatively, urban political machines, which built their political power base in working-class ethnic communities (Shefter 1985). When such studies made reference to the national political context, it was mainly done in order to demonstrate how urban politics shaped national policies rather than the other way round (Mollenkopf 1975).

Alongside the expansion of the urban welfare state, there was an upsurge of social unrest in the city, which soon gave a critical inflection to the analysis of urban politics (see Chapter 6, this volume). The context was the Civil Rights movement in the USA and similar struggles for democratic rights occurring throughout Europe in the 1960s. Rioting in Los Angeles (1965) and Detroit (1967), and student protests in Paris (1968), exposed deep racial, class and gender divisions in capitalist society at large. Inspired by Marxist analysis, critical urban scholars began to consider how the social tensions and contradictions of capitalism were being played out within the urban political arena. The urban sociologist Manuel Castells argued that political struggles in the city manifested a failure on the part of the state to deliver services collectively consumed by the urban poor and the working class (Castells 1977). Indeed to many critical urban scholars, it seemed that something politically important was happening within and around urban space. Not only were urban politics becoming separated from national debates about economic production but in addition, and crucially, a new set of concepts was needed to explain those sets of social relations and political tensions which were distinctively 'urban' in appearance and substance (Saunders 1986).

In Europe, critical attention focussed upon how struggles between different consumption groupings in the city gave rise to a dual state in which social welfare and economic development were allocated between, respectively, local and national branches of the state (Cockburn 1977; see Chapter 9, this volume). If there were spatial variations in the state provision of social welfare and housing to the cities, these could be explained in terms of local social structures, politics, and gender relations (Dickens *at al* 1985). In the USA, by way of comparison, the focus was much more on the urban fiscal crisis and how the tension between capital accumulation and the provision of social welfare led to reforms in local political arrangements inside the city. Infrastructural and economic development functions tended to be located within special purpose districts, which were separated institutionally and fiscally from general-purpose municipal functions and services (i.e. those functions garnering electoral votes) (Piven and Friedland 1984). Growing demands for welfare and services put upward pressure on urban social expenditures; but cities could not generate sufficient revenues from economic development to meet such demands, which forced many, including New York City in 1975, to declare a fiscal crisis.

Critical urban scholars soon recognized that there were limits to urban government's capacity to forge stable and enduring political alliances around accumulation and social redistribution. For David Harvey, the issues highlighted a growing territorial discrepancy between urban politics and the workings of capitalism at large:

> The political processes at work in civil society are much broader and deeper than local government's particular compass. Indeed, there are many facets that make it ill-suited to the task of coalition-building. Its boundaries do not coincide with the fluid zones of urban labor and commodity markets or infrastructural formation; and their adjustment through annexation, local government reorganization, and metropolitan-wide cooperation is cumbersome, though often of great long-run significance. Local jurisdictions frequently divide rather than unify the urban region, thus emphasizing the segmentations (such as that between city and suburb) rather than the tendency towards structured coherence and class-alliance formation.
>
> *(Harvey 1985: 153)*

Meanwhile, radical geographers in the UK, such as Doreen Massey, made connections between the decline of the inner city and national decisions influencing the location of different branches of production (Massey 1984). Massey observed a shift away from redistributive urban policies towards inter-regional competition. Those urban areas experiencing economic decline and mounting social problems were forced to compete for regional development grants and public social assistance.

In summary, by the end of the 1980s urban scholars on both sides of the Atlantic sensed that the nature of capitalism was changing and so too was the role of cities within wider geographies of capital accumulation and state social provision. Less a site of struggle around collective consumption and social welfare, the city was being strategically and politically repositioned as a location for a new regime of capital accumulation based around new spatial divisions of production and consumption. This regime was characterized as entrepreneurial or market-driven (i.e., neoliberal) rather than redistributive or dependent on generous state social expenditures (see Chapters 11 and 18, this volume).

The New Urban Politics

In an influential paper published right at the end of the 1980s, David Harvey argued that the dominant *modus operandi* of urban government in Europe and North America was undergoing a decisive shift from social-welfare managerialism to entrepreneurial urban development (Harvey 1989a). It seemed to Harvey that urban political elites everywhere were engaged in a competition to attract new investment and consumption opportunities to their cities; an insight which was soon examined and endorsed by other critical urban scholars (Hall and Hubbard 1998). Typical investments sought might include the headquarters operations of major corporations, research and development facilities, international sporting and cultural events, creative industries, or a combination of all of these. City governments, in turn, were responding to such intensified inter-urban competition by offering all sorts of tax and regulatory incentives. The urban political arena was becoming populated by a host of new institutions better to serve the interests of private capital, such as public-private partnerships and quasi-public redevelopment authorities, as well as designated spaces designed to attract inward investment, such as docklands, marinas, museum quarters, and cultural districts.

Marxist scholars have come to see these developments as indicative of the emergence of a New Urban Politics (NUP) (see Cox 1991; and Case study 20.2). The rise of the NUP is set against the threat of capital mobility, placing private business (which is presumed to be more mobile than workers and cities) in a position of strength with respect to urban government, thereby allowing the private sector to exact various concessions in the form of tax incentives and favourable planning. Consequently, it has become more difficult for working class and populist social movements to secure a foothold in urban government.

Another group of urban scholars known as urban regime theorists likewise argued that urban governance has become more entrepreneurial. However, they suggest that urban entrepreneurialism reflects how economic restructuring has shaped the trajectory of struggles inside urban government for the control of increasingly scarce fiscal resources (Stone and Sanders 1987). Cities differ in terms of the balance of political forces within them, and so internal struggles tend to

---

CASE STUDY 20.2 THE NEW URBAN POLITICS (NUP)

NUP describes a body of literature about urban politics, which claims that economic development has become a priority for urban governments but at the expense of its traditional welfare and public service functions. This priority is a reaction to the increased levels of capital mobility across local jurisdictional boundaries. Urban governments, which are not so mobile and are therefore dependent on local resources, must compete in order to attract inward investment and generate local revenues. Moreover, urban competition diverts scarce public resources away from the needs of ordinary urban residents (for example, tax incentives used to attract inward investment reduce the overall levels of funding for other services such as schools). However, critics argue that the NUP underestimates the degree to which capital itself is dependent on urban localities (Cox 1993). Consequently, the literature has a tendency to portray urban authorities and residents as hapless victims of economic restructuring rather than as active political participants in processes of urban development.

generate different types of urban regime, which can vary from those that are aggressively entrepreneurial to others that are more redistributive in intent (e.g., managerialist regimes). For example, some urban regime theorists have argued that conditions in European cities are not yet ripe for the emergence of growth regimes along the lines of those typically found in US cities (for example, Harding 1994). Nevertheless, these same scholars point to the growing importance of structural forces, such as globalization and state intervention, in shaping the trajectory of urban politics across different national contexts.

Kevin Cox has taken the discussion further, suggesting that scholars of the NUP often fail to distinguish between necessary social structures and contingent conditions (Cox 1991, 1993). He argues that a necessary condition for the formation of local political interests in urban economic development is the problem of local dependence: a condition, moreover, that arises from the tension between capital mobility and spatial fixity. This problem creates different material stakes in how a city or metropolitan area – and particular spaces within these, such as the downtown or the suburbs – develop (Cox and Mair 1988). For locally dependent firms, urban economic development might be important for growing the local market for products and services, or for realizing gains on capital invested in the built environment. Workers, on the other hand, might have a different stake in how cities develop; perhaps for them affordable housing, training and education might be important considerations. For the local state and locally elected officials, the key concerns tend to be expanding the local tax base, servicing local residents, and securing votes. To the extent that such different interests tend to converge around the same urban locality, locally dependent actors might set aside their ideological, social or class differences and enter into political coalitions. Some coalitions form around the further development of the urban locality; others might be opposed to economic growth – it all depends on how local coalition participants stand to gain or lose materially.

Development-driven coalitions resemble the 'growth coalitions' that Logan and Molotch (1987) have examined and described (see Case study 20.3 and Figure 20.2). Such growth coalitions aggressively pursue all sorts of opportunities to attract inward investment and maximize profits from rental income associated with the redevelopment, improvement, and intensification of land uses in the city. In some circumstances, they face organized opposition and may be forced to compromise, ensuring that the distributional benefits of urban economic growth are spread more evenly across the locality (for instance, by securing adequate funding for local schools or negotiating additional services from developers). Crucially, the state is not a neutral player in these coalition-building processes; in fact, it is quite the opposite. Urban politics are partly about organizing structures of the (local) state that are more accessible to local interests in urban development, on the one hand, and yet also manage local interests in collective consumption, on the other.

Cox and Jonas (1993) add a further dimension to this scenario when suggesting that efforts to manage the tension between urban development and collective consumption can result in the organization of new territorial structures of the state (i.e. new spatial units of government like a regional transit authority) within a city

CASE STUDY 20.3 THE GROWTH COALITION

The growth coalition is a local political alliance of businesses and local government officials, which has a collective stake in the development and expansion of the urban economy. According to Logan and Molotch (1987), a growth coalition is comprised of developers, banks, corporate investors, and other interests in the development of urban land and property. Members of the growth coalition are motivated primarily by gains associated with the exchange of land. Hence securing access to the instruments of local government that determine land use, zoning and services is critical to these interests. However, growth coalitions frequently come into conflict with political interests seeking to defend use values in the urban living place. For example, local residents might be opposed to a change in land use associated with neighbourhood gentrification because it alters the character of the neighbourhood. Moreover, existing residents on low incomes might not be able to afford the new housing on offer (see Figure 20.2).

The concept of the growth coalition works quite well as a way of explaining the politics of urban development in the USA, where land-use planning, elections and taxes are based around local structures of government. It is a less convincing argument in contexts where the structures of local government finance and land use planning are more centralized as is the case, for example, in some European countries such as the United Kingdom (Cox and Mair 1989; Jonas and Wilson 1999).

FIGURE 20.2   Notice of a proposed local land use action in the South Lake Union district of Seattle, WA, USA. This neighbourhood has been undergoing gentrification, resulting in opposition from local residents concerned that the character of the area is being changed through decisions about land use and zoning, which are occurring beyond their control (Source: Andrew Jonas).

or metropolitan area. For example, locally dependent business interests might want to consolidate local government into a larger metropolitan or regional structure in order to realize economies of scale in the delivery and funding of transportation infrastructure or to generate a new source of revenue to support regional economic development (Jonas *et al* 2014). Conversely, developers and local firms often prefer to work with local jurisdictions in order to exploit niche markets for housing and local schools or to provide goods and services exclusively to homeowners living in master-planned residential communities (McGuirk and Dowling 2011). If there has been a general drift towards entrepreneurial urban development, local distributional outcomes often vary quite markedly within metropolitan areas, and the ensuing struggles over territorial distribution influence and mould urban politics in ways that can still produce unique local outcomes.

Regulating urban politics

Although critical urban scholars remain interested in political processes occurring within cities, these days they strive to situate such processes in a wider state territorial context. This is exemplified by the rapid growth of critical urban scholarship on the regulatory transition of the state from spatial Keynesianism to its present competitive, neo-liberal form (Lauria 1997). Traditional conservatives have subscribed to the view that competition forces urban government to be more fiscally efficient and publicly accountable. Yet not even the most die-hard neoliberals could have anticipated the extent to which urban government has become not just consumer-oriented but also thoroughly privatized. Indeed it seems that any semblance of an urban politics fought around welfare and social redistribution has disappeared as local government has been 'hollowed out' from above by the state and economic globalization (Cochrane 1993).

This is not to say that the state's efforts to neoliberalize urban politics have not been contested from below. Rather it is more the case that the degree to which this has happened varies from place to place, reflecting a widening and deepening of processes of uneven urban and regional development (MacLeod and Goodwin 1999). Thus, for instance, attempts to make local government in the UK more responsive to the private market have proved problematic due to the absence of organized urban growth coalitions operating in British cities. Instead, the UK state has often stepped in, establishing new urban institutions with the necessary powers and resources for working with the private sector, raising local revenues from non-traditional sources (such as bonds and tax increment financing), and attracting inward investment into the city. For example, a number of British cities have set up Business Improvement Districts (BIDs), which are modelled on BIDs in the USA (Ward 2012). A BID enables retailers and landowners based in a downtown retail district to provide additional services, such as street-landscaping and security patrols, and also to market the district.

Given the proliferation of new institutional and governance spaces in the city with ties to global capital, critical scholars are increasingly critical of work that continues to find inspiration from studies of place-based politics. For instance, when accounting for the 'retooling of the urban growth machine' in the UK, Jessop *et al* (1999: 159) make the forceful argument that urban politics are becoming systematically structured by wider social relations and political forces at large in the global economy. Given this limited scope for local political agency, Jessop *et al* believe that it is necessary to situate the analysis of urban development politics in relation to broader regulatory processes, a challenge that many critical scholars are now taking up. A case-in-point involves those who look at how cities respond to new global environmental agendas imposed from above, such as climate change and attempts to reduce emissions of carbon dioxide (CO_2). What has been called the New Environmental Politics of Urban Development (NUPED) seems to be leading to new political alliances and compromises between interests in, respectively, urban economic development and collective consumption (Jonas *et al* 2011).

At the same time, the territorial scope of urban political analysis is undergoing a process of rescaling as city-regions are geopolitically repositioned as territorial drivers of accumulation and redistribution in the global economy (Brenner 2004). The politics of rescaling refers to how powers and resources pertaining to urban and regional development have been redistributed among the different territorial structures and hierarchical levels of the state. For example, the rescaling of city-regional governance has been accompanied by an upsurge of state spending on urban infrastructural projects, and corresponding efforts to channel resources for collective provision to those city-regions that are better able to compete by having suitable regional partnerships and growth coalitions already in place (Jonas 2013).

Notwithstanding the availability of state expenditures on infrastructure, urban development increasingly happens in a context of austerity (especially since the global financial crisis of 2007–08). For some critical urban scholars, austerity urbanism represents a political objective that is deliberately orchestrated by the neoliberal state in order to downsize the public sector and cutback on welfare and other services as a prelude for restoring suitable conditions for private investment in the city. As Peck (2014b: 20) argues, '[a]usterity politics...involve pushing the costs, risks and burdens of economic failure onto subordinate classes, social groups and branches of government...it is about making 'others' pay'. If urban politics in the 1960s reflected struggles around redistribution to ordinary people, today's austerity urbanism manifests how far the state is prepared to go in orchestrating spatial redistribution *back* from ordinary people *to* private capital.

Conclusion

As the above discussion suggests, critical urban theory has been informed in various different ways by thinking about how social relations and structures stretch beyond the jurisdictional limits of the city. This so-called 'relational' approach to urban politics can be contrasted with the more traditional 'territorial' readings, which hitherto had dominated urban political analysis (McCann and

Ward 2011). Critics of the territorial approach highlight the difficulty of separating conceptually those aspects of social relations that are seemingly local – as in 'local dependence' – from those that are regarded to be more spatially extensive in their origins and effects. As Allan Cochrane (1999: 120) explains, '[n]o doubt some social relations are indeed locally dependent – in the medium term at least – but for many others, businesses as well as social groups, definitions of place (or locality) are constructed far beyond the places in which they are located'. The challenge, then, is to decide whether the drivers of urban politics are primarily local (i.e. occurring within the city), global (i.e. social relations occurring at large including processes occurring across many cities), or perhaps a combination of the local *and* the global working together.

As the emphasis shifts from examining social relations and political processes inside cities to those extending between them, scholars acknowledge that urban political actors are increasingly outward-looking or 'extrospective' (McCann 2013). For example, urban authorities bidding for major events like the Olympic Games often refer to other places which have bid successfully in the past, drawing lessons for how to put into effect similar changes within their own urban jurisdiction in terms of political institutions and processes (Cochrane *et al* 1996). Such urban authorities are more favourably disposed (at least, in this narrow political sense) to approve projects, policies and plans, which mimic successful projects, policies and plans pursued elsewhere. This, in turn, has led to a replication of urban political processes and institutions across a range of cities, such as in the case of BIDs mentioned earlier (Ward 2012). The rapid nature of inter-urban policy transfer is generating a certain degree of uniformity in the political look and feel of cities, making the distinction between 'here' and 'there' (and perhaps even 'local' and 'global') intellectually redundant (see Rodgers *et al* 2014).

In summary, and for a variety of different reasons, the analysis of urban politics has progressively lost elements of its 'localness'. This has important implications for how you might want to examine the 'urban' as a particular spatial manifestation of politics. It suggests that the territorial scope of urban politics should not be restricted to the city itself. You should not assume that the analysis of urban politics ends at the jurisdictional limits of the city. Instead, politics assume an urban form by virtue of the fact that various social relations and political processes – some of which originate and extend far beyond the city – tend to converge around and within urban political boundaries. It is the task of the critical urban scholar to map this process of territorial convergence against the particular spaces (and scales) that make up the urban political arena. Sometimes this does involve mapping the reorganization of the territorial arrangements of the city itself (as in Case study 20.1); in other cases it involves thinking about the flows of information, power structures and social relations through which urban jurisdictions are connected to wider territories (and scales) of the state and economy at large. In each instance, the trick is to use relational thinking to animate conceptual understandings of urban politics and their contestation around and within particular urban territories.

Key reading

Cochrane, A. (1999) 'Redefining urban politics for the twenty-first century', in A. E. G. Jonas and D. Wilson (Eds.) *The Urban Growth Machine: Critical Perspectives Two Decades Later*, Albany: State University Press of New York.

This chapter is an excellent overview of the state of critical urban political theory at the dawn of a new millennium. It anticipates many of the issues covered in the present chapter.

Cox, K. R. (1993) 'The local and the global in the New Urban Politics: a critical view', *Environment and Planning D: Society and Space*, 11: 433–48.

An important critical assessment of the literature on the New Urban Politics.

Logan, J. R. and Molotch, H. (1987) *Urban Fortunes: The Political Economy of Place*, Berkeley and Los Angeles: University of California Press.

The definitive monograph about urban growth coalitions.

Rodgers, S., Barnett, C. and Cochrane, A. (2014) 'Where is urban politics?' *International Journal of Urban and Regional Research*, 38: 1551–1560.

A useful introduction to a recent journal special issue on urban politics, which discusses the emergence and implications of the relational approach.

21

RHYTHM

Tim Edensor

Introduction

This chapter investigates the multiple rhythms of the city, the repetitions, routines and habitual movements that focus on places in myriad combinations. I will endeavour to identify key agents in the production of urban rhythms. I provide a contrast between the looser organization of rhythms in pre-modern places with the more disciplinary rhythms that were forged in the nineteenth-century industrial metropolis to accord with the needs of bureaucracy and capital. However, I emphasize that though they are organized to synchronize the regular working, leisure and mobile patterns of city dwellers, urban rhythms are dynamic, often contested and continually liable to change. New habits of consumption, work and communication, as well as resistance to fast rhythms manifest in the cultivation of slow rhythms, testify to the rhythmic changes solicited by the increasingly globalized world. The chapter concludes by emphasizing that these diverse urban rhythms vary according to geographical contexts and that they are always partly co-constituted by the non-human agents with which we share the city.

Place and rhythm

In *Rhythmanalysis*, Henri Lefebvre (2004: 15) emphasizes that '(E)verywhere where there is interaction between a place, a time, and an expenditure of energy, there is *rhythm*'. Lefebvre identifies repetitive movements and actions, phases of growth and decline, and the distinction between linear and cyclical rhythms. In identifying the rhythmic specificities of place, he contends that 'every rhythm implies the relation of a time with space, a localized time, or if one wishes, a temporalized place' (1996: 230). In thinking about the distinctiveness of cities, we can, according to Lefebvre, investigate how rhythmic processes and practices interweave to compose a particular 'polyrhythmic ensemble' (Crang 2001).

In theorizing time as a social force, Barbara Adam (1995: 66) draws attention to the ways in which how the 'when, how often, how long, in what order and at what speed' are governed by a host of 'norms, habits and conventions' about temporality that regulate social life and space. In this chapter, I investigate how urban rhythms are imposed by norms laid down by commercial, bureaucratic and regulatory imperatives that shape the regular beats of the city. Yet I will also discuss how such authoritative and commonplace rhythms constitute a normative temporal framing from which people can deviate in creating their own rhythms in place.

I will also look at how such rhythms change, can be disrupted, become volatile or descend into arrhythmias. I argue that urban rhythms are becoming more varied and contested as work schedules, leisure pursuits and cultural events become more flexible, transcending the rigid rhythms of yesteryear, with the persistence of fixed opening hours, clocking-on times and stricter divisions between day and night. In addition, I explore how complex urban systems break down to produce arrhythmias and acknowledge the non-human rhythms that supplement those made by humans to co-constitute the city and urban experience. Though my discussion primarily focuses on cities in the UK, I include a brief discussion of the very different rhythms in an Indian city.

In considering the tension between the desire for rhythmic order and its disruption, it is important to acknowledge the ongoing tension between being and becoming – in other words, while regular rhythms can certainly be identified, this does not mean that they are immune to change or challenge. Lefebvre (2004: 6) is explicit that there is no 'rhythm without repetition in time and space, without reprises, without returns, in short, without measure', but he is equally insistent that that 'there is no identical absolute repetition indefinitely … there is always something new and unforeseen that introduces itself into the repetitive'. Accordingly, any discussion of rhythms must take account of their regularity and mutability.

Lefebvre exemplifies urban rhythms by gazing out from his third-floor Parisian window, observing the changing rhythms over the course of the day. From this vantage point, all seems somewhat chaotic, but after some hours he is able to identify the changing rhythms of traffic and pedestrians, the regular beats of the traffic lights, and the distinctive phases of quiet and business over time. This method of overlooking a site and observing the comings and goings of workers, pedestrians cars, delivery vehicles and schoolchildren, rush hours and eating times can be undertaken anywhere and offers a means to track the distinctive rhythms of any place (Edensor 2010). To demonstrate this, I spent some hours at the busy Piccadilly railway station in Manchester.

CASE STUDY 21.1 THE RHYTHMS OF MANCHESTER PICCADILLY TRAIN STATION

The most obvious rhythm at Manchester Piccadilly train station (UK) is that imposed by the train timetable, which regulates the arrival and departures throughout the day. Travellers need to synchronize their own personal timetables to ensure that they time their own arrivals and departures from the station. They carry out a variety of regular rituals in the station before boarding their train: buying tickets and perhaps reading material or food. Yet *arrhythmia* results when trains fail to arrive, creating visible frustration for those who depend on trains leaving on time so they can make appointments that synchronize with the train's arrival at their destination. The process of scheduling requires a variety of personnel who perform rhythmic practices to ensure smooth operation: tickets are checked, trains stocked with food and drink, taxis are ready to transport passengers to their final destinations, floors are swept to remove potential encumbrances and hazards, and train toilets are emptied. The daily rhythms vary throughout the day, with hordes of commuters walking swiftly through the station in the morning as they make their way to work, and then returning home during late afternoon. Some rush and others linger. Catalyzing the energies of the station, they contribute to the sensory rhythms that saturate the site, adding to the rhythms of the soundscape with their hurried bustle and augmenting the regular punctuations of announcements. The trains themselves enact a particular mobile rhythm that slows when they enter the station but eventuates in rapid, rhythmic progress when they commence their journeys. Yet increasingly, the station is not only a centre for mere travel but for a growing range of other practices. Mobile phones and laptops have made contact with others possible during periods of waiting. One platform has been converted into a gallery at which people causally browse the photographs on display. A range of bars, cafés and fast food outlets offer opportunities to eat and drink at varying speeds, and train-spotters still congregate on platforms, methodically noting the makes of trains during their long vigils. Most obviously, people shop at the large range of retail outlets that occupy the station premises, seeking clothes, food for the evening meal, greetings cards or medicine, performing everyday rhythms of consumption at a site at which they would previously have been unable.

Pre-industrial, pre-urban rhythms

It is difficult for urban inhabitants, habitually steeped in particular routine, common sense practices, to apprehend that the kinds of regular rhythms they experience and follow are of relatively recent origin. In the pre-industrial rural past, the rhythms of life followed the seasons and the changing quantity of daylight that influenced the agricultural timetable. In Northern Europe, daily working rhythms were largely conditioned by the long summer hours that made lengthy spells labouring in the fields profitable, in contrast to the short days of winter when much work was not possible. People tended to rise with the sun and move indoors after nightfall, which meant that some days were filled with arduous labour while others required less work to be undertaken. This lack of a regular, regulated work schedule impacted upon sleeping patterns, along with the absence of artificial illumination. After performing the daily ritual of shutting in – bolting and securing windows and doors from potential intrusion – people would go to bed and sleep but would typically awake after four hours or so for a period in the middle of the night when they might perform domestic chores, converse, have sex or meditate. After this spell, they would enter a second, lighter sleep before rising at dawn. Dependence on seasonal variations also contributed to distinctive annual rhythms, with periods of abstinence enforced by the dearth of provisions during winter months. Yet the seasonal imperatives to work also allowed greater range for festivities; besides those organized around summer and winter solstices, it is estimated that in pre-industrial Britain, an average of 44 holidays occurred, shaped by religious and folk customs as well as agricultural processes. Such a rhythmic structure was inimical to the time discipline required for large-scale industrial production in the towns and cities that emerged in the eighteen and nineteenth centuries.

Modern, ordering rhythms: industrialization and bureaucracy

The ordering of the city's temporality is as important as its spatial regulation. The emergence of the modern city is synonymous with the rise of clock time, the key temporal instrument through which the regulation of the industrial city was organized, primarily though the systematic regulation of labour and production. Mass industrial production required that workers turned up at predictable, regular times to ensure an even and predictable output, and systems of clocking in were vital to discipline workers who had previously followed the agricultural rhythms discussed above. To ensure continuous manufacture, fixed shifts were introduced and workers were expected to synchronize all other activities around these working hours. Accordingly, they intersected with other rhythmic routines, including pub opening hours and leisure pursuits, scheduled at times that would not damage production.

Over the second half of the nineteenth and into the twentieth century, weekly and annual urban rhythms gradually transformed from being dominated by long working days and minimal leisure time; as social reform and prosperity increased, workers were allowed more time off at weekends and provided with longer holidays. In cities in the UK, the weekly routine thus became shaped by several institutionalized events oriented around leisure and religion: church on Sunday, drinking and attendance at mass entertainments such as music hall and cinema, and for men, presence at mass sporting events on Saturday afternoons. Annual rhythms were shaped by the institutionalization of bank holidays and workers in particular industries were typically given a week's holiday, often taken as part of synchronized visits to particular seaside resorts by these same workers en masse, adding to the shared, communal rhythm of the year: thousands of Lancashire mill workers descended on Blackpool for a week, and Yorkshire woollen labourers and their families would visit Morecambe or Scarborough.

Despite these temporal reforms, the control of workers on Fordist production lines was intensified through Taylorism, a system of devising a repetitive, optimal sequence of events from shop floor to loading bay, strictly timed to impose highly ordered rhythms. Here, the optimum time it should take to complete a mechanical task was carefully calibrated and ever-shorter periods were carved up into exact segments. Importantly, the body of the worker was itself conceived as a machine, trained to perform with precise regularity in accordance with the imperative to produce measurable outputs. This training of the body to perform regular, rhythmic actions epitomizes what Lefebvre identifies as 'dressage', akin to the ways in which horses are trained to perform particular and precise movements in competition (see Chapter 5, this volume). Though these working bodies may have been dulled through this rhythmic compulsion, this should not blind us into thinking that dressage simply turns workers into robotic creatures, unable to act and think for themselves. For a host of more pleasurable, though equally disciplined rhythmic pursuits have pervaded the modern city, including sports, physical exercise and dancing – practices that may be conceived as providing a release from the rhythmically conditioned industrial urban body (see Chapter 19, this volume).

These forms of dressage are part of a broader biopolitical context in which power is expressed through imposing rhythmic conformity and consistency. Rather than through compulsion, other forms of rhythmic regulation depend on disseminating rhythmic conventions that come to be unreflexively accepted as common-sense ways of doing things. As Lefebvre (2004: 68) claims, power 'knows how to utilize and manipulate time, dates, time-tables'. Such rhythmic norms include the scheduling of school hours, regulations about commercial opening times and the sale of alcohol, the confinement of late-opening clubs to limited areas of the city, broader social conventions about the time when 'noise' is labelled 'antisocial', along with many other beats that we are expected to follow. These 'good habits' are often established and enforced by local and national state bodies, but others are simply temporal norms that are widely followed.

There are believed to be 'productive' rhythms that privilege certain kinds of regular commercial endeavours and 'hard work', and those that are regarded as 'unproductive' and deviant. Though usually not subject to conscious reflection, rhythmic conventions are evident in comments about the supposedly 'wasteful' rhythms of the long-term unemployed or the 'workshy'; yet, while students are acknowledged to spend hours and days in unproductive, often hedonistic pursuits, this is not subject to similar judgment, since this lifestyle is usually conceived as acceptable within the span of study. Deviant rhythms may be identified in the movements of the homeless person or asylum seeker who wanders the city aimlessly, unable to find work or any place beyond institutional accommodation to relax, and the rhythms of begging and wandering the streets after midnight are subject to policing and regulation, conceived as suspicious and unsocial (Hall 2010).

Besides these weekly and daily scales, rhythmic, temporal ideals value the different timings and the durations of lifecycle stages, or the chronological age at which life-course transitions take place, subjecting those who fail to conform to scrutiny. For instance, a man suffering from 'mid-life crisis' might be admonished if he embarks on a relationship with a younger woman, or wears clothes that are conceived to be more suitable for younger males. We may see how such norms also produce conventional social practices and identities in the city. Children are not supposed to be out in public space after dark and most people would express surprise if they saw a group of people of pensionable age entering a nightclub at a late hour.

Finally, perhaps the most neglected yet most crucial ordering urban ordering rhythms are the ceaseless practices of maintenance and repair. The installation of regular programmes for continuous upkeep are essential if the city is not to become clogged with refuse, its material fabric imperilled by decay and its transport and social services in ruins. Shops must be restocked each day to operate smoothly. Without regular checks by ticket inspectors many passengers are likely to refrain from payment, endangering the economic viability of the transport system. In the absence of the daily routines of street sweepers and cleaning vehicles, piles of detritus would pile up on roads and pavements, impairing mobility and enabling pests to prosper. Different systematic procedures of urban maintenance are organized at different rhythmic scales and temporal frequency. For instance, while busy commercial streets in urban centres tend to be cleaned each night, Church of England structures undergo quinquennial inspections through which designated architects survey churches and recommend necessary repairs (Edensor 2011a).

Synchronizing rhythms in the city

In organizing our daily and weekly schedules, we impose a rhythmic structure on an enormous range of mundane practices across work, consumption, leisure, sleep and relaxation. Those temporal units through which organize our lives – the day, the week, the month and the year – are thus ordered to produce a regular, structured pattern. Indeed, everyday life is constituted out of a multitude of habits, schedules and routines that lend to it a comforting predictability and security, and when this is disrupted or suspended, displacement and disorientation can result. Without such rhythmic and routine patterns, life would constitute a series of random events and passages that would challenge the very basis of identity and a sense of place. On the other hand, a rigid and unchanging routine might render life monotonous, tedious and unstimulating, and it is useful to speculate about how we might achieve a balance between unending rhythmic repetition and the importance of happenstance, surprise, difference and serendipity.

A moment's reflection highlights how our everyday experience is fundamentally grounded in rhythmic habits. Consider the rhythms through which we rise and sleep, brush our teeth and arrange regular checkups at the dentist, catch particular forms of transport, organize weekly forms of social engagement, shop, exercise, watch television, and eat and drink. Not only do such banal practices become unreflexive and habitual forms of common sense, they are often underpinned by particular cultural values. The maxim 'early to bed, and early to rise, makes a man wealthy, healthy and wise', captures one of the culturally esteemed ways of organizing habit. We might also draw attention to the advice meted out by experts and authorities about the need for regular exercise and eating habits. Such conventions about 'good' habits suffuse social life.

This is pertinent to a consideration of the rhythms of the city because not only do individuals engage in predictable, timetabled rhythmic activity, but social life operates according to the ways in which individuals *synchronise* their activities with others. As Frykman and Löfgren (1996: 10-11) emphasize, 'cultural community is often established by people together tackling the world around them with familiar manoeuvres'. These synchronicities shape the rhythmic practice and experience of the city – as friends, work colleagues and family share habitual routines through which they eat, watch television, play sport, take tea and coffee breaks, go on holiday, attend cultural events and worship, and so on – inscribing patterns of gathering and moving across urban space. This is augmented by the larger synchronic rhythms that shape patterns of commuting, going to and from school, working, shopping, clubbing and dining.

These patterns are marked by regular paths and points of spatial and temporal intersection which routinize action in space and collectively constitute the 'time-geographies' within which people's trajectories separate and cross in regular ways, as described by Torsten Hagerstrand (1977). Shops, bars, cafes and garages are meeting points at which individual paths congregate, providing geographies of communality and continuity within which social activities are coordinated. These regular trajectories and points of convergence vary throughout the day and week but give a rhythmic consistency to places. Throughout weekdays, cities are characterized, for instance, by early morning sporadic deliveries to shops and households and the passage of cleaners and postal workers through the streets, and this is superseded by a surge of commuting and school runs. This may subside into a mid-morning lull, followed by the business of lunchtime, only to recede once more into the quiescence of mid-afternoon. Subsequently, this is then successively followed by the return journeys of commuters and schoolchildren, the subdued early evening, the excitable congregations of revellers, theatre-goers, music fans and drinkers – who particularly energize the city on Friday and Saturday evenings – and the slide into nocturnal quiet, punctuated by a few late merrymakers. These quotidian, ordinary rhythmic processes may also be accompanied by more infrequent, but nevertheless regular, urban festivals, processions and holidays that are themselves etched into the annual rhythms of the city.

More and more multiple rhythms

While these synchronized habits constitute some of the multiple rhythms that characterize cities, the rhythmic quality of place is increasingly typified by more diverse and contesting rhythms. Global flows of migrants, tourists, workers and business people who live and move through cities bring different practices of eating, worshiping and socializing, adding to the urban rhythmic mix. Moreover, whereas workers were likely to follow schedules governed by the clock, now flexible patterns of working are emerging, and where once Sunday was formerly organized as a day of prayer and rest during which shops were closed, it is now a time for work, consumption and an ever-expanding range of leisure pursuits. The opening hours of shops, pubs and restaurants increasingly extend across the day and night to produce a 24-hour city. To this rhythmic complexity, we can add the online rhythmic patterns through which people interact with the internet and mobile phones, as everyday urban life becomes punctuated by multiple, brief conversations and texts, as well as longer passages surfing the web. Perhaps the disruption of familiar rhythms has become more commonplace and formerly rigid routines and timetables have become more flexible and open to improvisation.

It is crucial to recognize how new capitalist arrangements are producing novel rhythmic patterns across urban space that supplement and supersede the earlier arrangements of work and leisure discussed above. In urban financial districts, share- and currency-trading ceaselessly takes place throughout the day and night, and we can see how the life cycles of products and fashion shorten, producing an accelerated obsolescence through which commodities rapidly become outmoded and are replaced by trendier items. These strategies to feed the desire for distinction mean that rhythms of marketing, advertising (Cronin 2006) and acquisition have sped up for many. Another way in which the rhythms of consumption become inscribed in place is detailed by Mattias Kärrholm (2009), who shows how retailers in Malmo, Sweden, attempt to synchronize commercial rhythms with everyday urban rhythms and mobilities. Accordingly, in pedestrian precincts, shopping malls, museums, stations and airports, shopping rhythms become aligned with those of cultural events, festivals, movement and a range of leisure practices.

However, though it may seem that such speeded up, intensified rhythmic processes are increasingly dominating urban life, there are also rhythmic practices that resist them. In fact, Ben Highmore insists that there have always been 'slower, more sedentary and stagnant aspects of modern life' (Highmore 2002: 172), citing the work-to-rule and the 'go slow', to which we may add 'chilling out' and hedonistic drug-taking as rhythms that oppose modern speed. In recent years, the 'slow food' movement and 'slow tourism' have arisen to challenge the culture of fast food and scheduled tours respectively. Sarah Pink has written about how the *Cittàslow* movement involves urbanites in engaging more slowly 'in everyday practice in particular ways at a routine, personal, individual level' (Pink 2007: 63), and we are seeing an escalation of the long-standing practice of temporarily escaping the pace of urban life for the countryside. In addition, there are practices that disrupt the rhythmic flows of urban life such as Critical Mass cycling events, flash mobs and convoys that slow down traffic.

These practices that challenge the rhythms imposed by bureaucracy and capitalism are supplemented by the ways in which the increasingly complex and speedier rhythms of the city require technological infrastructures that themselves break down. Despite sophisticated traffic-flow systems, congestion has become a feature of the mobile experience of many urban commuters as trains and buses break down, accidents block streets and roadworks must be carried out. Rhythmic predictability is also threatened by power cuts and blackouts, droughts, terrorist acts and riots, flooding and high winds. Such occasions produce *arrhythmias*, where experience devolves into unexpected stasis or disruption, just as the habitual rhythms of our bodies are disrupted when we become sick or are jetlagged (Edensor 2011b).

Different cities

In addition to the increasingly complex intersecting array of rhythms that characterize contemporary urban life, it is vital to acknowledge that so far, I have primarily focused on cities in the UK. Yet the rhythms of cities outside the UK form widely divergent temporal patterns to those I have described. Indeed, the cultural specificity of the performance and apprehension of urban rhythms that typifies unreflexive, everyday experience can be put into perspective when one moves to a city that lies outside one's usual quotidian orbit. For instance, daily life in cities in which Muslims form a majority of the population are inescapably characterized by the sonic rhythm of the *muezzin* calling the faithful to prayer five times each day. In North African cities, large numbers of people crowd the streets between 6 and 8 pm, whereas the central districts of many American cities are virtually deserted after office and shop workers return home. In Spanish cities, although increasingly modified by global business norms, the siesta breaks up the day for a number of hours, lengthening the working day and extending the social life of the streets through late evening as families pour into the streets to dine and socialize. In other settings, rhythms diverge more markedly from UK cities, as I now describe with reference to a busy bazaar area of an Indian city (see Edensor 2000).

CASE STUDY 21.2 THE RHYTHMS OF THE INDIAN BAZAAR

The bazaar typically constitutes an unenclosed realm that provides a meeting point for a variety of people and multiple activities, mixing together small businesses, shops, street vendors, public and private institutions and domestic housing. Accordingly, the street is often a site for social activities such as loitering, sitting, chatting and observing, and domestic activities such as collecting water, washing clothes, cooking and child-minding, activities characterized by different rhythms that occur adjacent to each other. Some linger for several hours, some rush through while others take their time, ambling along the street, stopping to shop and chat. Bazaar space also comprises a variety of overlapping spaces – alleys, niches, awnings and offshoots – composing a labyrinthine structure with numerous openings and passages, enabling a flow of different bodies and vehicles to crisscross the street in multi-directional patterns. Pedestrians must weave a path by negotiating obstacles, and take account of the other people and traffic that cross their path at a variety of speeds. Walking is thus likely to be a sequence of interruptions and encounters that disrupt smooth passage, and jostling amidst the crowd involves continuous physical contact with other bodies. The confrontation with multiple textures alongside the body and underfoot further compounds the experience of diverse tactile sensations. Visually, the maintenance of a cultivated appearance through control or theming is rarely imposed; an unplanned bricolage of structures is infested with ad hoc signs, contingent and personalized embellishments, and unkempt surfaces and facades, and a lack of visual rhythmic order persists. This is supplemented by numerous distractions and diversions offered by heterogeneous activities and sights, and an ever-shifting series of juxtapositions and assemblages of diverse static and moving elements, which can provide surprising scenes. Moreover, the medley of noises generated by numerous human activities, animals, forms of transport and performed and recorded music, creates a changing symphony of diverse pitches, volumes and tones; an incessant composition of clashing and variegated rhythmic sounds.

The rhythms of the non-human

As Lefebvre (2004: 17) says: '(There is) nothing inert in the *world*'. He illustrates this with examples of the seemingly tranquil garden that is suffused with the polyrhythms of 'trees, flowers, birds and insects' and the forest, which 'moves in innumerable ways'. All places are always in the process of becoming, and humans are particular elements in spaces that seethe with intersecting trajectories and temporalities. Besides the multiple human rhythms discussed above that contribute to the ongoing experience and reproduction of the city, it is also crucial to acknowledge the equally multiple, ever-present rhythms of the non-human agents that shape urban environments. There are the obvious rhythmic patterns of weather, ranging from annual monsoons, heatwaves and dry spells, and periods where ice and snow cover the landscape, as well as the tidal surges that contribute to the experience of coastal cities, disastrously in the recent case of New Orleans. Rivers continuously flow through cities as does wind. Seasonal rhythms shape plant growth, notably in temperate cities where deciduous trees are stripped bare in winter in contrast to thick summer foliage, and at certain times of the year, spores float through the atmosphere, causing allergies. Flowerbeds in parks and gardens produce summer hues at variance to the more muted tones of winter. Birds also follow seasonal patterns, adding to the rhythmic experience of the city: in autumn, large flocks of starlings might congregate in city centres; in spring, birds seek nesting materials in suburban gardens, and swallows arrive to signify the advent of the new season. At a diurnal scale, the suburban spring dawn chorus can exceed the thrum of traffic in the early morning soundscape, and the almost ever-present, but usually unseen, foxes, bats and hedgehogs emerge after nightfall.

In considering the co-production of the city by non-humans, it is useful to return to the crucial rhythms of repair and maintenance discussed above, for it is easy to see what happens when these practices cease at derelict or ruined sites. As soon as a building is abandoned, and especially if visitors break windows and scavengers steal tiles and lead from roofs, non-human agents move in and the building is subject to a different series of rhythms shaped by the agencies of water, the nest-building and roosting of birds, the burrowing of woodworms and the lifecycles of insects (Edensor 2015). Cracks are rapidly colonized and subject to the rhythms of plant and fungal growth, and processes of decay mean that the erosive agents that come to pervade the building, transform its materiality according to destructive rhythms that are wholly different to those which held them at bay through maintenance and repair.

Conclusion

In this chapter, I have explored how the study of urban rhythms can highlight the contested nature of the city, as the powerful try to impose schedules and timetables that are often resisted or bypassed. Besides highlighting these ongoing tensions and conflicts, a focus on rhythms and their multiplicity can help us to investigate more thoroughly the ways in which space is ordered through time, the processes that flow through the city, transforming space and experience, and the different agents that contribute to the endless reordering of the built environment. Rhythmanalysis can also reveal the distinctive qualities of particular cities, helping to avoid over-generalizations, and disclose the dynamic global flows and increasingly complex practices through which contemporary cities are constituted.

Key reading

Edensor, T. (Ed.) *Geographies of Rhythm: Nature, Place, Mobilities and Bodies,* Aldershot: Ashgate.
   Published to capture the response among scholars to the publication of Lefebvre's *Rhythmanalysis,* this volume includes essays that focus on different kinds of mobility, embodied rhythms, the distinctive rhythms that circulate around place, and various non-human rhythms.
Lefebvre, H. (2004) *Rhythmanalysis: Space, Time and Everyday Life,* London: Continuum.
   This was Lefebvre's brief introduction to the study of rhythms, a project he had wished to expand; sadly his death prevented further analysis. Its translation into English has prompted increased interest in rhythm.
Smith, R. and Hetherington, K. (Eds.) (2013) *Urban Rhythms: Mobilities, Space and Interaction in the Contemporary City,* Oxford: Wiley-Blackwell.
   A collection of chapters that explore the diverse rhythms of a range of contemporary cities, including Manchester, Jakarta, Shanghai, Rio de Janeiro and Lisbon.

22

RIGHTS

Joaquin Villanueva

Introduction

> One only has to open one's eyes to understand the daily life of the one who runs from his dwelling to the station, near or far away, to the packed underground train, the office or the factory, to return the same way in the evening and come home to recuperate enough to start again the next day.
>
> *(Lefebvre 1996: 159)*

> Ron: There are times when I had to pee in the street.
>
> *(Duneier 1999: 175)*

Reproduction in the capitalist city can be tiring for some and unjust for many. The opening quotes highlight the obstacles that a worker (living in a distant suburb, an isolated housing project, or a satellite city) and a homeless person must circumvent on a daily basis to reproduce their labour power *and* survive in the capitalist city. Both quotes seek to highlight the extent to which the form of the city - its property regime (Blomley 2004) as well as its vastness – shapes our daily social and spatial practices. For Henri Lefebvre (1996: 159), the routine of the worker reveals the 'untragic misery of the inhabitant'. The homeless, we can add, exhibits the 'tragic misery of the inhabitant', for Ron's ability to satisfy a basic human need was impaired by the systematic closure of public bathrooms across New York City (Duneier 1999). Do we, as critical urban scholars, possess the conceptual framework and moral language to evaluate the above conditions? In this chapter I argue that the language of rights, understood as something to which one has a just claim, can provide us the conceptual framework and moral grounds to denounce unjust spatial practices and transform the social and spatial conditions that shape our (tragic and untragic) urban experiences.

Within urban studies, recent debates about rights have figured most prominently in discussions on the right to the city (Lefebvre 1996) – from what that right looks like, to the kind of right it refers to. Many of the criticisms of the right to the city within urban studies parallel criticisms of liberal rights more generally (Attoh 2011; Blomley 2008; Merrifield 2011). That is to say, rights come under attack for their indeterminacy, their abstraction and individualism. Despite these critiques, I argue that rights and the right to the city remain useful tools in the fight for social and spatial justice. The first part of the chapter looks at the general criticisms of liberal rights. I then discuss the right to the city as sketched by Henri Lefebvre. In the last section I organize the right to the city scholarship into three thematic categories: first, the right to the city as right to space; second, a transformative process; and third, an all-encompassing slogan. In the conclusion I invite urban scholars to continue fighting for the right to the city.

Liberal rights and the city

I discuss the example of property rights to demonstrate how rights can conceptually frame our understanding of cities and morally guide our analysis. Property, legal geographer Nicolas Blomley (2004) reminded us, has been almost exclusively conceived around the 'ownership model' whereby property is a thing to be owned by private individuals or a public entity (the state). Because property is something to be owned, it is often seen as a 'finite space'. Liberalism, a political philosophy fostered by the Enlightenment project and adopted by modern democratic and capitalist states, has been historically conceived around the autonomous, property-owner individual – an individual that possesses inalienable rights that preserve his (and later her) freedom. As Michael Walzer (1984: 315) noted, liberals mastered the 'art of separation', which consequently helped draw a map of the spaces where rights best operated. The archetypal subject *and* space of liberal-capitalist states is therefore the property owner and his/her 'finite space', a space where rights are almost always guaranteed.

Liberal and individual rights bring about the paradoxical situation according to which the property owner can justly exclude anyone from his/her property, thus denying anyone the right to be in their 'finite space'. The privatization of urban space and services, a trend that has intensified under the current political and economic constellation known as neoliberalism (see Chapter 18, this volume), has further expanded the spaces for individual property rights-claimants; claimants that can *justly* expel Ron from using *their* bathroom. Unfortunately 'we live in a world', decries David Harvey (2012: 3), 'where the rights of private property and the profit rate trump all other notions of rights one can think of'. By not owning property or being able to afford rent, the homeless is by consequence denied his/her 'right to sleep, defecate, eat or relax *somewhere*' (Mitchell 2003: 28; emphasis in original). Because liberal rights protect and, at times, perpetuate the conditions of inequality that deny Ron a 'finite space' for his reproduction, some progressive and radical scholars have mistakenly concluded that rights are not an effective weapon to fight 'economic injustice' (Rorty 1996). Moreover, since rights can simultaneously protect the weak *and* the powerful, critical scholars resist fighting for rights precisely because of their indeterminacy (Tushnet 1984).

Geraldine Pratt (2004) proposes that we conceive liberal rights and their claims to universal values as an 'empty space'. Pratt maintains that 'groups compete to define what counts as universal, a competition that requires in democratic societies the difficult labour of *translation* across competing claims' (2004: 85, emphasis added). Dismissing rights for their indeterminacy blinds us from seeing 'universal rights as a kind of empty space that we measure inequalities against' (Pratt 2004: 115). Rights are a lens through which 'the behaviour of the state, capital, and other powerful actors must be measured – and held accountable' (Mitchell 2003: 25). The institution of rights established by the liberal state has historically provided minority groups a discourse and a material basis to articulate their exclusion and alienation vis-à-vis equal rights-bearing subjects. As Dworkin (1977: 205) put it, the institution of rights 'represents the majority's promise to the minorities that their dignity and equality will be respected'. Modern citizenship rights' 'institutionally defined rules' (Young 1990: 25) protected and upheld by the nation-state include civil, political and social rights that over time have enabled women, the unemployed, retirees, dissenters, and minorities to be heard, to be insured, and to protest (see Chapters 6 and 23, this volume). However, these rights were not given away by the liberal state, but fought for – battles that have opened up new spaces of rights that have helped translate competing universal claims. 'The adjudication of rights', says Mitchell (2003: 27) 'is a function of force, a result of *political* action' (emphasis in original). It is precisely political action that determines whether right-claiming spaces remain open or close down.

Drawing on these expanded understandings of rights and their spatiality, urban scholars began to explore the connection between rights and the city, a critical arena for 'claiming, expanding or losing rights' (Isin 2000: 5). More impactful for urban scholars was perhaps the publication of the first English translation of Henri Lefebvre's *The Right to the City* in 1996. Since then, geographers, planners, architects and activists have adopted the right to the city as a rallying cry to denounce unjust urban practices and to organize their political struggles.

The right to the city

French sociologist and philosopher, Henri Lefebvre, wrote *Le Droit à la Ville* (*The Right to the City* hereafter) in 1967 (published in 1968) to commemorate the 100th anniversary of the publication of Karl Marx's *Capital*. If the latter was intended as a critique of industrial society (and political economy) and its various impacts on the working class, the former was meant as a critique of the capitalist city and its impacts on the working class. Lefebvre's urban historical analysis highlighted what the capitalist city had displaced and destroyed. In regards to medieval cities of Western Europe, Lefebvre (1996: 66) noted that these tended to accumulate wealth, but also 'knowledge (*connaissances*), techniques, and *oeuvres* (works of art, monuments)'. In fact, the medieval city was itself an *oeuvre*, a work of art. The main purpose of the city-*oeuvre*, Lefebvre insisted, was to use the streets, squares, monuments, and symbols of the city for the pleasure of its inhabitants. 'The eminent use of the city', he further added, 'is *la Fête* (a celebration which consumes unproductively, without other advantage but pleasure and prestige and enormous riches in money and object)' (Lefebvre 1996: 66). The arrival of industrialization and its obsession with commodity production and exchange would eventually erode this essential aspect of cities – the *oeuvre*. As Kofman and Lebas (1996: 20) put it, '[c]apitalism and modern statism have both crushed the creative capacity of the *oeuvre*'.

The capitalist city instead 'created the centre of consumption' (Lefebvre 1996: 169) where tourists and shoppers flood the shopping malls, boutiques and coffee shops that now dominate urban space, with the sole purpose of consuming the latest and most 'authentic' consumer goods (see Chapter 9, this volume). Lefebvre scorned urban planners, state bureaucrats, and private investors for transforming the traditional city into a consumer centre at the expense of the 'working class, victim of segregation and expelled from the traditional city, deprived of a present or possible urban life' (Lefebvre 1996: 146). The segregation of the poor deepened as exchange value gradually mediated urban relations and displaced the use value of the city. Central to Lefebvre's political project was the need to restore the use value of the city, to collectively *create* the city-*oeuvre* for the pleasure of all, and not just a few. Lefebvre adopted the language of rights in order to denounce unjust urban practices and to urge urban inhabitants to claim and fight for a city open for *all*, regardless of class, gender, race, ethnicity, or sexuality.

The right to the city is 'like a cry and a demand' (Lefebvre 1996: 158). It is a cry against unjust urban practices that have generalized marginalization and segregation, and a demand for a better and more just city. The right to the city, Lefebvre asserts, 'can only be formulated as a transformed and renewed *right to urban life*' (1996: 158; emphasis in original). It is the right to meet, to gather, and enjoy the city and its spaces without the mediation (and constraints) of exchange value. This renewed urban life also implies our 'right to the *oeuvre*, to participation and *appropriation* (clearly distinct from the right to property)' (Lefebvre 1996: 174; emphasis in original). In other words, the right to the city demands the opening up of right-claiming spaces that allow urban inhabitants to collectively create *their* city-*oeuvre*.

The right to the city is a *radical concept* that calls for the complete transformation of the institutions and economic relations responsible for the production of the contemporary city. In negative terms, it is the right of everyone *not* to be excluded (Lefebvre 1996: 195). As a positive term, it is the right of everyone to *be* 'an integral part of city' and to *be* a co-creator of the city-*oeuvre* (Mitchell and Villanueva 2010: 671). For Marcuse (2009: 193), the right to the city is a 'moral claim' for 'a better system in which the demands [of the oppressed and alienated] can be fully and entirely met'. Committing to the right to the city entails rejecting the existing institutions and the economic, political, cultural, and social processes that perpetuate oppression and domination. Mitchell and Heynen (2009: 616) argued that the value of the right to the city is its 'capaciousness', because it encompasses a variety of rights and demands that seek to satisfy ever-changing urban needs. Although this was Lefebvre's intention, to open up a concept for us to fill in its contents (just like the 'empty space' of universal rights), it is this very capaciousness that has opened the door for trivialization, as the concept is appropriated by UN agencies and urban scholars that have unfortunately stripped it of its emancipatory and revolutionizing potential (Souza 2010). It is to these problems that we now turn.

Appropriating the right to the city in the neoliberal city

The right to the city gained popularity thanks in large part to the publication of edited volumes on the topic. We can cite (not exhaustively): Holston (1999); Isin (2000); *GeoJournal* special issue the 'Right to the City' (Staeheli and Dowler 2002); *Cities for People, Not for Profit* (Brenner *et al* 2011); and a special issue on the 'Right to the City' on *Architectural Theory Review* (Stickells 2011). These constitute some of the major contributions that analyse the right to the city in the 'Global North'. Recently, the concept has also drawn attention from urban scholars critically assessing urbanization processes in the 'Global South' (Samara *et al* 2013). Similarly, the right to the city is on the agenda of non-government organizations and policymakers that came into contact with the slogan through international conferences. The first such conference was held in Porto Alegre, Brazil in 2001 by the World Social Forum and later conventions led to the creation of a World Charter for the Right to the City (Kuymulu 2013). The United Nations has similarly adopted the language of the right to the city when UN-HABITAT organized the first World Urban Forum in 2002. Since then, UNESCO and UN-HABITAT have brought together government officials and policymakers to discuss possible ways of implementing the right to the city at a global scale. Grassroots activists, such as the Right to the City Alliance in New York City and the Abahlali baseMjondolo movement in Durban, South Africa (shack dwellers fighting against eviction), have adopted the right to the city for organizing their political agendas (Huchzermeyer 2014). In Brazil, urban social movements since the 1970s pressed legislators to pass the Statute of the City in 2001 under which the right to the city officially became federal law, albeit unevenly enforced and less radical than Lefebvre's original cry (Friendly 2013).

The right to the city has clearly become an extremely popular concept for urban scholars, social activists, and policymakers alike. In particular, critics of neoliberal urbanization have found the right to the city a useful paradigm because it provides them a moral ground from which to measure the privatization of urban space, the transformations of urban governance regimes, and the incessant criminalization and marginalization of the poor that often accompanies these processes (Brenner *et al* 2011; Harvey 2012; Mitchell 2003). I will briefly consider these critiques alongside three broader categories that capture the ways in which the right to the city has been theoretically discussed by urban scholars.

Right to space

Lefebvre suggested that the right to the city entailed the right to 'participation and appropriation' (1996: 174). Following this statement, urban scholars have emphasized the appropriation of urban space as the basis for participation in the creation of the *oeuvre*. Dikeç (2001: 1790), for instance, invited us to conceive the right to the city not only as a 'participatory right' or a right to 'urban space, but to a *political space*' (emphasis added) Similarly, Mitchell (2003: 82) documented how struggles over free speech, particularly among marginalized groups, have almost always been struggles 'for an appropriate *place* to speak' (emphasis in original).

These accounts highlight the physical presence of groups and individuals as the foundation for participating in the decision-making process of the city and for articulating urban inhabitants' just claims to the city. Purcell (2002: 102) argues that 'the right to appropriation' of urban space is a necessary precondition for 'shifting control away from capital and the state and toward urban inhabitants'. Iveson's (2013: 946) research on 'do-it-yourself' urbanism such as graffiti artists and cultural jammers similarly highlights how urban space is appropriated to challenge authority claims (see Case study 22.1). Laam Hae (2011), on the other hand, explored the regulation and closure of spaces for 'social dancing', an expressive and political activity, in gentrifying neighbourhoods across New York City. Its banning, she argues, seriously trumps 'urban rights' – 'rights to spaces in which [inhabitants] can legitimately enjoy diverse urban life' (Hae 2011: 140). Tovi Fenster (2005: 229) stressed the diversity of urban life to advocate for the 'right to the gendered city', that is 'the right to use and the right to participate' to challenge 'patriarchal power relations'. For these authors, the right to space is the basis for any political engagement in the *making* of the city of difference (Dikeç 2001).

CASE STUDY 22.1 ABDEL'S URBAN POLITICS OF INHABITING

I first encountered Abdel, a 23 year-old French-Tunisian, as he wandered the streets of Paris on a hot summer night. He engaged me in conversation, and, curious about my accent, decided to keep the conversation alive for the next couple of hours. Abdel was unemployed at the time, despite repeated attempts to find a job. There was a sense of desperation and anger in his voice that oddly mingled with his charming personality. We walked the streets of north-eastern Paris for two hours as Abdel recounted his family history – having originally migrated to Paris, where Abdel was born, they moved back to Tunisia before returning to France eleven years prior. They now reside in a crowded apartment in a social housing project on the outskirts of Paris.

Abdel felt that his citizenship rights (he is, after all, a French citizen) were being trumped by his ethnic origins and physical appearance – hence his inability to find a job. That night Abdel wandered the streets of Paris to reclaim his right to the city – a city he felt denied him access to a job and decent housing. Abdel insisted on getting something back from France. To white males, Abdel would ask for a cigarette, despite having a full pack in his pocket. At times he would ask for a light, despite having a lighter in his other pocket. To police officers Abdel defiantly asked them for the time, even when I had told him the time two seconds ago. Fire, cigarette, time: these are basic elements easily available in the streets. They allow for easy and sporadic encounters with people. These are, moreover, some of the use values that Abdel found in the city; their usage (and appropriation) enabled Abdel to enjoy the city. Abdel's daily spatial practices are a prime example of the 'urban politics of the inhabitant' (Purcell 2002), whereby the mere presence and practice of inhabiting (through the appropriation of urban space) can constitute an act of resistance – a way for inhabitants to challenge 'authority claims' (Iveson 2013). Abdel's spatial practices, more importantly, *temporarily* opened up a space where he could validate his right to *be* – without being excluded from the city's daily encounters.

Transformative process

As Lefebvre conceived it, the right to the city cannot be implemented under current institutional and economic conditions. A new society should be constructed instead. While discussing the World Charter on the Right to the City – the first attempt to define the rights in the right to the city – Mayer (2009: 369) correctly argued that 'this institutionalized set of rights boils down to claims for inclusion in the current system as it exists, it does not aim at transforming the existing system'. The right to the city is a popular slogan partly because many narrowly understand it as a juridical right of inclusion to the current system. This has prompted critical urban scholars to highlight the radicalness of the concept – its transformative potential.

For David Harvey (2008: 40), the right to the city is a 'working slogan and political ideal' that could potentially unite social movements around a common cause: establishing 'greater democratic control over the production and utilization of the surplus' that circulates and is generated in cities (2008: 37). For Harvey, controlling the surplus produced in cities is a just claim to the 'right to change ourselves by changing the city' (Harvey 2008: 23). Mayer (2009: 367) similarly noted that 'movements' around the world are already 'building on the Lefebvrian conception where urbanization stands for a transformation of society and everyday life through capital'. For these authors, the right to the city presupposes a radical transformation of the institutions that control the surplus, the institutions that grant and enforce the rights of citizens, and the institutions that perpetuate oppression and inequality. The institutionalization of the right to the city, in other words, requires a new set of institutions that respond primarily to the *social needs* expressed by urban inhabitants through 'social and political action' (Mayer 2009: 367) (see Case study 22.2).

CASE STUDY 22.2 RIGHTS *IN* THE CITY AND RIGHTS *TO* THE CITY

Guest workers from France's ex-colonies partly satisfied the need for labour in the postwar period. Their arrival along with natural population growth exacerbated the housing crisis in France. In the 1950s, the central government allied with the private sector to initiate a 'social project' to house postwar France (Cupers 2014). Satellite cities, bedroom communities, New Towns, and high-rise social housing estates mushroom in the suburbs of traditional cities. Lefebvre critiqued the new landscapes for having replaced the practice of 'to inhabit' (i.e. Abdel's ability to appropriate the use values of the city) with that of 'habitat' – an abstract and functional dwelling that prevents residents from appropriating and making their daily spaces (Lefebvre 1996: 78–80).

By the 1970s, housing estates experienced a demographic shift as many of the French working and middle classes acquired their own private properties, while an increasing number of families from immigrant extraction moved in. Since then, unemployment and geographical exclusion have isolated suburban housing estates. The history of revolts – often triggered by a police action gone wrong – has further isolated these communities, which the authorities and the media have labeled *zones-de-non-droit* (lawless places). A lawless area is, by definition, a place *closed* for claiming, expanding, or respecting the rights of inhabitants (Villanueva 2013).

The Minister of Justice has since the 1990s campaigned to, literally, *open* up spaces where residents of the infamous *zones-de-non-droit* can access and claim their rights. A network of 'local courts' has been erected to increase the judiciary's presence across the urban landscape. Judicial presence in 'lawless urban areas' should, in principle, 'contribute to the prevention of delinquency, to the assistance of victims, and access to rights' (*Code de l'Organisation Judiciare*).

One of these courts, the *Maisons de Justice et du Droit* (MJD, Houses of Justice and Law), serves not as a means for urban inhabitants to claim their right *to* the city or to transform their urban conditions, but rather as a place for the construction of model citizens of the French Republic. One of the central missions of the MJD is to remind young people from immigrant extraction 'that citizenship requires respect, of the fact that if we have rights we also have duties' (MJD Judge, personal interview).

MJDs are sites for claiming rights in the city – residents of the Paris region can claim their rights in 33 different locations – while most of their clients remain excluded from participation in the processes that produce the city. In this context, the right to the city entails opening up spaces of rights across the urban landscape, spaces created (through struggle and conflict) and administered by urban inhabitants and not the Minister of Justice. The right to *political* spaces – 'empty spaces' where the translation of competing claims takes place – is a precondition for collectively claiming the right to transform ourselves by transforming the city.

An all-encompassing slogan

Andy Merrifield has recently proposed we 'ditch' and give the concept of the right to the city to the 'enemy' because it has lost all its utility (2011). Merrifield's argument is twofold. First, he argues that the 'city' no longer exists – it has been absorbed along with the countryside by the 'urban fabric' (Merrifield 2011: 474). Second, if the city no longer exists, then we must ask ourselves, a right to *what* city? (ibid: 475). In his words (ibid: 478), 'the right to the city quite simply isn't the *right* right that needs articulating. It's too vast because the scale of the city is out of reach for most people living at street level'. Given the 'capaciousness' of rights *and* the city, Merrifield proposes we recalibrate our political motivations around the 'politics of the encounter', an abstract notion that, as he says, 'utters no rights, voices no claims' (ibid: 479). How a politics of the encounter more concrete than real-life claims raised by, say, a group of individuals fighting against eviction (i.e. Abahlali baseMjondolo) is never fully explained by the author.

Merrifield, however, is correct in pointing out that the right to the city is too encompassing and as such is losing its utility, as the concept is appropriated by social movements, UN agencies, and urban scholars that do not critically engage with its content and meaning. Nevertheless, ditching the concept seems a bit premature. Merrifield could have instead theoretically engaged with the literature of rights and their geographies. This is a critique that Nicholas Blomley (2004, 2008) has consistently launched against the right to the city literature, which he labels as 'interesting', and yet 'remains under-theorized, failing to engage the broader critical literature on rights, their politics and their geographies' (Blomley 2008: 159). Kafui Attoh (2011), similarly, took to task answering the important question: 'What kind of right is the right to the city?'. Attoh observed that 'Lefebvre's notion of rights was sketchy at best, and, worse, the growing scholarship on the "right to the city" offers little clarification' (Attoh 2011: 674). While welcoming the 'capaciousness' of the concept, Attoh (ibid: 674) noted that the 'right to the city' scholarship has failed to assess occasions when 'rights not only collide but are incommensurable' (ibid: 674). A possible solution to this conundrum is offered at the conclusion of the essay when the author claims that we 'integrate the right to the city into a general theory of social justice or substantive democracy' (ibid: 678) – that is, to wrestle with the 'empty space' of universality.

Conclusion

Rights constitute an important arena from which we can resolve 'political struggles' (Blomley 2008: 158). It is imperative, however, that urban scholars engage critically with the literature on rights and their geographies. The 'right to the city' literature, for instance, needs to define and clarify how the right to the city fits within and sets itself apart from the geographies of liberalism. It must likewise engage in the tedious task of mapping the spaces of right-claiming and ask how the right to the city can bring about a more inclusive geography of rights. Despite its all-encompassing capacity and recent trivialization, we should not 'ditch' the right to the city because we have yet to fully theorize or even imagine its full potential and mobilizing power.

In fact, I suggest that the right to the city provides urban scholars an important conceptual framework to measure the urban experiences of workers commuting hours long to reproduce their labour, Ron's daily struggles to satisfy his biological needs in an increasingly privatized city, Abdel's creative tactics to be a part of the exclusive city, and our inability to control the surpluses generated by the city. Seriously engaging with the right to the city provides us the capacity to imagine an alternative urban daily experience. More so, fighting for the right to the city is to commit ourselves to a collective struggle to transform our current urban conditions: a transformation that will do away with the narrow liberal and individualized conception of rights – and its exclusionary geographies – in favour of a set of collective rights that respond to ever-changing urban needs. Ultimately, Lefebvre would insist, the right to the city is an invitation to fight against our tragic and untragic urban experiences and to collectively create a pleasurable and festive urban life.

Key reading

Attoh, K. (2011) 'What kind of right is the right to the city?', *Progress in Human Geography*, 35: 669–85.

In the now vast scholarship of the right to the city, this article stands alone, as it is one of the few pieces that have seriously wrestled with theories of rights in order to enrich the politics and debates around the right to the city.

Lefebvre, H. (1996) 'The right to the city' (translated by E. Kofman, and E. Lebas) in *Writing on Cities*, Oxford: Blackwell.

Lefebvre's classic essay is a must read for anyone interested in the right to the city. The text can be challenging at times but it will surely provide much-needed theoretical and political inspiration.

Mitchell, D. (2003) *The Right to the City: Social Justice and the Fight for Public Space*, New York: Guilford Press.

This text should be read alongside Lefebvre's essay, as Mitchell powerfully uses the 'right to the city' conceptual framework to denounce unjust urban practices in the neoliberal city and to illustrate that a more just city is possible through the appropriation and construction of democratic public spaces.

Pratt, G. (2004) *Working Feminism*, Philadelphia: Temple University Press.

This text provides a sophisticated feminist critique of liberalism and its associated geographies. With empirical evidence to support her theoretical interventions, Pratt explores the place of Filipina domestic workers and Asian youth in Canada's liberal rights landscape.

23

SEXUALITY

Jon Binnie

Introduction

This chapter examines how sexuality contributes towards our understanding of urban life. Specifically, I wish to focus my discussion on queer theory and politics and to consider how queer can contribute towards urban theory. The chapter examines four key themes, which I suggest are key in progressing understanding of the relationship between sexualities and urban life; first, the institutional politics of urban studies that govern the possibilities for urban research on sexualities; second, the potential of queer politics and theory to transform understanding of the city; third, the value of the material and economic in understanding the queer politics of urban space; and fourth, recognition of the value of transnational approaches to the study of the sexual politics of the city and the limitations of 'methdodological nationalism' in studies of sexualities and urban space (see Chapter 8, this volume). I do not mean to suggest that these are the only issues that are central to thinking through the contemporary state of the field. How one narrates or maps the trajectory of this burgeoning field inevitably means rendering some bodies of work central, and others peripheral or marginal, and this brief examination of the field is far from exhaustive. However, in discussing these four themes, I draw on Bell's (2001) critique of attempts to frame discussions of queer's impact on urban geography. Informed by Hemmings's (2011) assessment of what is at stake in the re-telling, re-imagination and re-narration of the past within contemporary feminism, Bell argues against seeing the development of geographies of sexualities that have been heavily influenced by queer theory purely in terms of progress from absence and repression to success, acceptance and institutionalization.

The chapter starts with a discussion of the institutional politics of researching the sexual politics of urban space. Such a discussion is necessary, I argue in order to understand the status and position of sexualities research within urban studies and theory. This leads into a discussion of queer politics and theory, where I provide an overview of the multiple and contested understandings of queer. Queer politics and theory have been central in contributing to theoretical understanding of the relationship between sexuality, power and knowledge and contesting the exclusion and marginalization of sexualities from social theory. I then move on to discuss how queer theory can help us think about urban life differently – how it can contribute towards an understanding of urban life that recognizes the value of 'non-normative' gendered and sexual subjectivities and lives. In particular, I argue that queer can provide a more holistic understanding of sexuality and sexual desire as both a key human need, but also a key locus of power within cities. It can contribute to a more comprehensive and complete understanding of how certain bodies are marked as having greater value within the city (see Chapter 5, this volume). Issues of class, race, gender and mobility are central to the value that certain bodies are accorded within the contemporary city; the final part of the chapter examines how we might think of the transnational dimensions of queer urbanism – beyond 'methodological nationalism'. In order to ground my discussion of these themes and issues, I will use the case studies of De Waterkant 'gay village' in Cape Town, South Africa, and the March for Equality in Poznan, Poland. These case studies demonstrate that place matters (Monk 1994) and geographical context is core to theorizing the relationships between sexuality and urban space and the particular value of knowledge produced outside of Anglo-American contexts in challenging the assumptions and geographical imaginations of scholars of sexualities rooted in Anglo-American geographical traditions (Brown et al 2010; Silva and Viera 2014).

'Dirty work': sexuality and urban theory

Despite the fact that sexuality is a fundamental dimension of urban experience, institutional support for work on sexuality in urban studies varies widely, even within geographical centres of academic knowledge production of scholarship on the sexual politics of urban life. Here, it is worth considering why sexuality remains marginal within urban scholarship. In a passionate and politically engaged essay, Irvine (2014) has argued that sexuality research is systematically devalued and often stigmatized within the academy. Writing about the status of sexuality research within sociology in the United States, Irvine argues (2014: 636) that this field of research is constructed as 'dirty work', suggesting that: 'The persistence of stigma towards sexuality research is striking, coterminous with public demand and any professional benefits that might accrue to those studying a culturally fraught and therefor sensationalized topic. "Dirty work" conceptually captures this paradox'''. Moreover, Irvine notes that progress narratives about the liberalization of US culture and the growth of academic research on sexuality are simplistic, for they mask a more complex and paradoxical picture of institutional regulation of sexuality research. Irvine notes that: 'On the one hand, venues for academic research have expanded over the last decades, many people are eager for the knowledge that sexuality researchers produce, and in some circles the field is respected, even trendy.

On the other hands, sexuality researchers have attempted for over a century to establish academic legitimacy in the face of deep cultural anxieties about their object of study' (Irvine 2014: 633). I contend that a similar paradox to that outlined by Irvine exists within urban theory. Work on the relationship between sexualities, sexual desire and urban space has certainly blossomed in the past two decades. The theoretical and geographical scope of such work has expanded dramatically in the past twenty years, and this topic has received much more critical attention than has previously been the case, enough for Hubbard (2012) to be able to write a major overview of the relationships between cities and sexualities. However, it would be misguided to interpret this welcome development as a simple linear progress narrative, and to celebrate this as a case of unbridled progress. Recalling Irvine's discussion of the institutional politics of research on sexuality within sociology in the United States, we need to reflect more critically and rigorously on the state of knowledge production in this area. While research on the sexual politics of urban life has expanded of late, we should acknowledge the fact that mainstream accounts of urban life still tend to neglect questions of sexual politics, and specifically the role of sexual desire within urban life. Despite the growth of theoretical engagement within urban theory on the non-representational; affect and embodiment it is still rare to find accounts engage directly and explicitly within issues of sexual desire and the erotic politics of urban life (though see Brown 2008 for a notable exception). These issues are of urgent importance for scholars engaged in gender and sexual politics, for urban scholars informed by queer politics and theory, and for those concerned with questions of the complex relationships between sexual acts, identities, subcultures and communities considered outside of the socially normative.

Futhermore, I contend that political economic accounts of urban inequalities, and research focussing on affect and embodiment are constrained to the extent that they overlook sexuality as a central locus of urban power that should play a much more significant role within urban theory than at present (see Chapters 2 and 11, this volume). As Phil Hubbard (2012: 209) argues, sexuality should not be seen simply as a private or personal matter, but in fact has profound consequences for how cities function socially and economically: 'Sexuality matters in city life not just because it is a question of personal choice that matters to the individual, but because it is deeply implicated in the ways that cities *work* as sites of social and economic reproduction.' It is this role played by sexuality in the operation and functioning of urban life that is often neglected within urban theory. For instance, sexuality rarely figures within accounts of the everyday or the mundane within urban theory; as if sexual desire was outside of the everyday; rather than an important dimension of everyday urban life. Moreover, I would contend that sexuality is strongly bound up with economics: certain desires are valued and recognized; others are stigmatized. Having examined the institutional constraints that shape and govern the possibilities for research on sexualities and urban space, I wish to turn my attention to queer theory and politics, which have been prominent in challenging the neglect of sexualities within social theory. I suggest that queer is therefore of central importance in helping us to theorize the relationships between sexuality and urban space.

What is 'queer'?

Queer theory and politics are central to theoretical attempts to understand the relationship between cities and sexualities. However, queer is a notoriously slippery concept that escapes easy definition. To offer a prescriptive definition of queer, to pin down, or to judge what can be described or labelled as queer inevitably leads to exclusions. There are multiple and different uses of queer and the ways in which it can be understood and it is important to reflect on these before we can examine how queer can inform our understanding of the city. Despite the fact that there is a politics of contention associated with these multiple understandings and uses of queer, I would like to retain an open-ended, holistic sense of the term to maintain some sense of its complexity, rather than authorizing a 'correct' use of the term. What can count as queer and what may be seen as queer will also vary across time and in space. Moreover, different theoretical vocabulary exists to examine the politics of sexuality in different geographical contexts; for instance, in their discussion of research on sexualities in Brazilian geography, Silva and Viera (2014) note that 'queer' has not been used widely in Brazilian geographies of sexualities. To offer a genealogy of queer inevitably anchors one geographical context as being foundational, and others as peripheral or 'regional'. Queer emerged as a critique of assimilationist strategies of mainstream, established gay organizations; specifically as a counter to the narrowing down of political claims of such organizations to issues such as gay marriage; acceptance in the military and the embrace of the market. Queer was associated with the rejection of labels and fixed identity categories such as 'gay' and 'lesbian' and has been associated with a politics of collation and solidarity around social and sexual justice across and beyond fixed category categories. Edelman (2004: 17) has suggested that: 'queerness can never define an identity; it can only ever disturb one'. Despite Edelman's assertion, a number of people have identified with queer because it affirms their cultural political practice and values. For instance, many bisexual activists have been attracted to queer because it promised a more bi-friendly space for activism and scholarship than had hitherto been the case within some lesbian and gay groups, political organizations and academic camps. The adoption of queer as an identity to describe political actions and activism was significant in a generational narrative (Plummer 2010) that can be seen to distinguish activists emerging in the 1990s from the previous generation above. A key feature of the generational narrative was the focus on coalitional work – not just coalitions across boundaries of identity categories such as gay and lesbian, or heterosexual – but also across other differences including race and class (Gould 2009).

Queer is sometimes used as an umbrella term – for non-normative sexualities – yet this can be problematic, particularly if it is used as a synonym for 'gay' as it can render identity categories such as transgender, intersex, lesbian and bisexual invisible. However, for the purposes of this chapter I would like to use queer as shorthand here, in a similar way to feminist. While there are many types of feminism, each with their own political agenda and critical lens, which may in some instances be in opposition to one another; it is also the case that queer contains within it different values, subject positions, and political priorities. There is no singular understanding of queer – it is a travelling concept that changes its shape and meaning from one geographical context to another. This can make it difficult when discussing queer across different contexts; and across different times and generational viewpoints. For O'Rourke (2011: xv): 'queer is a term that brings problems of translation, transmission, transport and dissemination with it as it travels across borders'. While certain political possibilities and understandings of queer may get lost in translation from locality to locality, others may be opened up as activists articulate new visions of queer appropriate to their political struggles. Queer is thereby subject to being transformed as well as being transformative and challenging. A utopian vision of queer articulated by Munoz sees it as means of attempting to imagine and forge new more socially just intimate worlds (Munoz 2009), about not accepting social injustices. Here, queer is not simply deconstructive and resistant, but can be envisioned as also about forging collectivities in difference. Queer is therefore not simply about challenging and unsettling norms, but forging new more socially just collectivities and solidarities.

Queer and the city

A sense of queer's rejection of fixed identity categories is captured in this statement from Hubbard describing in a holistic way the extent to which sexualities impact on urban experience *regardless* of one's identity categories:

> Whether we identify as lesbian straight, gay or bisexual (or perhaps none of these categories), sexuality has a profound impact on our experience of the city given urban space can open up, or close down, different spaces for sexual expression.
>
> *(Hubbard 2012: 204)*

This statement hints at what a queer 'take' on urban life may entail. It means always questioning the assumptions and norms that we may be prone to make about the urban experience. It means being open to the fluidity of sexualities, and challenging attempts to categorize and fix sexual identity categories into place. Having discussed the complex and unruly nature of queer, how can we apply a queer perspective on urban life? What would a queer city look like? In an essay thinking through the relationship between queer and the city, it is significant that Bell (2001) uses the term 'fragments for a queer city', which suggests an understanding of queer as something insurgent, ephemeral and unsettling. Manalansan IV (2015: 567) has spoken about queer in terms of value, of those whose lives are rendered as being without value within the capitalist city: '[where] Queer is also about the productive possibilities of people who are left out, displaced, or dispossessed because of their position within the landscapes of the normal.' This important distinction marks out queer as distinct from 'gay' and 'lesbian'. Manalansan's discussion of working class

Filipino gay men in the USA, in terms of their being queer, is that queer is about shaping a life in the context of exclusion from mainstream urban life because of their gender and sexual identity as well as their ethnicity and class position. Manalansan draws attention to two critical issues – the social and economic basis of queer; and queer as a relationship with notions of normality and normativity. In this regard, we need to be attentive to challenging what we mean about normativity, and how queer critique can centre, reproduce, esssentialize and fix 'normativity' while seeking to challenge it. Manalansan's research shows the contribution of queer in bringing critical attention to those whose sexual subjectivities contribute towards their marginal social and economic status within the city. It brings critical attention to questions of value and the importance of sexuality, and the need to live a rewarding and non-oppressive intimate life as a human need alongside other basic needs such as shelter, water, food, and meaningful work.

In a landmark essay on the relationship between sexual desire, urban space and power, Califia (1994: 205) has spoken of the city as 'a map of the hierarchy of desire, from the valorized to the stigmatized. It is divided into zones dictated by the way its citizens value or denigrate their needs'. Califia goes on to argue that: 'In the city there are zones of commerce, of transit, of residence. But some zones of the city cannot be so matter-of-fact about the purposes they serve. These are the sex zones–called red-light districts, combat zones, and gay ghettoes… [T]here is no city in the world which does not have them. In part because of these zones, the city has become a sign of desire: promiscuity, perversity, prostitution, sex across the lines of age, gender, class, and race.' (Califia 1994: 205). It is this boundary-crossing that is seen as particularly troubling and troublesome, as unsettling of the hierarchies of urban life. This makes an intersectional approach to urban life as essential to any thinking through of the relationships between sexuality and urban space, even though I recognize criticisms of intersectionality as a concept (Erel *et al* 2008; Puar 2004). These include that intersectionality reproduces a narrow grid of identities that runs counter to the messiness of queer's resistance to fixed identity categories and can be used as problematically diversity management framework (Puar 2004) and that intersectional critique can be depoliticizing (Erel *at al* 2008). Sexual desire can be unruly, and seen as a negative force by authoritarian regimes, particularly those obsessed with spatial containment and separation such as in an apartheid city, where boundaries and social-spatial segregation were central to the city's continued operation (Tucker 2009; Visser 2003b).

Re-reading Califia's essay on the hierarchy of desires in urban life makes me think immediately of recent scandals in the United Kingdom around the sexual exploitation of vulnerable young working class women in northern English towns such as Rotherham and Rochdale, which have been the focus of widespread and intense media attention and scrutiny. Here we must recognize that the media attention on these cases of Rotherham and Rochdale can be partially explained by the fascination and fear about boundary-crossing that such cases generate. Such cases reveal much about whose bodies and whose sexualities are constituted as being normative and whose are marked as deviant: with the vilification of white working class girls and women, and the hypersexualization of British Asian men of

Pakistani descent. They are also significant for reproducing an intersectional politics of fear – the threat of contagion and pollution, and anxieties about 'race', class and gender. Through the Rotherham and Rochdale child sexual exploitation scandals, we can also see the ways in which sexuality can be significant in the negative place imagery of certain cities. Within media reports on these scandals, the name of the town often stands in for the child sexual exploitation scandal so that each town's name becomes synonymous with child sexual exploitation and used as media shorthand for discussions of child sexual exploitation. While such issues have generally been under-studied by urban geographers, there has been some work on child sex abuse and the representation of place – for instance Cream's (1992) work on the 1987 Cleveland (UK) child sex-abuse scandal, which focussed on the actions of social services in removing a number of children from the families under suspicion of child sex abuse in the north-east of England. Cream's pioneering essay remains one of the very few in human geography to explicitly discuss child sex abuse work in relation to the politics of place, and to understand the gendering of media discourses on the case and the way that social workers and health professionals were represented by the media. Willis *et al* (2015) have written movingly and persuasively of what they see as human geography's silence around child sex abuse (CSA), arguing that 'CSA is a "present absence" in human geography' (2015: 4). They go on to suggest that 'there is a very strong likelihood that being sexually abused as a child will impact upon a person's access to, use of, behaviour in and perceptions of place and space'. Willis *et al*'s paper attests to the amount of work still to be done to investigate the sexual politics of urban space, despite the growth of the field and the proliferation of research in this area. It also demonstrates the limitations of current theorizing of the sexual politics of urban life and shows the necessity of ongoing critique of the institutional constraints that govern the possibilities for sexualities research in human geography and urban studies (see also Silva and Viera 2010).

While urban geography has done much to focus on the sex zones, there is still much to do as Bell (2001) has argued to examine the sexual politics of urban life beyond these sex zones. Debates still circulate about the place of 'gay villages' and neighbourhoods within research on urban sexual geographies (Brown 2014) and whether such a conceptual lens, renders invisible other ways of seeing the relationships between sexuality and urban life. For instance, Cattan and Vanolo (2013) argue in their essay on queer clubbing spaces in Paris and Turin that they can be characterized by a 'shared characteristic of ephemerality'. Here, I do not wish to legislate against certain types of research, or certain objects of research, though it is important to recognize the importance of research on gay and lesbian neighbourhoods in helping to secure the legitimacy of sexuality as a topic within urban studies.

Queer, value and urban life

Discussing academic critiques of what he terms the de-queering of 'gay space', Hubbard concludes his study of the relationship between cities and sexualities by arguing that: 'cities may be getting sexier, but only *profitable* expressions, representations and performances are being encouraged' (Hubbard 2012: 208–209). Hubbard thereby draws critical attention to notions of value: whose sexualities are valued within urban space? Whose sexualities can be recognized and celebrated? Whose sexualities are rendered invisible and are represented as a threat to the urban norm?

While it is important to recognize the way in which certain more-respectable invocations of gay space – the most profitable, least threatening to the mainstream – have in some instances been folded into city-branding discourses that promote the city as diverse and tolerant, it is also important to note that 'not all cities are equally open to sexual diversity' (Hubbard 2012: 206). Marches against Pride events in some cities, for instance Budapest (Renkin 2015) or the banning of Pride marches in some cities e.g. Poznan in Poland (Gruszczynska 2009) demonstrate that the relationship between cities and sexualities are complex and multiple. We also need to think critically beyond simply thinking about these issues in terms of for and against sexual diversity.

While some cities may be encouraging the development of gay commercial districts and hallmark events such as Pride to enhance the city's place brand as being cosmopolitan and as welcoming difference (Bell and Binnie 2004), it is evident from the Poznan case study that this is not the case everywhere. Moreover, as my earlier discussion of the Rotherham and Rochdale child sex abuse scandals demonstrates, we must also recognize how sexual politics can impact negatively on a city's place image. Given the high media profile of child sexual exploitation scandals in shaping public discourse about these urban areas, it is surprising that urban studies and urban theory generally remain silent on these issues. Within research on sexualities and urban space there are similarly silences and omissions – as Willis *et al*'s (2015) discussion of CSA in Human Geography demonstrates. Another pressing issue related to my earlier discussion of geographical context and place in the discussion of work on sexualities and urban space, is the need to examine the transnational dimensions of the urban experience and sexualities (a largely neglected issue within studies of cities and sexualities), which leads on to the final substantive section of this chapter.

CASE STUDY 23.1 DE WATERKANT 'GAY VILLAGE', CAPE TOWN

In the Apartheid period, which ran from 1948–1994, the regulation of sexualities and sexual practices was fundamental to the operation of he Apartheid state. For instance the Mixed Marriages Act 1949 outlawed marriage across the racialized categories established by the Apartheid state. Moreover, male same-sex activities were restricted. Section 9(3) of the post-Apartheid constitution (1996) explicitly prohibited discrimination on the basis of race, gender, and sexual orientation. South Africa thereby became the first state in the world to enshrine anti-discrimination on the basis of sexual orientation into its constitution. As Visser (2003a: 175) states:

> In just one-and-a-half decades, the country has gone from persecuting and arresting individuals with same-sex dispositions to allowing them to marry and adopt children. No country has so radically changed its position towards homosexual individuals or the perception of itself in such a short period of time.

De Waterkant is a gay commercial space in Cape Town that emerged in the post-Apartheid period. Its emergence became symbolic of the recognition of diversity in post-Apartheid South Africa. However, its sexual geography also reproduces a geography of racialized, gendered and classed exclusion. Gustav Visser (2003b) argues that while the Waterkant district has affirmed the identity of affluent, white gay men, this has been at the cost of racialized, gendered and classed exclusions. As Tucker (2009: 61) states: 'the popular representations of Cape Town as a liberated space for queers, with a culture similar to that represented in spaces in the West, does not reflect the diversity that exists within the city'. The commercial spaces are primarily targeted towards gay men and there are fewer lesbian-orientated venues (which is not specific to Cape Town). De Waterkant reproduces social and spatial segregation from the Apartheid period. This means for instance, that the district remains both physically and economically inaccessible to those black townships and Coloured areas. At the same time, the space emerged as international tourism to South Africa grew in the post-Apartheid period, with the end of the country's international isolation. This facilitated the growth of De Waterkant, with the return of expats who had experienced gay-oriented commercial spaces in Europe and North America. Visser highlights the importance of transnational connections for the development of De Waterkant. He also argues that the gentrification of the neighbourhood reproduced social and economic exclusions. We can see that approaching the geographies of inclusion and exclusion within De Waterkant can help us to think through the queer possibilities for the study of the city. On the one hand we can argue that focusing on the social economic and broader gendered, racialized and classed politics of sexualities in an urban space such as De Waterkant is queer in the sense of Manalansan IV – it means putting under a critical lens questions of social and economic justice relating to sexual justice. However, it could be argued that focussing on De Waterkant as a sex zone runs counter to the spirit of what I argued earlier in this chapter. In focusing on this zone it renders invisible different models of same-sex intimacies and identification in different urban spaces. Tucker's (2009) work has suggested that De Waterkant remains a separate world for queer black men within Cape Town's townships. We also need to recognize the importance of women's prisons for same-sex intimacies; and the spaces for same-sex intimacy afforded by migrant labour practices in the gold-mining industry. It also brings our questions of recognition to the fore: whose subjectivities, whose agency is recognized, or rendered intelligibly 'queer'? Or, in other words, how is queer seen, or unseen.

Transnational queer urbanism

In this section, I suggest that adoption of a transnational critical lens can be productive for developing our understanding of the relationship between sexualities and urban space. As Vertovec (1999: 447) has argued: '"Transnationalism" broadly refers to multiple ties and interactions linking people or institutions across the borders of nation-states.' Adopting a transnational critical lens on queer urbanism means being attentive to the importance of the circuits of flows of people, ideas and commodities across national borders. It means challenging the uncritical taking-for-granted of the nation-state as the starting point for analysis of social processes. It also means foregrounding of the migrant experience within a sometimes-abstract discussion of globalization. Hubbard has argued that global or world cities have a distinctive sexual geography, suggesting that 'world cities are not merely major markets for sexual consumption, pornography and prostitution but are the hubs of a global network of sexual commerce around which images, bodies and desires circulate voraciously'. (Hubbard 2012: 176). This suggests that such cities play a disproportionate role in the commercialization of sexualities: in constituting profitability from sex. For instance, Hubbard argues that sex workers in world cities are disproportionately from a migrant background: 'The extent to which migrant workers dominate sex markets varies massively, but there is certainly plentiful evidence to suggest that non-native and/or illegal migrant workers make up the majority of sex workers in major world cities.' (Hubbard 2012: 190–191). Such accounts are significant for drawing critical attention to migrant lives, subjectivities and agency and shaping urban life. Manalansan argues that for working class queer Filipino men in New York, their identities are structured transnationally. Moreover, he argues that their lives are precarious and characterized by disorder and messiness:

> What arises in these stories is a powerful vernacular form of queer world-making amidst precarious conditions... this very sense of impossibility or untenable chaos lies at the heart of a queer worlding. Their lives exemplify the ways in which their lives embody a queer and wayward art of being global. The tangled and untidy nature of their lives and experiences precisely positions them in a queer location outside the realm of the normative, the possible, the desirable and the orderly.
>
> *(Manalansan IV 2015: 576–577)*

Manalansan's work attests to working-class Filipino queer men and he frames his discussion of their subjectivities, and practices of queer world-making – their experiences – as a 'messy urban vernacular'. A focus on queer as a process of world-making, an orientation towards the global, can help bring to the fore queer lives that perhaps would be overlooked by accounts that centre on self-identified gay men. A focus on practices of 'queer worlding' need not be restricted to so-called 'world cities' such as New York and Manila that are the focus of Manalansan's study. A transnational critical lens can also be appropriate and productive in the study of sexualities in 'regional' or small cities (Myrdahl 2013). A narrow focus on 'world cities' can also obscure how we can think more holistically about the transnational and international connections between a wider range and typology of cities. We can recognize the importance of transnational circuits of queer mobility in a range of cities. Focusing on queer transnational urbanism means drawing on more mundane, less visible dimensions of queer lives – for instance the mundane actions of activists in offering support and solidarity for queer lives threatened or endangered elsewhere. It means recognizing sexuality as a significant dimension of urban life, not as a spectacular rupturing of the mundane.

CASE STUDY 23.2 SPACES OF QUEER SOLIDARITY: THE MARCH FOR EQUALITY

This case study discusses how challenging the heteronormativity of urban public space can sometimes be met with violent opposition. It draws attention to the fact that city authorities may use violence to banish those contesting the heteronormativity of urban space. In 2005 a proposed March for Equality was banned by city authorities in Poznan, Poland – a ban that was later deemed unlawful. Sixty-eight activists marching in defiance of the ban were arrested by the police. Grusczynska (2009) has argued that this event led to solidarity actions across a number of Polish cities. The march constitutes a queer march for solidarity, which has roots in feminist organization. The march is a constellation of actors and activists each worth their own agendas, that go way beyond gay or lesbian identity politics. These include secularist activists, environmental issues, and those focusing on issues of fat activism and ageism. Such events constitute queer event-based solidarities that articulate a vision of queer urban life in Poznan that represents a challenge to gender-normativity and heteronormativity. This march with its strong focus on coalitions represents one key feature of queer solidarities in difference (see Binnie and Klesse 2013 for further discussion). The banning of the Poznan equality march, as well as equality marches in other Polish cities such as Warsaw and Krakow, led to significant mobilization among pro-LGBT and queer activists across Poland, but also abroad – particularly within the European Union. Some activists in The Netherlands and Germany made use of existing twinning arrangements as a platform for transnational civic activism around the issue of gender and sexual politics in Poland (Binnie 2014). This shows the limitations of 'methodological nationalism' uncritically taking the nation-state as the starting point for the generation of theory about social movements and activism. These events demonstrate the value of investigating the transnational connections and flows of ideas, values, people and finance across national borders, and their place in shaping struggles around sexual politics in different urban political contexts. The Poznan case also attests to the significance of activism around sexual politics in articulating the right to the city of those whose gender identity and sexuality does not conform to societal norms.

Conclusion

In this brief discussion of sexuality and urban theory, I have outlined four key themes, which I argue, can be particularly productive in understanding the relationships between sexualities and urban space. First, I have argued that highlighting the legitimacy of knowledge production and how sexuality research is recognized and valued are of central importance in challenging the relative marginal status of sexualities research in urban theory. Moreover, I have suggested that questions of place and geographical context are key to developing theoretical understandings of the sexual politics and geographies of the city. Second, I have suggested that queer politics and theory can contribute towards a more holistic view of sexualities, beyond fixed identity categories. Third, I have argued that the issue of value can help us to think through an integration of the sexual and the economic to examine how non-normative gender identity and sexuality renders them socially and economically marginal and vulnerable within urban space. Finally, I called for greater critical attention to be paid to the *transnational* politics of queer urbanism. Discussing Mananlansan's study of working class queer Filipino men in New York and Manila, I have suggested that Manalansan's notion of 'queer worlding' can help us to understand queer urban space beyond the national.

In outlining these four themes, I have sought to identify recent trends within research on sexuality and the city, and signpost possible future directions. To develop the field further, and to expand and develop theoretical understandings of the relationships between sexuality and urban space, there is a pressing need for more comparative research across a range of geographical contexts (see Chapter 8, this volume) – and for more joint and collaborative work between scholars examining the sexual politics of the city in different urban contexts (such as by Nash and Gorman-Murray 2014). However, in calling for such a trajectory it is imperative to recognize the institutional and other power relations that govern and shape the possibilities for producing such research.

Key reading

Collins, A. (Ed.) (2007) *Cities of Pleasure: Sex and the Urban Socialscape*, London: Routledge.
   A reprint of a special issue of *Urban Studies* on sexualities and urban space, which contains papers examining the regulation of sexualities in a wide range of urban contexts; this is still one of the few journal special issues that concentrates on the theme of sexuality and urban space.

Hubbard, P. (2012) *Cities and Sexualities*, London: Routledge.
   This is the most authoritative single point of reference for research on sexuality and the city. It is both comprehensive and accessible.

Tucker, A. (2009) *Queer Visibilities: Space, Identity and Interaction in Cape Town*, Chichester: Wiley-Blackwell.
   Tucker's monograph provides a theoretically informed, empirically rich study of the urban geographies of queer men in Cape Town and how they are fractured by the politics of race and class.

24

SUBURBAN

Roger Keil

Introduction

Urban theory has traditionally been an exercise of explaining the city from the inside out. For the time being, when we say 'suburban', we imply a simple definition: Suburban refers to suburbanization as the movement of people or economic activity to the peripheries of existing cities and the simultaneous expansion of built (and used) urban space; it also relates to suburbanism as a particular way of life.

From the first thinkers in urban theory, for example, Weber (1921), the Chicago School (Park *et al* 1925), their Marxist challengers in the 1970s (Castells 1977), the city has been conceptualized from the centre outward. There have been counter-tendencies, for example Lefebvre (2003), and the Los Angeles School (Scott and Soja 1996; Soja 1996; Sieverts 2003; for a recent overview see Judd and Simpson 2011), yet generally the peripheral, or suburban experience has been treated as derivative, or even deviant. Urbanization is still mostly imagined as a concentric expansion of centrality – of space, functions, economies, people. In turn, suburban studies (a large and growing field) has usually reproduced the centre-periphery split. Their disregard of the centre has been equally problematic (for recent overviews see Harris 2010; Forsyth 2012; Jauhiainen 2013).

We are being told that we have entered the urban century. More than half of us are now living in what by all manner of statistical criteria would be considered 'urban' environments. Surely, we can safely assume that most inhabitants of the planet have now become city or town dwellers, i.e. they live in a built and social environment that both the ordinary observer and the statistician would recognize as urban. More importantly, many of us have subscribed to a theory, first pioneered by Henri Lefebvre in the 1960s, that there is a 'virtual object' we can call 'urban society' in which even those areas that are not obviously urban in form are part of a process of 'complete urbanization' (Lefebvre 2003). For Lefebvre, this emergence of urban society is linked to a series of dialectical processes of implosion and explosion (see 2003; Brenner 2014). In this chapter, I will argue that not enough attention in this intellectual tradition has been paid to the constituting properties of the explosive antithesis of the implosive thesis. My argument has two aspects. I will first posit that the urban revolution is really a suburban revolution and (urban) society today is characterized by a horizontality that has its roots in global suburbanisms (Quinby 2011). Most urbanization today is suburbanization; urbanism as a way of life has increasingly been overshadowed by multiple suburbanisms as a way of life. I will secondly claim that the – quite undialectical – nuclearphysical notion of explosion and implosion leaves the point of origin in the centre intact; by contrast, I will argue that the generalized horizontality of global suburbanisms forces us to rethink, at least to a degree, the core beliefs of urban studies: we need to abandon the centripetal-centrifugal orientations of urban theory (that allow explosions only as a movement away from an imagined centre) and take Lefebvre's virtual object seriously as a generalized condition of post-agrarian and post-industrial possibility: urban society as a set of multiple centralities that are neither geographically nor functionally linked to a pre-existing traditional core.

Suburbanization was considered part of the American dream. And so has been its crisis. From Brazil to India, from South Africa to China, American suburbanization provides an important model for the global urban middle-classes and elites (Hamel

FIGURE 24.1   Southern California Cloverleaf (Source: Roger Keil).

and Keil 2015; Keil 2007). But the imagination of a globally scaled and themed suburbanity now comes from elsewhere, not from the American heartland. As Ananya Roy has reminded us, non-central worlding practices now include strategic suburbanizations as part of a multilogue of interreferencing in the global south itself (Roy 2009, 2015). Roy notes accordingly: 'the edges of metropolitan regions are a patchwork of valorized and devalorized spaces that constitute a volatile frontier of accumulation, capitalist expansion, gentrification, and displacement' (2015: 342). Let us, for a moment, contemplate the state of the suburban world.

The suburban prospect: observations on global suburbanization

Land covered by urban uses will have tripled in just thirty years between 2000 and 2030. This ubiquitous trend will imply significant consequences for climate change, biodiversity, and so forth (Seto *et al* 2012). Two aspects stand out in this perspective. First, this projected urbanization will be extremely unequal, with China and Africa absorbing the lion's share of global urbanization during the next generation. Second, extrapolating from current trends, we can expect that the majority of the urban expansion we face in the next generation or two will not mirror the current trend towards re-urbanization, widely celebrated in the urban North, but will continue to be sprawling in nature. This will take wildly different forms in places such as China or Turkey, where more dense, high-rise type suburban developments are driven by large scale state sponsored programs, and (most of) Africa or India, where we see continued and continuous lower density suburbanization prevail (Bloch 2015; Gururani 2013; Mabin 2013; Wu and Shen 2015; Wu 2013). At the same time, there will also be additional suburban extension of cities in North America, Europe and Australia, where half-hearted growth controls can barely withstand the tide of further sprawl, often now driven by aggressive infrastructure development, including airports, private motorized transportation and public transit that now reaches the far corners of the commutershed and extends the urban region (Addie and Keil 2015).

In line with the argument put forth in this chapter, I am therefore assuming that the majority of the new urban forms and processes brought on by this tremendous wave of urbanity will be suburban in nature. There will be some massing of built form and concentration of urban process but given the uneven distribution (Africa and China) and specific path dependencies of those urbanities, the majority of urban settlement will not just be geographically peri-urban, outside of the current perimeter of urban areas in those areas but also – at least in Africa – mostly low to middle density in nature. Even in China's massive tower neighbourhoods that are pushed outward concentrically, ringroad after ringroad (Fleischer 2010), the majority of new and existing urbanites will live in suburban constellations (Keil 2013).

That is, under the conditions of current trends in technology, capital accumulation, land development and urban governance, the expected global urbanization will necessarily be largely *sub*urbanization. The rise of the suburb(s) is a global phenomenon but those peri-urban extensions to existing urban form and structure are locally built and re-built. There is a large diversity in process and outcomes of suburbanization. This diversity in form and structure is reflected in diversity of concepts in a 'world of suburbs' (Harris 2010). Suburbs also must not be considered an organic product of aggregate consumer choice but the composite outcome of planned and regulated interactions of public and private actors in very different systems of land economies and governance (Hamel and Keil 2015).

Instead of speaking about an Urban Age (with all its problematic uses, see Gleeson 2014), or an urban revolution (Lefebvre 2003), we need to at least consider the prevalence of processes and forms that warrant deploying the notion of a '(sub) urban revolution'. If, as Harvey (2007) has noted, 'the planet as building site' collides with the 'planet of slums', suburbanization in its myriad forms, and that includes all manner of post-suburbanizing processes in existing and emerging peripheries, will be the defining characteristic of the shared human experience in the remainder of this century. We can therefore speak of global suburbanization and the proliferation of global suburbanisms – that is suburban forms of life (Drummond and Labbé 2013).

This conjuncture also denotes a break in the history of urbanization: from increasingly dense and mostly industrial Manchester or Chicago, or finance capitalist New York, London or Tokyo, we gravitate to suburban expansion everywhere. As this 'exploding' urbanization (to use Lefebvre's (2003) space age, astrophysical language) proceeds, 'imploding' centralizations will continue to take place, leading to a kaleidoscopic global (sub)urban landscape. We see a great multiplicity of form in this rapidly suburbanizing world. But there is also much blurring and bleeding among and between the different world regions. In a post-colonial, post-suburban world, the forms, functions, relations, etc. of one suburban tradition get easily merged, refracted and fully displaced in and by others elsewhere, near or far.

Suburbanization appears now as original and constitutive, not a derivative element of urbanization. To return to Harvey's turn of phrase that he based on Mike Davis, we can speak now of a 'planet of suburbs'. This means that the epistemic lens through which we can know the urban changes in significant ways. There is a growing need to put Lefebvre (2003) – who had bemoaned the lack of urbanity in the periphery in the 1960s – from his head onto his feet: now the periphery is not just about deficit of urbanity anymore – albeit there are some that are linked to the impoverishment of older, 'inner ring' suburbs in North America and Europe, for example (Charmes and Keil 2015). The geographical periphery, the outskirts, the peri-urban claim new kinds of centralities (Lefebvre 2003).

In this new landscape, spatial peripheralization goes along with social marginalization and/or the sequestration of privileges both in classical gated communities and in newer forms of segregation, such as condominium complexes even in suburban hubs. Suburbanization as a general phenomenon now has to be understood on a continuum that stretches from crisis management – Keynesianism, *crisis switching* (Harvey 1982; 2003), or in other words, '*the suburban solution*' (Walker 1981) – to being recognized as a cause of crisis (subprime crisis, etc.). While most of the talk by well-known urbanists is about the creative cores of the urban revolution, suburbanization is now an increasingly dominant arena for the socio-material process of the production of urban space.

We need to ask now whether, once all is suburban, it still makes sense to speak about suburbanization as a process, suburbs as a place and suburbanism as a distinct way of life. Or should we begin talking about a postsuburban world? These questions are not rhetorical and deserve an honest and productive answer. There are fundamentally two ways to respond. First, in this dramatic period of peripheral urbanization that spans the world, we are seeing a ramping up in scale and intensity of the process of suburbanization, daily expansions of actual suburban areas and a world wide spread of distinctly suburban forms of everyday life (Drummond and Labbé 2013). In that sense, at least, we can continue to speak of suburbanization.

The second answer is a bit more evasive but still true and useful, as it shifts our perspective and course of action in urban theory-building to a process of pervasive postsuburbanization. Nick Phelps and Fulong Wu have argued that postsuburbia is a 'composite picture' because of its global manifestations, divergences and mixing of land uses, less predictable geographic forms, new politics, new work-residence relations, and discordant land use (Wu and Phelps 2008). We now see a more reflective process that consists of both the retrofitting of existing suburbs and the continuing emergence of 'original' suburbanization: we can call this postsuburbanization, as it points beyond the traditional form of linear peripheral development.

In parallel with post-suburbanization, we can also observe a shift in the *meaning* of the peripheral suburban form. Suburbanization, long seen as the material process of Keynesian-Fordist economics with its virtuous circles of programmed mass consumption, has now become a prime terrain of neoliberalization. Paul Knox has noted the emergence of a particularly vulgar form of capitalist spatial fix which, over the past two decades, has given neoliberal US-society a specific (sub)urban form of ostentatious consumption of space, nature and resources through large-lot monsterhomes, often in secluded treed domains (Knox 2008). Jamie Peck (2011) has pictured suburbs in the US as privileged sites for the roll out of actually existing forms of neoliberal governance. These observations fit neatly with the notion that the urban periphery has been the 'dream space' of post-Fordist restructuring. The emergence of clusters of high tech or other production, logistics and commercial complexes in wider urban regions is now visible all across the globe, but was first noted in places like Silicon Valley or Orange County in Southern California (Soja 1996). Similarly, the global city economy, tightly linked to the global post-Keynesian evolution, is 'going up the country' in search for space and function. An increasingly 'oscillating' growth dynamics between the financial and business service core and the sprawling office park hinterland has been the trademark of that period of urbanization (Keil and Ronneberger 1994).

Suburbs are now a global phenomenon, not just the specifically American spatial fix. Ethnoburban developments are one of many forms of that development. The competence and institutional arrangements of integrating and differentiating populations normally identified with the inner city has now moved to the suburbs

(Saunders 2010). But these often-informal peripheries don't follow a pre-trodden path towards some more formal, fixed and central. While they are 'cities-in-waiting' as they gel into something more permanent, this permanence will not be anything like the outcome of industrialization-based urbanization in its twentieth-century western form (McGuirk 2014; Keil 2011). Yet, the global periphery is now not seen as the periphery of globalization anymore since its 'worlding' has been part of the overall shift of the geographies of theory towards a post-western historical geography of urbanization (Roy 2009; 2015).

It is difficult for urban studies to concede that most urbanization is now suburbanization which creates a theoretical conundrum. Through textbooks and in introductory university courses in urban studies, we have consistently fostered the notion that 'normal' urbanization tends to produce centrality, and even the sprawling metropolitan areas of the twentieth century have been imagined as agglomerations first and foremost. The suburbs have been largely written into a theoretical space of derivative and second-order quality. They were either considered on their way of becoming urban eventually (look at the history of streetcar suburbs, for example); or worse: unchangeable and timeless in their geographical and semantic location. Suburbanization rebels against us urban intellectuals and our sense of self, as we cannot imagine the suburban to be part of our personal lives or worthy of serious investigation: they lack the centrality from where meaningful discourse springs. They are the colony to the centre from where we usually construct our narratives and theorizations. The suburbs in the urban imagination appear largely as terra incognita: an unknown world, a colonial space. In the worst versions of this mental displacement in urban theory, they become the equivalent of Tolkien's 'Mordor': the place where the horrors of industrial and urban society are played out in the peripheries of the mind. The suburbs have been made extraordinary and pathological, always distant from but ideally on their way to a normalized urbanism of centrality. They are not part of a dominant geography of theory (Robinson 2006; Roy 2009).

Part of the problem of the theoretical colonization of the suburban in critical urban studies stems from a certain reading of Henri Lefebvre's (1968; 1996) 'right to the city' as a claim to the 'inner' city (see Chapter 22, this volume). This results on one hand from the conjoining of the perceived deficits of suburban life with the conceived superiority of urban (mostly European) life. On the other hand, critical urban theory has sided naturally and rightfully in the past with the disadvantaged and oppressed masses of core urban populations whose precarity was set in one with the crisis of the city. Manuel Castells (1976a) is exemplary here as he cites suburbanization as a strategic element of the US model of economic dualism, class domination and spatial segregation, built on the historical articulation of metropolitanization, suburbanization and social-political fragmentation but then proceeds to focus solely on the oppressed inner city neighbourhoods of the 'wild city'. Today, we know better than confusing urbanity with geographical centrality.

Progressive and critical urbanist thinking was further biased against taking suburbs seriously by what one can term 'the tragedy of suburbia', by which I mean the expulsion of the working class from the centre of the city, for which Baron Haussmann's Paris has largely stood in as the prime example (Lefebvre 2003: 109–10). From this shift arose a disdain for the suburbs on the side of the political Left and progressive scholars (Harvey 2013; Davis 1990). But the mainstream urbanist tendency, especially in the tradition of Jane Jacobs (1961) and Richard Florida (2002), to celebrate the creative centrality of cities over what has been largely perceived as a barren suburban field – almost enemy territory to urbanists – has been even more influential.

The need for a reorientation

In the light of this theoretical unease with all things suburban among much critical and mainstream urban theory, there is a dire need to abandon historically privileged spots for observing urbanization. This includes both the privilege of the urban centre and the privilege of the Global North. Many downtowns are now Disneyfied, predictable and uniform. They are often gentrified monocultures, filled with green infrastructures (New York's High Line), privileged spaces of recreation and highly stratified economic machines with creative workers in charge, and provided for by a precarious service class. Many suburbs, on the other hand, are raw, unpredictable and diverse. They are zones of vast differentiation in built environment (Charmes and Keil 2015), mobility access (Keil and Young 2014), politics (Young and Keil 2014) and socio-economic structure (Hulchanski 2010).

FIGURE 24.2 Southern Ontario Cloverleaf (Source: Roger Keil).

CASE STUDY 24.1 LOS ANGELES

The suburban archetype, never quite reached by any other western city, was twentieth-century Los Angeles, a city so expansive that it already dwarfed other urban centres across the world in size and automobile density early in the century. The 'Southland', called '19 suburbs in search of a metropolis' as early as the 1920s, seemed to lack any sense of centrality. Built on the grid of its extensive regional rail system, it was the ultimate auto city where freeways connected the various parts of the suburban American dream into a metropolitan whole, both logistically and conceptually (Keil 1998).

Yet, throughout the twentieth century, even in the rapidly suburbanizing United States, Los Angeles remained an exception. Only in the 1980s did scholars detect a marked shift. Not just in Los Angeles but also in other American urban regions, monocentrality gave way to a more polycentric pattern of development. This created the specific moment in the writing on cities that it was possible to call Los Angeles both the 'ultimate American city' and the 'capital of the twenty-first century'. Equipped with a mix of traditional political economy and 'postmodern' speculation, authors of the so-called LA School – a real and imagined joint project of Los Angeles scholarship and practice – turned our attention away from the standard centrality of the urban narrative to the suburban expanses of the Southland, and began to reassemble urban theory from the outside in. The enthusiasm for this eversion was only partial. Beyond the cultural turn towards the peripheral, the location of industries in sub- and exurban industrial districts, for example, was a matter of fact and hardly reason for exuberance on the urban theory front (Scott and Soja 1996). What's more, the suburbs were once again acknowledged mostly as the locus of real existing revanchism, right-wing populism and homeowner associations masquerading as grassroots democracy (Davis 1990). Yet, two scholars in particular took the empirical look through the 'sixty mile circle' around Los Angeles as the starting point of a more fundamental rethink of the urban world in which we live. UCLA's Edward Soja (1996: 239–239), clearly a pioneer of a postmodern take on urban geographies noted on the basis of his Southern California experience:

> Exopolis, the city without, to stress their oxymoronic ambiguity, their city-full non city-ness. These are not only exo-cities, orbiting outside; they are ex-cities as well, no longer what the city used to be. Ex-centrically perched beyond the vortex of the old agglomerative nodes, the Exopolis spins into new whorls of its own, turning the city inside-out and outside-in at the same time, unraveling in its paths the memories of more familiar urban fabrics, even where such older fabrics never existed in the first place.

Michael Dear, Soja's crosstown geographer colleague, then at USC, ultimately took the identification of Los Angeles with postmodernism furthest. Dear and Dahmann (2011: 74) approached the suburban in Los Angeles in typically deconstructivist manner: 'There is no longer such a thing as suburbanization, understood as a peripheral accretion in a center-dominated urban process'. Henri Lefebvre (1996: 208) confirmed its limitlessness in this famous observation: 'There is something stupendous and fascinating [about Los Angeles]. You are and yet are not in the city. You cross a series of mountains and you are still in the city, but you don't know when you are entering or leaving it.'

At the same time, what is true for the increasingly inappropriate and impossible use of Northern models to make theoretical sense of global urbanization patterns, is particularly visible in the suburban areas of the globe's emerging urban fields. Surely, the classical North American case remains relevant to our understanding of the history and geography of suburbanization and its theoretical implications. Not least Los Angeles is the gift that keeps on giving. Yet, perhaps contemporary Cape Town's Mitchell's Plain of squatter settlements more than Los Angeles's post-war middle class community of Lakewood is the true icon of contemporary suburbanization. In Asia (McGee 2015), Africa (Mabin 2013) and Latin America, 'the post-colonial suburb' (Roy 2015) is taking shape as a global form and they point back to those suburbs of the North, be they in Europe (France, the UK, Germany, Portugal, etc.) or Canada, where large numbers of immigrant communities have begun to settle on the fringes of the cities.

FIGURE 24.3   Barra da Tijuca, Rio de Janeiro (Source: Roger Keil).

A call for suburban studies?

Yet this plea for recognition of the planet's burgeoning peripheries as an object of study and source of theory that is not subordinate to the 'normal' centre, is not a call for a genre of 'suburban studies'. Instead, it is an argument for an urban studies as centrally concerned with processes of suburbanization and suburbanisms as non-derivative processes in the move towards 'complete urbanization' (Lefebvre 2003). Such an extended urban studies takes the suburban as a historically evolving human geography in which more questions are posed than answers given. If being urban is increasingly the shared condition of our humanity, for many if not most of us, this takes place in what we would recognize as a suburban space. So why, and how do we speak about suburbs and suburbanization today?

First and foremost we must avoid a replay of the western colonial bias in urban studies. New developments and dynamics originating from the margins and peripheries defy the traditional dependencies of outsides from insides, suburbs from cities, and expand our understanding of the dialectic of the urban process. Theorizing suburbia now needs resituating it in wider concerns for new geographies of urban theory and 'planetary' (Brenner and Schmid 2014; Brenner 2014), 'ordinary' (Robinson 2006), 'postcolonial' (Roy and Ong 2011) and comparative (IJURR symposium on urban theory, forthcoming) configurations (to mention a few).

We need to engage in an exercise of provincializing the (North American) suburb. It is not the only model from which theory springs. The inclusion of urbanization in the Global South in the debate on global suburbanism(s) is not a mere addition of more empirical cases to an existing script of peripheral expansion. This implies the acknowledgment that the script of urban theorizing has to be rewritten from scratch. The suburbs are a good place to start that intellectual journey. It is from the emerging geographies of non-European and non-American (sub)urbanity that the architectures of urban theory await rebuilding. Making the suburban central to urban theory in this manner is, then, a contribution to a 'reloaded urban studies' (Merrifield 2012) that creates openings beyond the traditional dichotomies of the field (see also recent work by Schafran 2013; and by Walks 2013).

Having thus moved the theoretical object of urban studies out from under the obscurity that covered its peripheries, we can return to the notion of an ongoing, and one might argue accelerating, proliferation of urban society (Lefebvre 2003). In the tradition of Lefebvre, urban society has conventionally been seen as having spread in an implosion-explosion dialectics (Lefebvre 2003: 14). In Lefebvre's nuclearphysical metaphor, explosion sees asteroids flee the exploding star into the far corners of an ever-expanding universe. This – figurative – movement from the centre outward poses a conceptual issue at the heart of the 'reloaded' urban studies: if the explosion leads to 'disjunct fragments', how are they held together? What are the processes and experiences they share? What is the causality that leads to their 'projection'? Lefebvre's image seems to suggest that the antithesis of the explosion is a consequence, not a simultaneous moment, of the implosion thesis. Viewing' these two issues together, it appears as if the 'numerous, disjunct fragments' of spatial expansion are, in fact, mere products (not producers), derivative (of central processes), and perhaps even disjointed from one another (rather than linked through processes of state action, (i.e. policies), capital accumulation through the production of space, and private authoritarian suburban ways of life (Ekers *et al* 2012). We need to imagine, then, urbanization as a process of myriad explosions, expansions and contractions, some of them recentering, some of them extending the urban into the general space of society.

Manifesto of the suburban revolution

Urban society is not a virtual object anymore as it has materialized many ways that were still emergent when Lefebvre formulated his original theory. Yet it has also obliterated our conceptual categories, often beyond recognition. There are no *essential* differences anymore between centres and suburbs. The suburban is not derivative. The politics of everyday life becomes central to the new materiality of the urban periphery. We can therefore also speak about the right to the suburbs (Carpio *et al* 2011). There is no more catching up to do. There is no deficit that has to be overcome: 'Within the postmetropolis, a reconfiguration of the meaning of urbanism is taking place. Centrality is increasingly reserved for immaterial networks of power and the physical assets that support them, while bodily existence within the postmetropolis is increasingly moved to the periphery' (Quinby 2011: 75). A 'suburban-like order of horizontality and dispersal' reflects the 'horizontal strategies of surveillance, dispersal, and consumption' that contextualize much of politics and governance in the post-suburban landscape of today's neoliberal capitalism (Quinby 2011).

We end up with a web (or we could say assemblages) of post-suburbanization processes (see Chapter 4, this volume). In their totality, they add up to a suburban involution. We witness not a complete dissolution but a reconsolidation of the urban fabric, even a balancing, and in any case a rejection of classical functional or conceptual dichotomies such as live-work. The process of post-suburbanization entails a profound re-scaling of the relationalities and modes of governance that have traditionally regulated the relationships between centre and periphery in the suburban model (Phelps and Wood 2011; Phelps and Wu 2011; Hamel and Keil 2015). The '(post)-suburban involution' has its own 'functional, spatial and rhythmical diversification of post-suburbanisms' (Charmes and Keil 2015); we are talking about 'a complexification and folding in of suburbanizing cultures and rationalities, as opposed to a linear process of centrifugal succession' (Peck *et al* 2014c: 389).

From this springs a wide-ranging field of research and practice, in these overlapping sections: physical form/built environment; social relationships/ governance; and (sub)urban political ecologies. These sections are roughly compatible with the classical sub/urban disciplinary approaches of planning, sociology/politics and geography/environmental studies. Three lenses lend further focus to these studies: governance, land and infrastructure: i.e. the dimensions through which suburbanization is driven forward, redefined, and through which globally diverse suburbanisms are engendered.

Conclusion

Studies of the suburban have always been present in the pursuit of knowledge about cities and regions. Yet, they have often been peripheral to the core themes in the field. The gaze of urban researchers has been sometimes distracted to stray across the suburban expanse, but it would invariably return to the centre of the city where insights on the essence of the urban were assumed to be had. This centre reflex, as we may call it, has eclipsed, for the most part, the formation of suburban constellations even in the second half of the twentieth century, when the production of space in the periphery was the driving force in the accumulation process in many capitalist societies. None the less, a strong continuity of urban studies that focuses specifically on the suburban now exists. It has already contributed to better understanding of metropolitan infrastructures, the production of land, regional governance and suburban ways of life. More recently, studies on the 'classical' suburban cases in Anglo-Saxon settler societies (USA, Canada, Australia) have been joined by broad examinations of the global variety of suburban constellations (Keil 2013). In this chapter, then, we have built the case for a theoretical reorientation in urban studies. Propelled by both empirical developments that see continued settlement of the world's urban peripheries and by a critical reading of a Lefebvre-inspired urban theory, I argued for a decentralization of urban thought and practice. The demand for an urban society, as hypothesized and predicted 'virtually' by Henri Lefebvre almost half a century ago, cannot be conceptualized anymore from the notion of the suburban deficit from which the 'right to the city' was developed. The centrality of the suburban form and society must be part of that reconsideration. The urban revolution, in this sense, does not have a central orientation but opens the city to (sub)urban society!

Key reading

Hamel, P. and Keil, R. (Eds) (2015) *Suburban Governance: A Global View*, Toronto: University of Toronto Press.
  This edited book owes its selection to its recent publication and innovative perspective. It offers the first comprehensive overview of suburbanization worldwide, especially under the perspective of suburban governance. Contributors to this volume are among the leading researchers on global suburbanization and their chapters offer an encyclopedic overview of suburbanization today.

Harris, R. (2010). 'Meaningful types in a world of suburbs', *Suburbanization in Global Society (Research in Urban Sociology)*, 10: 15–47.
  Like papers by Forsyth (2012) and Jauhiainen (2013), Harris's overview presents the current state of the art in suburban research. The Canadian geographer presents an analytic perspective on the existing world of suburbanization, the various conceptual

approaches that make sense of it, and the possible futures in globally suburbanized landscape that we might expect.

Lefebvre, H. (2003) *The Urban Revolution*, Minneapolis: University of Minnesota Press.

This theoretical masterpiece by the great French urban thinker Henri Lefebvre offers the key stimulant behind the current lively debate on what he called 'extensions of the city', which he presents as part of a dialectic of 'implosion' and 'explosion' through which 'urban society' emerges. *The Urban Revolution* is not a specialized text on the suburban, but the key to the understanding of its theoretical significance.

25

SUSTAINABILITY

Rob Krueger

Introduction

In recent years we have witnessed, yet again, a major rupture in the confidence of urban theory to explain 'urbanization' (Brenner and Schmid 2014; Wachsmuth 2012). It seems apropos here to paraphrase Manuel Castells (1976a: 59) who, a generation ago, wrote: 'the one subject that remains unexplored in urban theory is its subject matter'. More recently, at a conference that brought together representatives from the so-called Los Angeles, Chicago, and New York schools of urban theory, Dear and Dahmann (2011: 77) remarked that, 'we urgently need to revise our obsolete theoretical and analytical apparatuses'. Calling for a more ambitious move, Neil Brenner (2009) argues that we need to excavate twentieth-century urban theory for alternative cartographic and conceptual frameworks. This is to say we need to focus our analysis on the *transformative* potential brought by the current social formation, but also engage in a focused analysis of the attendant systematic exclusions, oppressions, and injustices wrought by them.

In an effort to make my own modest contribution toward these goals, in this chapter I want to begin developing the following argument: sustainable development is a 'still frame' of urbanization, a snapshot in a larger and longer-term process of urbanization. Following Wachsmuth (2012: 82), I argue that cities are not so much things as processes. Wachsmuth writes, 'Stop the process at any point in time, and the discreet spaces you observe are the extent of urbanization.' For me, sustainable development can be conceptualized as one of these 'still frames'.

For some time, urban theorists have ignored urban sustainable development or exiled it to the margins of raw empiricism – sometimes for good reason. The way scholars of urban sustainable development have conceptualized their subject has often been 'journalistic' reportage, green boosterism, or examined uncritically as a set of best practices or 'techniques'. Indeed, there are some excellent analyses on key factors that shape urban sustainable development practices or what forms of governance are critical to 'success' but these are outside broader theoretical debates about urbanization and urban theory. In this chapter, I will outline the main points of urban sustainable development literature to better align it with important urban questions of our time, including well-distributed economic prosperity, ecological integrity, and social equity.

To develop my argument I present a 'reading' of urban sustainable development as a policy discourse and set of ideological practices that position it as a coherent historical moment, an active and coherent set of social and material relations, and a process of creative destruction for overcoming contradictions of capitalism. Put simply, understanding urban sustainable development as a 'still frame' of the process of capitalist urbanization may enable us to more clearly elucidate the practices that bring about social injustice in the city. Why? Because urban sustainable development is one of a few, if not the only, policy discourse that explicitly calls for social equity to be built into the decision making process (cf. Krueger and Buckingham 2012). I develop this argument by first examining, in general, some of the key aspects of sustainable development, and then, in particular, urban sustainable development, including how it developed as a concept and grew from a set of discreet practices to a vision of urban form. I go on to reflect on a body of work – 'critical sustainability studies' – that has offered a rejoinder to the ambitious, if not overly optimistic, claims of success in the urban sustainable development arena. The point of this discussion is to establish urban sustainable development as a coherent set of social relations that have come together to shape urban form. In the next section, I reflect on how urban theory has evolved to examine urbanization as a process of contradictory, periodized, and paradigmatic social relationships, and how urban sustainable development could offer new insights into the urbanization of injustice.

Contextualizing [urban] sustainable development

In contrast to many of the other concepts presented in this book, the origins of urban sustainable development are not found in, nor motivated by, urban concerns, *per se*. Urban sustainable development came out of a broader discourse of sustainable development – at a global scale – and must first be understood in the larger context of human-environment relationships, specifically to those outlined in the scarcity and growth debates of the 1970s and 80s. For example, the International Union of Conservation of Natural Resources (IUCN) first coined the phrase 'sustainable development' in the 1980s, but conceptualized it not in terms of development as we know it, but in terms of (non-human) species conservation and biodiversity. It was in 1987, when *Our Common Future* (or 'The Brundtland Report') was published that linkages between economy, environment, and society were first expressed (see Figure 25.1). In particular, Brundtland focused on intergenerational equity and the notion that the economy should be understood as, and even organized by, the larger context of human-environment and human-development relations. Figure 25.1 and this conceptualization of it have often been referred to as a 'three-legged

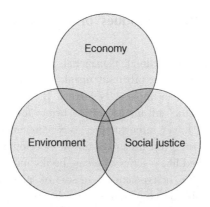

FIGURE 25.1   'Three-legged stool': linkages between economy, environment, and society (Source: Rob Krueger).

stool'. There is plenty written on this history so there is no need to dwell on this further here (c.f. Satterthwaite 1999).

For my argument it is important to note that the urban, or what was then called the 'local sustainability' agenda, emerged from the Earth Summit of 1992, where advocates introduced two core concepts to local development agendas. First, the concept of 'subsidiarity' was, they argued, a central feature of sustainability governance. Subsidiarity maintains that many decisions regarding a 'sustainable' society should be made at the scale where the most relevant decision makers are, which are those who are closest to the people. To address this a second set of practices, Local Agenda 21 (LA 21), was presented by Earth Summit architects. LA 21 principles recognized that local governments play a key role in bringing about sustainable development across different spatial scales. Here it was emphasized that local authorities needed to make considerable changes to their policymaking approaches in order to incorporate a broad-based, multi-stakeholder planning process focused on balancing economy, social equity and the environment. The approach's architects believed the success of LA 21 would also turn on the ability of a local authority to redirect its policies, laws and regulations to align with the principles that emerged from the planning efforts.

While rhetoric from sustainable development's architects suggested certain procedural interventions (e.g. LA 21), at sustainable development's conceptual core was an effort to address the tensions between the 'technological optimists' who argued that finite resources were not a barrier to growth, and the 'limits to growth' or 'steady state' accounts that suggested otherwise. As a result, urban sustainable development emerged from this moment as a set of procedural and managerial approaches for bringing into alignment urban development and limits to growth.

Best practice and the 'techniques' of urban sustainable development

Constructed as a set of procedural, managerial, and technical responses to the problem of 'unsustainability' the urban sustainable development literature emerged outside urban theory. For example, UN Habitat II, which took place in 1996, brought to the fore the advantages cities bring for addressing sustainable development: density, lower infrastructure costs, efficient recycling, and public transport. Scholarly writing on urban sustainable development in the 1990s and early 2000s often looked like a series of recipe books and 'how to' manuals, and included case studies for 'best practice' – a sort of 'pick-and-mix' toolbox for enacting urban sustainable development. As a result, urban sustainable development did not contribute to a body of theoretical literature; instead it amounted to a set of practices for delivering more sustainable urban forms and lifestyles. In this section, I will briefly and selectively develop this point and demonstrate my points with exemplars from the literature.

Identifying the problem of 'unsustainability': ecological footprint analysis

Ecological Footprint Analysis emerged in the 1990s and stirred broad interdisciplinary debates around the resource flows and waste sinks in cities. Today, any undergraduate student – or faculty member – can go online and in ten minutes learn the hectares of land required to support their lifestyle (see Case study 25.1 for an analysis of my own 'lifestyle' quantified this way). Twenty years ago the progenitors of this approach, William Rees and Mathias Wackernagel, showed that cities and industrial regions appropriate ecological carrying capacity of an area vastly larger than the areas that they physically occupy. Complicating this, was the continued rapid urbanization of modern society; people were moving from rural areas to cities. From these analyses other scientists and planners should determine what interventions could take place to reduce the human footprint on the land, especially in terms of raw materials like timber, food, and CO_2 absorption. Other issues involved consumption, the resource sinks, and other building materials like cement, and steel. As a result, these analyses brought to the fore key areas for intervention.

Policy responses: developing and advertising 'best practice'

In the mid-1990s, with the help of ICLEI, an international non-government organization dedicated to urban sustainability, many cities embarked on LA 21 processes. It started in 1994 with the publicity around the Aalborg Charter and subsequent high profile events. At this time thousands of cities and towns, through ICLEI's 'European Cities and Towns Network' began to organize, deliberate on, and implement LA 21 processes and plans. Still others were taking urban sustainability as defined by the material input-output model. Solid waste programs were instituted to 'redirect' waste, trams were constructed, bicycle infrastructure

CASE STUDY 25.1 MY ECOLOGICAL FOOTPRINT

My Ecological Footprint - Quiz Results

If everyone on the planet lived my lifestyle, we would need:

= 2.57 Earths

Petitions by Change.org| Start a Petition »

Reduce your footprint

MY FOOTPRINT IN GLOBAL HECTARES BY CONSUMPTION CATEGORY

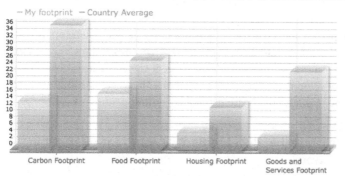

Total: 40.32

created and/or improved, urban greening was promoted, sewer overflows were counted, and municipal expenditures on sustainability initiatives were tallied. To measure these successes cities developed indicator projects – typically part of the LA 21 process. To assist local authorities in their sustainability endeavors, knowledge networks were funded by the European Union (e.g., Local Authorities Self Assessment of Local Agenda 21 or LASALA). ICLEI-US initiated the 'Cities for Climate Change' campaign and network, and ICLEI-Europe developed 'Eco-Budget', a 'financial' accounting system for a municipality's environmental expenditures that was piloted in Europe and disseminated to Asia and the US. Academic planner Tim Beatley (1999) wrote volumes on best [technical] practices for urban greening and sustainability. The point here is that the urban sustainability agenda evolved as a set of discreet practices that, if implemented, could help reduce resource inputs and mitigate the externalities of urban life, i.e. create more sustainable cities. Following the discourse of ecological modernization, these interventions when implemented amounted to a series of technological fixes. Moreover, it became a discourse of the up-and-coming or 'progressive' city: city boosters wanted be part of the movement (Krueger and Gibbs 2007).

From function to form

In the late 1990s and 2000s the urban sustainability agenda moved beyond these discreet fixes and sought to redefine the urban form. No longer was it enough to redirect a preponderance of waste from municipal landfills, urban space had to be reorganized to support the new sustainable development imperative. Concepts of 'compact urban development', 'smart growth', 'urban greening', 'transit oriented development', 'eco-cities', and others emerged in the planning literature and policy documents in the US and Europe. 'Starchitects', economic development and planning gurus, and others seized on consulting opportunities to recreate sustainable quarters in cities: The 'Vauban' in Freiburg, Germany; Fells Point in Baltimore, Maryland, USA; Wilhelminaplein in Rotterdam, the Netherlands; the Gowanas Canal in Brooklyn, New York; and London's East End are a few exemplars. Indeed, urban sustainable development evolved beyond a set of discreet techno-interventions that supported the goals of sustainable development while offering a vision of urbanization that extended beyond material throughputs to something more comprehensive with utopian ideals. Indeed, these practices coalesced into broad urban redevelopment strategies designed to address certain social and ecological and economic problems that were defined by various political forces.

Urban sustainable development: a coherent moment in urbanization?

Over the past dozen or so years a growing body of literature has offered trenchant critiques of the urban sustainability literature and practices, especially as it moved from discreet interventions to quarter- and city-wide development strategies. For example, Gibbs (2002) showed that many LA 21 processes, which were supposed to examine and develop comprehensive goals for the sustainable development agenda, were in many cases focused on the economic development aspects of sustainable development. Similarly, environmental justice scholar Julian Agyeman has shown in his work that justice and equity remain absent–urban sustainable development plans (Agyeman 2005). In other words, urban sustainable development, in general, and urban natures, in particular, are fabrications of white, upper-class ideologies. Finally, in their edited volume Krueger and Gibbs (2007) bring a variety of essays together to show how intertwined urban sustainable development initiatives are with contemporary capitalist social relations (see Case study 25.2). They note that the 'sustainable city' had emerged as a coherent set of policies designed to promote the stability of the 'three-legged stool', but as a matter of practice the 'sustainability stool' acted more as an engine of local and regional economic growth (see Figure 25.2).

CASE STUDY 25.2 URBAN SUSTAINABLE DEVELOPMENT IN BOSTON

In 2003, Boston Mayor Thomas Menino set out on an ambitious track to 'green' the city. GIS maps show how all neighborhoods are subsequently enjoying more tree cover, green spaces, and bicycle lanes. The air is demonstrably cleaner now, with the introduction of hybrid taxis, city fleet vehicles, and, especially LPG and hybrid buses. Wind-power and natural gas have replaced coal-fired power plants, too. Menino really set himself apart from other US mayors by promoting green roofs and establishing the first municipal green building standard – Article 37 of Boston's municipal zoning standards. He also provided incentives to redevelop South Boston and East Boston, areas where derelict buildings stood to create innovation districts for the new economy. These mixed-use, highly dense developments rebuilt and rebranded communities. But, being green is much more than all of these practices. According to a presentation by Menino in 2005, this agenda meant: 'stimulating our growing clean tech economy'. Urban sustainable development, as a collection of practices, results in a distinct form of urbanization.

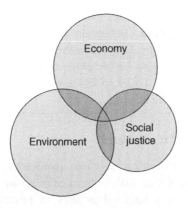

FIGURE 25.2   'Sustainability stool': linkages between economy, environment, and society (Source: Rob Krueger).

Over the past few years this growing body of work has evolved into a coherent literature often referred to as 'critical sustainability studies'. The references above are key ones, but other important discussions are ongoing and address many of the themes in this book: such as privatization, the commons, infrastructure, neoliberal urbanism, right to the city, and the like (see key reading at the end of this chapter). This proliferation of work, the themes it addresses, and the way commentators frame it suggests that urban sustainable development and urban theory may have reached a point where a discussion is now not only feasible, but both desirable and extremely useful; I will get to why shortly.

To summarize: in this first part of the chapter I have covered the history and evolution of the urban sustainable development literature. In the early days of the research and practice urban sustainable development responded directly to its historical roots: limits to growth. As a result, it emerged as a procedural and managerial approach to social change that existed largely outside the literature on social change, especially urban social change. However, as sustainable development evolved as a worldwide phenomenon, the urban sustainability literature and the political economy literature collided in ways that have provided some fruitful insights. For example, we can see how neoliberal visions of economy influence (or creatively destroy), rather than eschew, the progressive potential of the urban sustainable development discourse. In fact, this theme is prominent and well developed with rich conceptual and empirical accounts in the critical sustainability studies literature. I think it's time we go further and bring these two threads together in a more fruitful and comprehensive way: we need to understand sustainable urban development as a 'stop frame' in capitalist urbanization that has shaped urban form and the process of urbanization around the world.

Urban sustainable development and the narrative of a 'transformative' urban theory

In this second part of the chapter I offer a rather selective 'reading' of urban theory and its historical development. My intent is to capture some key conceptual moments to support my points about urban theory's current condition and where urban sustainable development can be plugged in and incorporated as a set of paradigmatic social relations. To set up and sustain the argument I reflect on some current thinking around 'the urban'.

The study of 'the city' is obsolete?

'The city' as a unit of observation came into question in the in the 1970s. Urban studies scholars redirected their analyses from cities as containers of social relations to questions of *how* cities operated as sites of production, circulation, and consumption of commodities. In particular, they focused their attention on the socio-spatial organization, governance systems, and patterns of resistance and conflict in and around the urban. These ideas challenged the assumptions of previous accounts of 'the urban' in that they questioned the trans-historical assumptions and metaphors earlier urban scholars relied upon. For this coterie of scholars, cities were decidedly capitalist forms, and, as such, were, like capitalism itself, fraught with ephemeral cycles, both virtuous and vicious, that at one moment brought synchronous movement and stability and, later, dissolved to challenge the once-dominant social arrangements and their attendant material forms. Let's explore some of these points a bit further.

Cities are inherently capitalist forms...

To invoke a phrase from Harvey's *Social Justice and the City*, these conceptualizations of 'the urban' sought to 'chart the path' for a rigorous line of conceptual and empirical argument and analysis that brought to the fore the intrinsically *social* nature of urban development and economic restructuring. Where their predecessors had relied on modernist notions of 'the subject' and on determinist metaphors regarding 'competition' borrowed from biology, these scholars sought to reveal these framings and the ideology of that they supported. In a paper published in *Economic Geography* on population (Malthus) and resources (Ricardo), Harvey (1974: 256) states:

> I am arguing that the use of a particular scientific method is of necessity founded in ideology, and that any claim to be ideologically free is of necessity an ideological claim. The results of enquiry based on a particular version of scientific method cannot consequently claim to be immune from ideological assault.

Invoking Marx – and not as an ideologically neutral position, quite the opposite – illustrates the point that Malthus's scientific analysis, which was 'derived' from natural law, was actually apologia for the extant state of affairs in England at the time. Similarly, he argued that Ricardo's effort was to support the release of the productive forces championed by the bourgeoisie. Thus, if these activities have social origins – origins that serve ideological ends, and not the outcome of some biological or chemical predisposition of people, there could be a way forward – a new path revealed – not only to a more just urban environment, but to a 'just' society.

Bringing this point – that ideological 'stop frames' of urban issues and the way development occurs – to cities, I argue that cities could no longer be understood as 'hosts' of activity; in fact, they were made in the image of capitalist forces that were exerted from within and outside urban 'boundaries'. For Brenner *et al* (2011: 3) cites were themselves viewed as intensively commodified: 'from buildings and the built environment, to land-use systems, networks of production and exchange, and metropolitan-wide infrastructural arrangements [were] sculpted and continually reorganized in order to enhance the profit making capacity of capital'. Wachsmuth (2012) recognizes this 'sculpting and reorganization' but insists we understand them as processes that are experienced by actors in everyday life.

Imagining, sculpting, organizing, and reorganizing the 'urban'

In contrast to Louis Wirth, who defined the city as a way of life built on three pillars – size, density, and heterogeneity – later urban scholars refused to define the city as an object or specific unit. Rather, they invoked a dialectical approach to explore the contradiction and tension of capitalist societies. Indeed, one key concern was how capitalism, despite these structural contradictions and tensions, continued to survive. From these analyses, *meso*-level concepts such as periodization emerged. The concept of periodization suggests that there is the capacity for multiple *capitalisms*: capitalist urban forms will continue to mutate as their attendant social relations seek to overcome the contradictions and tensions from previous periods of growth. Here, Lefebvre's notion of the city as *process* not form is important. For Lefebvre urban form, and therefore cities, reflects dominant, yet ephemeral, ideological representations and thus cities must be understood as historical categories that are constantly changing through the process of urbanization.

Relational approaches to 'the urban'

Most recently, urban scholars have begun to develop relational theories to express the process of capitalist urbanization. For Massey (2012: 4) this means that, 'places are what they are in part precisely as a result of their history of and the present participation in relations with elsewhere'. The concept of policy mobilities that has been advanced by the work of McCann and Ward (2012) illustrates the idea of relationality well: 'The concept derives from Deleuze and Guattari's work and speaks not to the static arrangement of set parts, whether organized under some logic or collected randomly, but the 'process of arranging, organizing, fitting together where an assemblage is a whole of some sort that expresses some identify and claims a territory' (McCann and Ward 2012: xv). Assemblages are inventions (see Chapters 4 and 15, this volume). They are made and unmade. They are affixed then disassembled. Here, the concept of assemblages, with its most rigorous application in the policy mobility literature seeks to understand 'how those involved in urban politics and policy making act beyond their own cities to practice or perform urban globalness and to articulate their cities in the world' (McCann and Ward 2012: xvii). Thus, cities are not objects, they are assemblages of objects, created by people to serve different purposes in different places and times. Yet, even though we invoke the 'assemblage' metaphor, we cannot lose sight of the 'stop frame' view of the city, the ideological representation of capitalist social relations that the built environment can reveal. My discussion now turns to how urban sustainable development can be one of those vantage points.

Urban sustainable development as a transformative account?

> I have not been able to find a single source that is against 'sustainability'. Greenpeace is in favour, George Bush Jr. and Sr. are, the World Bank and its chairman (a prime war monger in Iraq) are, the pope is, my son Arno is, the rubber tappers in the Brazilian Amazon are, Bill Gates is, the labour unions are.
>
> *(Swyngedouw: 2007)*

I like Swyngedouw's dry commentary on sustainability because it captures well the notion that urban sustainable development has become a powerful discursive and material process of capitalist urbanization over the past two decades. Indeed, outside a few hundred members of the US Tea Party, who is against urban sustainable development? Thousands of cities around the world, on every continent, have adopted the process of Local Agenda 21, so-called 'smart growth' or 'new urbanist' philosophies and practices, engaged in sustainability-visioning and planning, branded themselves as 'green', or just aligned themselves informally with the urban sustainability agenda. Like the information city, the ordinary city, naked city, the creative city, the fill-in-the-blank city, the sustainable city has emerged as a paradigmatic set of political economic relationships; that is it represents a coherent agenda, suite of policies and practices, and exists in both material and representational ways on the landscape (see Case study 25.2). It is a paradigmatic set of social relationships, not a paradigmatic city. Sure Freiburg, Portland (Oregon), or Växjö (Sweden) may be viewed as paradigmatic cities. However, the forms these cities have taken are the result of assemblages of ideas and practices that flowed across space and were re-territorialized in place.

As different types of cities across the world system are being repositioned within increasingly volatile, financialized circuits of capital accumulation, it is increasingly urgent to understand the implications of these for urbanization. Given the pervasiveness, and perversity, of the 'sustainable city' it seems unwise for urban scholars to ignore the concepts of and the material practices that have brought these urban forms into existence. It is true that some scholars have examined the effects of the financial crisis on sustainable cities. In these analyses, however, the causal arrow points one way: toward the effects on the sustainable city from the financial crisis wrought by neoliberal policies and institutions. In contrast, the approach I propose here would explore the possible theoretical interventions that come from turning the arrow toward the other direction. What crisis did urban sustainable development respond to? How did the 'rolling out' of urban sustainable development fuel or foil the normative vision that propelled it to the international policy agenda? In other words, what opportunities did urban sustainable development seek and how did their assembly, disassembly, and reassembly promote and constrain the alternative vision imagined by urban sustainable development?

Conclusion

> The whole is the untrue.
>
> *(Adorno 1951: 30)*

And this brings us back to where we started: to charting the alternative cartographies and creating the theoretical apparatus. If we leave urbanization, in general, and sustainable urban development, in particular, to the experts we will end up confident, yet our focus will be targeted on the 'broken windows' (see Case study 25.3). Sustainable urban development represents an ideal metropolis and yet has been bastardized through the process of implementation. Perhaps a close

---

CASE STUDY 25.3 MISSING THE 'METROPOLITAN' EXPERIENCE

[U]rbanization destroys individuality; and yet, out of this destruction, new forms of individuality, ones with more open-mindedness to ideas and cultures, especially radical ideas and cultures emerge. With urbanization, the windows are smashed and the modern person can breathe in the open air. What unfolds in this open air is the metropolitan experience itself, the experience of urbanism.

(Merrifield 2002: 10)

Where is this urban experience and why are we missing it? An exploration of urban sustainable development can render transparent the process through which alienation occurs.

For example, in the opening lines of his excellent book *Dialectical Urbanism* (2002), from which the above quotation is taken, Andy Merrifield utilizes the imagery of F. Scott Fitzgerald's essay 'My Lost City'. Merrifield uses Fitzgerald's prose and protagonist to explore changing urban realities and the implications of how people adapt to them. Fitzgerald's protagonist, Bunny, adapts well to the city, 'no longer the shy little scholar'. The city is dynamic, and, later, when the author next sees his protagonist, the city has changed, and Bunny's confidence had increased; but, so too, has his alienation.

Let's fast forward Fitzgerald's Bunny some eighty years into the future and place him in the context of the sustainable city. We could ask him: what makes your electric car sustainable? How is your state-of-the-art solar settlement sustainable? How is the tram extension sustainable? Where is the rush of new air that allows us to see beyond these objects of 'sustainable urbanism'?

---

examination of these processes and their material manifestations will reveal a new metropolitan ideal? And through this process we must not only examine what is out there, but what is, as experts in us. We hope through this analysis we can open up these various perspectives to allow for a sustained critique of capitalist urbanization and not give up on the multiple useful perspectives available to create a praxis of change.

Another feature of urban sustainable development is the explicit attention, in theory at least, to the social dimensions of urban development. As a conceptual framework, then, urban sustainable development could be used to show the injustice of development, with a particular focus on various forms of environmental injustice. Because cities often comprehensively adopt the three pillars of sustainable development while leaving out the social component, this framework is useful to hold decision makers accountable for their choices. If, as I argued above, that urban sustainable development is a process of urbanization that represents a coherent historical moment – an active and coherent set of social and material relations, a process of creative destruction for overcoming contradictions of capitalism – it can, then, I argue, complement existing accounts of urban injustice because the very goals of urban sustainable development explicitly include justice and equity in them. Historically, urban development schemes have been discursively supported by the fuzzy, yet deterministic, logics of conventional economic theory. It's one thing, for example, to employ Pareto Optimality and show the injustice that ensues from abstract winners compensating real losers; it's another to take as a starting point the concept of social justice or equity, for example, and show it is redirected in ways that seem legitimate, as in the case of urban sustainable development. For a generation, urbanists such as Smith, Merrifield, Massey, Lees, and Robinson, to name just a few, have been interested in the relationship between 'urbanization and injustice' (see Chapters 8 and 12, this volume). These scholars, as well as many others, have done excellent conceptual and empirical work to elucidate the processes and implications of capitalist urbanization. I hope the analysis that follows will add some small dimension to this excellent urban scholarship.

Key reading

Brenner, N., Marcuse, P., and Mayer, M. (Eds) (2011) *Cities for People, Not for Profit: Critical Urban Theory and the Right to the City*, London: Routledge.
 This book explores new ways of examining 'the urban' as an emancipatory space, a way of seeing positive social change, rather than merely an expression of the whims of capitalist accumulation.
Krueger, R. and Gibbs, D. (Eds) (2007) *The Sustainable Development Paradox: Urban Political Economy in the United States and Europe*, New York: Guilford Press.
 This book is the first to take on 'sustainable urbanism' from a political economy perspective. It shows the various forms of power, institutional relations, and economic logics that have brought urban sustainability from its fugitive status to the status quo.
Merrifield, A. (2002) *Metromarxism: A Marxist Tale of the City*, London: Routledge.
 This is an excellent book that explores different Marxian perspectives on the city. Each chapter is rich in description and theoretical nuance and is suitable for students of the city, no matter whether one is an undergraduate or an advanced scholar.

26

AFTERWORD

Mark Jayne and Kevin Ward

So, you've made it to the end of the book! Well done! Or, perhaps you have headed straight for the conclusion. We hope not! If this is the case, we suggest you pause and rewind. Go back to the introduction, and get a sense of the intellectual context for each of the chapters. Then, have a look over the chapters and see what each offers. Alternatively, if you have got to this point through more traditional means, what chapters did you read, and how did you read them? Did you choose a particular chapter because of an assignment, or in order to prepare for a lecture or presentation? Did you read across one, two or even more chapters because you have been synthesizing a collection of key thinking for your dissertation or thesis? Did you read some chapters in depth and focus on the detail of the arguments? Did you read any of the chapters more than once? Did you dip into a handful of chapters because the titles sounded interesting? Were any parts of the book just too difficult to understand…, or conversely did you get so engrossed that you have read this book from cover to cover? Well done if the latter is the case! However you have engaged with this book and its various chapters, we hope it has amazed, challenged, educated, excited, annoyed, frustrated, inspired and stimulated you.

Taken together, the contributions to this book offer you – the reader – an insight into the diverse and multiple intellectual agendas of contemporary urban theory. As you will hopefully have noted as you read the chapters, each is organized in a similar manner. In every one the author outlines the particular genealogy of approaches, ideas and understandings associated with a keyword. Together we hope the chapters have highlighted how arguments, debate and disagreements are the 'stuff' that fuels theoretical advancement, progression and refinement. We hope that you will agree that the amazing complexity and sophistication, depth and openness, multiplicity and variation of contemporary urban theory should be celebrated as a remarkable achievement. We believe that this collection offers an important statement of the most challenging, demanding, exciting, and innovative aspects of contemporary urban theory. This is despite the choice of keywords saying as much about the two of us as editors, as it does about the field that the chapter authors and ourselves have sought to represent.

For there is no field of urban theory pre-configured, no field waiting to be discovered. Rather, in as much as it is a bounded 'field', it is one whose boundaries are made and remade through books like these, for example, or through plenaries at academic conferences or lecture courses in universities around the world. And, over the years, its edges have been reworked, subject to contestation and refinement from inside and from outside. What has emerged is an inter-disciplinary, internally heterogeneous, multi-methods-based set of contributions under the auspices of urban theory. For some, this complexity, diversity and variation is a weakness. It is a problem to be solved. It is argued that there is now an 'urban theory of everything', diluting our capacity to understand and explain urban life. Of course, a competing body of thought suggests that such complexity and plurality is necessary. That is the disposition and orientation towards theoretical openness important for an urban theory that is alive to the making and remaking of a world of cities. We hope that the material presented in this book will help clarify your own thinking on this issue, or maybe like many others you have come to the conclusion that the critical tension at the heart of such debate is only ever going to be generative and productive, the intellectual impulse for the future reworking of urban theory and its boundaries.

Ask those who you believe have a passion for urban theory and they will certainly have felt varied emotions in and through their own engagement with critical perspectives on cities and urban life. In fact, we think that the contributors of this volume have done just that – and running through each chapter are the thoughts, ideas and reflections of theorists seeking to understand, make sense of and engage with their own research specialisms in order to give you a resource with which to make your own minds up about the importance and relevance of urban theory. In so doing, it appears that in writing their chapters authors have been able to remind themselves of how, why and what urban theory they do, and why they should celebrate their own contributions to a longer process of knowledge production.

We hope, of course, that for many of you the pages of this volume will inspire you to become the next generation of urban theorists. In time we hope it will simulate new thinking, to formulate new innovative ways of finding out about the city, and will in time influence diverse audiences in and beyond academia. In the introductory chapter we set the scene to help you understand the long traditions, trajectories of ideas, and strengths and weaknesses that have characterized the emergence and history of urban theoretical work. We also pointed to stubborn fault lines of contention and disagreement, while highlighting that argument and debate is often vital to sparking new fruitful avenues of theoretical innovation and new research agendas. We hope that you noted this explicit and also implicit terrain in the chapters you have read. If you have spent some time thinking through such ideas and arguments, then you now have the critical resources at your fingertips to begin to engage with the topics introduced in each of the chapters in more detail.

Please look at the key readings in detail and, if you are smitten by the urban-theory bug, in no time at all you will find yourself looking at the myriad sources found in the bibliography. Reading across the chapters and beginning to understand and engage with the way in which theory is produced and reproduced will be an enriching and rewarding experience. And if at this stage you don't think a professional academic life is for you, then do not fear, because engaging with urban theory is a challenge the benefits to which accrue not just in terms of directly benefiting your current studies. The analytical and critical skills that you have been developing, finessing and honing as you have read and reflected upon the chapters of this book constitute an on-going learning that will serve you well no matter what you do next with your life.

And for those of you looking forward to an academic career, we hope that this book has helped you see the intellectual returns to putting theory at the cornerstone of your own intellectual project. While empirical research might make your work easier to explain to your friends and family, or more accessible for diverse audiences beyond universities, we urge you not to shy away from trying to convince people of the importance of the power of ideas. Urban theory is our professional language that we draw upon not only to talk to our colleagues and peers, but is also the very thing that gives us a unique voice. Urban theory is the unique resource that underpins our response to the challenges that demand that academic thinking and rigor has relevance. In seeking to ensure our professional academic knowledge contributes to political, policy and popular debates and works towards developing progressive and socially just cities, we must work hard to ensure that the kinds of urban theory presented throughout this book are at the heart of such a project – within and beyond universities.

BIBLIOGRAPHY

Aalbers, M. B. (2016) 'The pre-histories of neoliberal urbanism in the United States', in C. Morel and G. Pinson (Eds) *Debating the Neoliberal City*, Aldershot: Ashgate.

Abler, R. J. S., Adams, R., and Gould, P. (1971) *Spatial Organization: The Geographer's View of the World*, Englewood Cliffs: Prentice-Hall.

Abrams, C. (1966) *Man's Struggle for Shelter in an Urbanizing World*, Massachusetts: M.I.T. Press.

Adam, B. (1995) *Timewatch: The Social Analysis of Time*, Cambridge: Polity Press.

Addie, J. P. D. and Keil, R. (2015) 'Real existing regionalism: the region between talk, territory and technology', *International Journal of Urban and Regional Research*, 39 (2): 407–417.

Adey, P. (2009) *Mobility*, London: Routledge.

Adey, P. (2008a) 'Airports, mobility and the calculative architecture of affective control', *Geoforum*, 39: 438–451.

Adey, P. (2008b) 'Aeromobilities: geographies, subjects and vision', *Geography Compass*, 2: 1318–36.

Adey, P. (2006) 'If mobility is everything then it is nothing: towards a relational politics of (im)mobilities', *Mobilities*, 1: 75–94.

Adey, P., Brayer, L., Masson, D., Murphey, P., Simpson, P. and Tixier, N. (2013) 'Pour votre tranquillité': ambiance, atmosphere, and surveillance', *Geoforum*, 49: 299–309.

Adorno, T. W. (1973) *The Jargon of Authenticity* (Transl. by K. Tarnowski and F. Will), Berkeley: Northwestern University Press.

Agard, O. (2008) 'La mélancolie urbaine selon Siegfried Kracauer', in S. Füzessery and P. Simay (Eds) *Le Choc des Métropoles: Simmel, Kracauer, Benjamin*, Paris: Editions de l'Eclat.

Agyeman, J. (2005) *Sustainable Communities and the Challenge of Environmental Justice,* New York: New York University Press.

Alderman, D. H., P. Kingsbury and Dwyer, O. J. (2013) 'Reexamining the Montgomery bus boycott: toward an empathetic pedagogy of the civil rights movement', *The Professional Geographer,* 65: 171–186.

Allen, J. (2011) 'Powerful assemblages?', *Area*, 43 (2): 154–157.

Allen, J. (2008) 'Powerful geographies: spatial shifts in the architecture of globalization', in S. Clegg and M. Haugaard (Eds) *The Handbook of Power*, Los Angeles: Sage.

Allen, J. (2006) 'Ambient power: Berlin's Potsdamer Platz and the seductive logic of public space', *Urban Studies*, 43: 441–455.

Allen, J. (2003) *Geographies of Power*, London: Blackwell.

Allen, J. and Cochrane, A. (2010) 'Assemblages of state power: topological shifts in the organization of government and politics, *Antipode*, 42: 1071–89.

Allport, G. W. (1954) *The Nature of Prejudice*, New York: Basic Books.

AlSayyad, N. (2004) 'Urban informality as a "new" way of life', in A. Roy and N. AlSayyad (Eds) *Urban Informality: Transnational Perspectives from the Middle East, Latin America and South Asia*, Lanham, USA: Lexington Books.

AlSayyad, N. and Roy, A. (Eds) (2004) *Informality: Transnational Perspectives from the Middle East, Latin America, and South Asia*, Lanham, USA: Lexington Books.

Amin, A. (2008) 'Collective culture and urban public space', *City*, 12: 5–24.

Amin, A. (2002) 'Ethnicity and the multicultural city: living with diversity', *Environment and Planning A*, 34: 959–980.

Amin, A. (2002) 'Spatialities of globalization', *Environment and Planning* A, 34: 385–399.

Amin, A. and Cohendet, P. (2004) *Architectures of Knowledge: Firms, Capabilities and Communities*, Oxford: Oxford University Press.

Amin, A. and Graham, S. (1997) 'The ordinary city', *Transactions of the Institute of British Geographers*, 22: 411–429.

Amin, A. and Roberts, J. (2008) *Community, Economic Creativity and Organization*, Oxford: Oxford University Press.

Amin, A. and Thrift, N. (2002) *Cities: Reimagining the Urban*, Cambridge: Polity Press.

Anderson, B. (2014) *Encountering Affect: Apparatuses, Encounters, Conditions*, Aldershot: Ashgate.

Anderson, K. J. (1991) *Vancouver's Chinatown: Racial Discourse in Canada, 1875–1980*, Kingston, ON and Montreal, PQ: McGill-Queen's Press-MQUP.

Anderson, N. (1961) *The Hobo: The Sociology of the Homeless Man* (First edition 1923), Chicago: University of Chicago Press.

Andreotti, L. and Costa, X. (Eds) (1996) *Theory of the Dérive and Other Situationist Writings on the City*, Barcelona: Museu d'Art Contemporani de Barcelona ACTAR.

Angell, E., Hammond, T. and Van Dobben-Schoon, D. (2014) 'Assembling Istanbul: buildings and bodies in a world city', *City*, 18 (6): 644–654.

Angelo, H. and Wachsmuth, D. (2014) 'Urbanizing urban political ecology: a critique of methodological cityism', *International Journal of Urban and Regional Research*, 39 (1): 16–27.

Angotti, T. (2013) 'Urban Latin America: violence, enclaves and struggles for land', *Latin American Perspectives*, 40 (2): 5–20.

Anholt, S. (2003) *Brand New Justice: The Upside of Global Marketing*, Oxford: Butterworth Heinemann.

Arantes, O, Vainer, C. and Maricato, E. (2000) *A Cidade de Pensamento Unico*, Rio de Janeiro: Vozes.

Arnstein, S. R. (1969) 'A ladder of citizen participation', *Journal of the American Institute of Planners*, 35 (4): 216–224.

Ascensao, E. (2015) 'Slum gentrification in Lisbon, Portugal: displacement and the imagined futures of an informal settlement', in L. Lees, H. Shin and E, López-Morales (Eds) *Global Gentrifications: Uneven Development and Displacement*, Bristol: Policy Press.

Askins, K. and Pain, R. (2011) 'Contact zones: participation, materiality, and the messiness of interaction', *Environment and Planning D: Society and Space*, 29: 803–821.

Attoh, K. (2011) 'What kind of right is the right to the city?', *Progress in Human Geography*, 35: 669–85.

Audier, S. (2012) *Neo-liberalisme(s): Une Archeologie Intellectuelle*, Paris: Grasset.

Augé, M. (1995) *Non-Places: Introduction to and Anthropology of Supermodernity*, London: Verso.

Autonomous Geographies Collective (2010) 'Beyond scholar activism: making strategic interventions inside and outside the neoliberal university', *ACME: An International E-Journal for Critical Geographies*, 9 (9): 245–275.

Auyero, J. (1999) '"From the client's point(s) of view": How poor people perceive and evaluate political clientelism', *Theory and Society*, 28 (2): 297–334.

Azuela, A. (1987) 'Low income settlements and the law in Mexico City', *International Journal of Urban and Regional Research*, 11 (4): 522–542.

Azuela, A. and Duhau, E. (1998) 'Tenure regularisation, private property and public order in Mexico', in E. Fernandes and A. Varley (Eds) *Illegal Cities: Law and Urban Change in Developing Countries*, London: Zed Books.

Back, L. (2007) *The Art of Listening*, Oxford: Berg.

Bakhtin, M. (1984) *Rabelais and his World*, Bloomington: Indiana University Press.

Barnes, T. and Duncan, J. (Eds) (1992) *Writing Worlds: Discourse, Text and Metaphor in the Representation of Landscape*, London: Routledge.

Barnett, C. (2008) 'Political affects in public space: normative blind-spots in non-representational ontologies', *Transactions of the Institute of British Geographers*, 33 (2): 186–200.

Bateson, G. (1956) 'The message "this is play"', in B. Schaffner (Ed.) *Group Processes: Transactions of the Second Conference*, New York: Josiah Macy Jr Foundation.

Baudrillard, J. (1983) *Symbolic Exchange and Death*, London: Verso.

Baxter, R. and Lees, L. (2008) 'The rebirth of high-rise living in London: towards a sustainable and liveable urban form?', in R. Imrie, L. Lees and M. Raco (Eds) *Regenerating London: Governance, Sustainability and Community*, London: Routledge.

Bayat, A. (2004) 'Globalization and the politics of the informal in the global South', in A. Roy and N. AlSayyad (Eds) *Urban Informality: Transnational Perspectives from the Middle East, Latin America and South Asia*, Lanham, USA: Lexington Books.

BBC (2012) 'Slough gets £200,000 to tackle shed homes'. BBC News, 31 August.

Beatley, T. (1999) *Green Urbanism: Learning from European Cities*, Washington, D.C.: Island Press.

Beaverstock, J. G., Lorimer, H., Smith, R. G., Taylor, P. J. and Walker D. R. F. (1999) 'A roster of world cities', *Cities*, 16, 445–458.

Bell, D. (2009) 'Winter wonderlands: public outdoor ice rinks, entrepreneurial display and festive socialities in UK cities', *Leisure Studies*, 28(1): 3–18.

Bell, D. (2001) 'Fragments for a queer city', in D. Bell, J. Binnie, R. Holliday, R. Longhurst and R. Peace (Eds) *Pleasure Zones: Bodies, Cities, Spaces*, Syracuse, NY: Syracuse University Press.

Bell, D. and Binnie, J. (2004) 'Authenticating queer space: citizenship, urbanism and governance', *Urban Studies*, 41 (9): 1807–1820.

Bell, D. and Jayne, M. (2010) 'Urban geography: urban order', in R. Kitchin and N, Thrift (Eds) *The International Encyclopedia of Human Geography*, London: Elsevier.

Bell, D. and Jayne, M. (2009) 'Small cities? Towards a research agenda', *International Journal of Urban and Regional Research*, 33 (3): 683–699.

Bell, D. and Jayne, M. (Eds) (2006) *Small Cities: Urban Experience Beyond the Metropolis,* London: Routledge.

Benjamin, S. (2014) 'Occupancy urbanism as political practice', in S. Parnell and S. Oldfield (Eds) *The Routledge Handbook on Cities of the Global South,* London: Routledge.

Benjamin, W. (2006) *Berlin Childhood Around 1900,* Cambridge, MA: Belknap Press.

Benjamin, W. (1999) *The Arcades Project,* (translated by H. Eiland), Harvard: Harvard University Press.

Bennett, J. (2010) *Lively Matter,* Durham, NC: Duke University Press.

Berlant, L. (2011) *Cruel Optimism,* Durham, NC, and London: Duke University Press.

Berman, M. (1992) *All That is Solid Melts Into Air: The Experiences of Modernity,* London: Verso.

Berry, B. (1985) 'Islands of renewal in seas of decay', in P. Peterson, (Ed.) *The New Urban Reality,* Washington, D.C.: The Brookings Institution.

Binnie, J. (2014) 'Relational comparison and LGBTQ activism in European cities', *International Journal of Urban and Regional Research,* 38 (3): 951–966.

Binnie, J. and Klesse, C. (2013) 'The politics of age and intergenerationality in transnational lesbian, gay, bisexual, transgender and queer activist networks', *Sociology,* 47 (3): 580–595.

Bissell, D. (2014) 'Encountering stressed bodies: slow creep transformations and tipping points of commuting mobilities', *Geoforum,* 51: 191–201.

Bissell, D., Adey, P. and Laurier, E. (2011) 'Introduction to the special issue on geographies of the passenger', *Journal of Transport Geography,* 19: 1007–1009.

Blanco, J., Bosoer, L. and Apaolaza, R. (2014) 'Gentrificación, movilidad y transporte: aproximaciones conceptuales y ejes de indagación', *Revista de Geografía Norte Grande,* 58, 1–23.

Bloch, R. (2015) 'Africa's new suburbs', in P. Hamel and R. Keil (Eds) *Suburban Governance: A Global View,* Toronto: University of Toronto Press.

Blok, A. (2013) 'Urban green assemblages: an ANT view on sustainable city building projects', *Science and Technology Studies,* 26 (1): 5–24.

Blomley, N. (2008) 'Making space for law', in K. Cox, M. Low, and J. Robinson (Eds) *Handbook of Political Geography,* London: Sage.

Blomley, N. (2004) *Unsettling the City: Urban Land and the Politics of Property,* New York: Blackwell.

Bloor, D. (1999) 'Anti-Latour', *Studies in the History and Philosophy of Science,* 30: 81–112.

Blyth, M. (2013) *Austerity: The History of a Dangerous Idea,* New York and Oxford: Oxford University Press.

Bondi, L. (1999) 'Between the woof and the weft: a response to Loretta Lees', *Environment and Planning D: Society and Space,* 17 (3): 253–255.

Bonefeld, W. (Ed) (2008) *Subverting the Present, Imagining the Future: Insurrection, Movement, Commons,* Brooklyn, NY: Autonomedia.

Borch, C. and Kornberger, M. (Eds) (2015) *Urban Commons: Rethinking the City,* London: Routledge.

Borden, I. (2001) *Skateboarding, Space and the City: Architecture and the Body,* Oxford: Berg.

Borden, I., Kerr, J., Rendell, J. and Pivaro, A. (Eds) (2001) *The Unknown City: Contesting Architecture and Social Space,* Cambridge, MA: MIT Press.

von Borries, F., Walz, S., and Bottger, M. (Eds) (2007) *Space Time Play: Synergies Between Computer Games, Architecture and Urbanism,* Basel: Birkhäuser.

Bourdieu, P. (1984) *Distinction: a Social Critique of the Judgement of Taste,* London: Routledge.

Braun, B. (2008) 'Environmental issues: inventive life', *Progress in Human Geography,* 32 (5): 667–679.

Brenner, N. (Ed.) (2014) *Implosions/Explosions: Towards a Study of Planetary Urbanization*, Berlin: Jovis Verlag.

Brenner, N. (2009) 'What is critical urban theory?', *City*, 13: 198–207.

Brenner, N. (2004) 'Urban governance and the production of new state spaces in Western Europe, 1960–2000', *Review of International Political Economy*, 11: 447–488.

Brenner, N. (1999) 'Globalisation as reterritorialisation: the re-scaling of urban governance in the European Union', *Urban studies*, 36 (3): 431–451.

Brenner, N., Madden, D. J. and Wachsmuth, D. (2011) 'Assemblage urbanism and the challenges of critical urban theory', *City*, 15 (2): 225–240.

Brenner, N., Marcuse, P. and Mayer, M. (Eds) (2011) *Cities for People, Not for Profit: Critical Urban Theory and the Right to the City*, London: Routledge.

Brenner, N., Peck, J. and Theodore, N. (2010) 'Variegated neoliberalization: geographies, modalities, pathways', *Global Networks*, 10: 182–222.

Brenner, N. and Schmid, C. (2014) 'The "urban age" in question', *International Journal of Urban and Regional Research*, 38: 731–755.

Brenner, N. and Theodore, N. (2002) 'Cities and the geographies of 'actually existing neoliberalism', *Antipode*, 33 (3): 349–379.

Brenner N. and Theodore N. (Eds) (2003) *Spaces of Neoliberalism: Urban Restructuring in North America and Western Europe*, Oxford: Blackwell.

Bresnihan P. and Byrne M. (2015) 'Escape into the city: everyday practices of commoning and the production of urban space in Dublin', *Antipode*, 47: 36–54.

Bridge, G., Butler, T., and Lees, L. (Eds) (2011) *Mixed Communities: Gentrification by Stealth?*, Bristol: Policy Press.

Bridge, G. and Watson, S. (Eds) *The Blackwell City Reader*, Oxford and Malden, MA: Wiley-Blackwell.

Bromley, R. (1978) 'The urban informal sector: why is it worth discussing?' *World Development*, 6 (9/10): 1033–1039.

Brown, G. (2008) 'Ceramics, clothing and other bodies: affective geographies of homoerotic cruising encounters', *Social and Cultural Geography*, 9 (8): 915–932.

Brown, G., Browne, K. Elmhirst, R, and Hutta, S. (2010) 'Sexualities in/of the Global South', *Geography Compass*, 4 (10): 1567–1579.

Brown, M. (2000) *Closet Space: Geographies of Metaphor from the Body to the Globe*, London: Routledge.

Brown, M. (2014) 'Gender and sexuality II: there goes the gaybourhood?', *Progress in Human Geography*, 38: 457–465.

Bunge, W. (1971) *Fitzgerald: Geography of a Revolution*, Cambridge, MA: Schenkman.

Bunnell, T., Goh, D., Lai, C-K. and Pow, C. P. (2012) 'Global urban frontiers: Asian cities in theory, practice and imagination', *Urban Studies*, 43 (13): 26–37.

Burgess, E. (1925) 'The growth of the city: an introduction to a research project', in R. Park, E. Burgess and E. Mckenzie (Eds) *The City*, Chicago: The University of Chicago Press.

Burgess, E. (1923) 'The growth of the city: an introduction to a research project', *Publications of the American Sociological Society*, 18: 86–97.

Burnett, K. (2014) 'Commodifying poverty: gentrification and consumption in Vancouver's Downtown Eastside', *Urban Geography*, 35 (2): 157–176.

Büscher, M., Urry, J. and Witchger, K. (2010) *Mobile Methods*, London: Routledge.

Butcher, M. (2011) 'Cultures of commuting: the mobile negotiation of space and subjectivity on Delhi's metro, *Mobilities*, 6: 237–54.

Butler, J. (2006) *Gender Trouble: Feminism and the Subversion of Identity*, London: Routledge.

Caillois, R. (1961) *Man, Play and Games*, New York: Free Press of Glencoe.

Califia, P. (1994) *Public Sex: The Culture of Radical Sex*, San Francisco: Cleis Press.

Callon, M. (2007) 'What does it mean to say that economics is performative?', in D. MacKenzie, F. Muniesa and L. Siu (Eds) *Do Economists Make Markets? On the Performativity of Economics*, Princeton: Princeton University Press.

Callon, M. (1986) 'Some elements of a sociology of translation: Domestication of the scallops and the fishermen of St-Brieuc Bay', in J. Law (Ed.) *Power, Action, and Belief: A New Sociology of Knowledge?*, London: Routledge and Kegan Paul.

Callon, M., Lascoumes, P. and Barthe, Y. (2009) *Acting in an Uncertain World: An Essay in Technical Democracy*, Cambridge, MA: MIT Press.

Campbell, C. (1995) 'The sociology of consumption', in D. Miller (Ed) *Acknowledging Consumption: A Review of New Studies*, London: Routledge.

Campkin, B. (2013) *Remaking London: Decline and Regeneration in Urban Culture*, London: I. B. Tauris.

Cantle, T. (2005) *Community Cohesion: A New Framework for Race Relations*, Basingstoke: Palgrave.

Carlsson, C. (2008) *Nowtopia*, London: AK Press.

Carpio, G., Irazabal, C. and Pulido, L. (2011) 'Right to the suburb? Rethinking Lefebvre and immigrant activism', *Journal of Urban Affairs*, 33 (2): 185–208.

Carr, J. (2012) 'Public input/elite privilege: The use of participatory planning to reinforce urban geographies of power in Seattle', *Urban Geography*, 33(3): 420–441.

Carver, T. (2010) 'Materializing the metaphors of global cities: Singapore and Silicon Valley', *Globalizations*, 7 (3): 383–393.

Castells, M. (2000) 'Materials for an exploratory theory of the network society', *British Journal of Sociology*, 51 (1): 1–24.

Castells, M. (1996) *The Rise of the Network Society*, Oxford: Blackwell.

Castells, M. (1977) *The Urban Question: A Marxist Approach*, London, Edward Arnold.

Castells, M. (1976a) 'Is there an urban sociology?', in C. Pickvance (Ed.), *Urban Sociology: Critical Essays*, London: Methuen.

Castells, M. (1976b) 'The wild city', *Kapitalistate*, 4–5: 2–30.

Castells, M. (1973) 'Movimiento de pobladores y lucha de clases en Chile', *EURE*, 3: 7: 9–35.

Cattan, N. and Vanolo, A. (2014) 'Gay and lesbian emotional geographies of clubbing: reflections from Paris and Turin', *Gender, Place and Culture: A Journal of Feminist Geography*, 21 (9) pp. 1158–1175.

Cetina, K., Schatzki, T. and Savigny, E. (2000) *The Practice Turn in Contemporary Theory*, London: Routledge.

Charmes, E. and Keil, R. (2015) 'The politics of postsuburban densification in Canada and France', *International Journal of Urban and Regional Research*, Volume 39, Issue 3: 581–602.

de Châtel, F. and Hunt, R. (2003) *Retailisation: The Here, There and Everywhere of Retail*, London: Europa.

Chatterton P. (2010a) 'The urban impossible: a eulogy for the unfinished city', *City*, 14 (3): 234–244.

Chatterton, P. (2010b) 'Seeking the urban common: furthering the debate on spatial justice', *City*, 14 (6): 625–628.

Chatterton, P. (2010c) 'So what does it mean to be anti-capitalist? Conversations with activists from urban social centres', *Urban Studies*, 476: 1205–1224.

Chatterton, P., Featherstone, D. and Routledge, P. (2012) 'Articulating climate justice in Copenhagen: antagonism the commons and solidarity', *Antipode*, 45 3: 602–620.

Chattopadhyay, S. (2012) *Unlearning the City: Infrastructure in a New Optical Field*, Minneapolis: University of Minnesota Press.

Chen, M. and Skinner, C. (2014) 'The urban informal economy', in S. Parnell and S. Oldfield (Eds) *The Routledge Handbook on Cities of the Global South*, London: Routledge.

Chen, M. and Vanek, J. (2013) 'Informal employment revisited: theories, data and policies', *The Indian Journal of Industrial Relations*, 48 (3): 390–401.

Chen, X. and Kanna, A. (2012) *Rethinking Global Urbanism: Comparative Insights from Secondary Cities*, London: Routledge.

Chicago Tribune (2012) 'Summit ends without giving Chicago a black eye', 22nd May: 9.

Christaller, W. (1933) *Central Places in Southern Germany*, London: Prentice Hall.

Cisney, V. W. and Morar, N. (Eds) (2015) *Biopower: Foucault and Beyond*. Chicago: University of Chicago Press.

Clarke, S. E. (1995) 'Institutional logics and local economic development: a comparative analysis of eight American cities', *International Journal of Urban and Regional Research*, 19 (4): 513–533.

Cloke, P., May, J. and Johnsen, S. (2011) *Swept up Lives: Re-envisioning the Homeless City*, Oxford: Wiley-Blackwell.

Clyde Waterfront Committee (2008) *Clyde Waterfront Regeneration*, www.clydewaterfront. com/home.aspx accessed 16th June 2015.

Cochrane, A. (1999) 'Redefining urban politics for the twenty-first century', in A. E. G. Jonas and D. Wilson (Eds) *The Urban Growth Machine: Critical Perspectives Two Decades Later*, Albany:, State University Press of New York.

Cochrane, A. (1993) *Whatever Happened to Local Government?*, Buckingham: Open University Press.

Cochrane, A. (1991) 'The changing state of local government: restructuring for the 1990s', *Public Administration*, 69: 281–302.

Cochrane, A. (1989) 'Restructuring the state: the case of local government', in A. Cochrane and J. Anderson (Eds) *Politics in Transition*, London: Sage.

Cochrane, A., Peck, J. and Tickell, A. (1996) 'Manchester plays games: exploring the local politics of globalisation', *Urban Studies*, 33: 1319–1336.

Cochrane, A., and Ward, K. (2012) 'Researching the geographies of policy mobility: confronting the methodological challenges'. *Environment and Planning A*, 44 (1): 5–12.

Cockburn, C. (1977) *The Local State: Management of Cities and People*, London: Pluto.

Colectivo Situaciones (2007) 'Something more on research militancy: footnotes on procedures and (in)decisions', in S. Shukaitis, and D. Graeber (Eds) *Constituent Imagination: Militant Investigations//Collective Theorization*, Oakland: AK Press.

Coleman, A. (1985) *Utopia on Trial: Vision and Reality in Planned Housing*, London: Hilary Shipman.

Collins, A. (Ed.) (2007) *Cities of Pleasure: Sex and the Urban Socialscape*, London: Routledge.

Colls, R. (2006) 'Outsize/outside: bodily bignesses and the emotional experiences of British women shopping for clothes', *Gender, Place and Culture*, 13: 529–45.

Comaroff, J. and Comaroff, J. L. (2000) 'Millennial capitalism: first thoughts on a second coming', *Public Culture*, 12: 291–343.

Connell, R. (2007) *Southern Theory: The Global Dynamics of Knowledge in Social Science*, Cambridge: Polity Press.

Corsín, A. and Estalella, A. (2014) 'Assembling neighbours: the city as archive, hardware, method', *Common Knowledge*, 20 (1): 150–171.

Cosgrove, D. and Daniels, S. (Eds) (1988) *The Iconography of Landscape: Essays on the Symbolic Representation, Design and Use of Past Environments*, Cambridge: Cambridge University Press.

Cox, K. R. (1998) 'Spaces of dependence, scales of engagement and the politics of scale, or: looking for local politics', *Political Geography*, 17: 1–23.

Cox, K. R. (1993) 'The local and the global in the new urban politics: a critical view', *Environment and Planning D: Society and Space*, 11: 433–48.

Cox, K. R. (1991) 'Questions of abstraction in studies in the new urban politics', *Journal of Urban Affairs*, 13: 267–280.

Cox, K. R. (1982) 'Housing tenure and neighbourhood activism', *Urban Affairs Review*, 18 (1): 107–129.

Cox, K. R. and Jonas, A. E. G. (1993) 'Urban development, collective consumption and the politics of metropolitan fragmentation', *Political Geography*, 12: 8–37.

Cox, K. and Mair, A. (1989) 'Urban growth machines and the politics of local economic development', *International Journal of Urban and Regional Research*, 13: 137–146.

Cox, K. R. and Mair, A. (1988) 'Locality and community in the politics of local economic development', *Annals of the Association of American Geographers*, 78: 307–25.

Crang, M. (2001) 'Rhythms of the city: temporalised space and motion', in J. May and N. Thrift (Eds) *Timespace: Geographies of Temporality*, London: Routledge.

Crary, J. (2013) *24/7: Late Capitalism and the Ends of Sleep*, London: Verso.

Cream, J. (1993) 'Child sex abuse and the symbolic geographies of Cleveland', *Environment and Planning D: Society and Space*, 11 (2): 231–246.

Cresswell, T. (2010) 'Towards a politics of mobility', *Environment and Planning D: Society and Space*: 28: 17–31.

Cresswell, T. (2006) *On the Move: Mobility in the Modern Western World*, London: Routledge.

Crewe, L. (2001) 'The besieged body: geographies of retailing and consumption', *Progress in Human Geography*, 25: 629–40.

Cronin, A. (2006) 'Advertising and the metabolism of the city: urban space, commodity rhythm', *Environment and Planning D: Society and Space*, 24: 615–632.

Cumbers, A. and MacKinnon, D. (2004) 'Clusters in urban and regional development'. *Urban Studies*, 41 (5): 1–23.

Cumbers, A. and MacKinnon, D. (Eds) (2006) *Clusters in Urban and Regional Development*, Routledge, London.

Cummings, J. (2015) 'Confronting favela chic: the gentrification of informal settlements in Rio de Janeiro, Brazil', in L. Lees, H. Shin and E. López-Morales (Eds) *Global Gentrifications: Uneven Development and Displacement*, Bristol: Policy Press.

Cupers, K. (2014) *The Social Project: Housing in Postwar France*, Minneapolis: University of Minneapolis.

Curry, M. R. (2005) 'Meta-theory/many theories', in N. Castree, A. Rogers and D. Sherman (Eds) *Questioning Geography: Fundamental Debates*, Oxford: Blackwell.

D'Andrea, A. (2009) *Global Nomads: Techno and New age as Transnational Cultures in Ibiza and Goa*, London: Routledge.

D'Eramo, M. (2014) 'UNESCOcide', *New Left Review*, 88: 47–53.

Dahl, R. A. (1961) *Who Governs?*, New Haven, CN: Yale University Press.

Daley, R. (2006) *Presentation at the Global Cities Forum*, University of Illinois at Chicago campus, 9th May.

Daley, R. (2005) *Presentation at the Global Cities Forum*, University of Illinois at Chicago campus, 4th May.

Darling, J. and Wilson, H. F. (Eds) (2016) *Encountering the City: Urban Encounters from Accra to New York*, London: Routledge.

Datta, A. (2012) *The Illegal City: Space, Law and Gender in a Delhi Squatter Settlement*, Aldershot: Ashgate.

Davidoff, P. (1965) 'Advocacy and pluralism in planning', *Journal of the American Institute of Planners*, 31(4): 331–338.

Davis, M. (2006) *Planet of Slums*, London: Verso.

Davis, M. (1990) *City of Quartz: Excavating the Future in Los Angeles*, London: Verso.

Davidson, M. and Iveson, K. (2014) 'Recovering the politics of the city: from the 'post-political city' to a 'method of equality' for critical urban geography', *Progress in Human Geography* 39 (5): 543–551.

Davidson, M. and Lees, L. (2005) 'New build 'gentrification' and London's riverside renaissance', *Environment and Planning A*, 37 (7): 1165–1190.

De Angelis, M. (2003) 'Reflections on alternatives, commons and communities' *The Commoner*, 6: 1–14.

De Angelis, M. (2007) *The Beginning of History: Value Struggles and Global Capital*, London: Pluto Press.

De Angelis, M. (2007) *The Beginning of history*, London: Pluto.

Dear, M. and Dahmann, N. (2011) 'Urban politics and the Los Angeles school of urbanism', in D. R. Judd and D. W. Simpson (Eds) *The City, Revisited: Urban Theory from Chicago, Los Angeles and New York*, Minneapolis: University of Minnesota Press.

De Boeck, F. (2011) 'Inhabiting ocular ground: Kinshasa's future in the light of Congo's spectral urban politics', *Cultural Anthropology,* 26: 263–286.

De Soto, H. (2000) *The Mystery of Capital: Why Capitalism Triumphs in the West and Fails Everywhere Else*, New York: Basic Books.

Degen, M. M. and Rose, G. (2012) 'The sensory experiencing of urban design: the role of walking and perceptual memory', *Urban Studies*, 49: 3271–3287.

DeLanda, M. (2006) *A New Philosophy of Society: Assemblage Theory and Social Complexity*, London, New York: Continuum.

DeLanda, M. (2000) *A Thousands Years of Nonlinear History*, New York: Swerve Editions.

Deleuze, G. and Guattari, F. (1987) *A Thousand Plateaus: Capitalism and Schizophrenia*, Minneapolis, London: University of Minnesota Press.

Deleuze, G. and Parnet, C. (1987) *Dialogues*, New York, NY: Columbia University Press.

Demant, J. and Landolt, S. (2014) 'Youth drinking in public places: the production of drinking spaces in and outside nightlife areas', *Urban Studies*, 51 (1): 170–184.

Denis, J. and Pontille, D. (2014) 'Maintenance work and the performativity of urban inscriptions: the case of Paris subway signs', *Environment and Planning D: Society and Space*, 32 (3): 404–416.

DeSilvey, C. and Edensor, T. (2013) 'Reckoning with ruins', *Progress in Human Geography*, 37: 465–485.

Dickens, P., Duncan, S., Goodwin, M. and Gray, F. (1985) *Housing, States and Localities*, London: Methuen.

Dikeç, M. (2006) *Badlands of the Republic: Space, Politics and Urban Policy*, Chichester: Wiley-Blackwell.

Dikeç, M. (2005) 'Space, politics, and the political', *Environment and Planning D: Society and Space*, 23: 171–188.

Dikeç, M. (2001) 'Justice and the spatial imagination', *Environment and Planning A, 33*: 1785–1805.

Domosh, M. (1989) 'A method for interpreting landscape: a case study of the New York World Building', *Area*, 21: 347–355.

Dorling, D. (2014) *Inequality and the 1%*, London: Verso.

Doshi, S. (2013) 'The politics of the evicted: redevelopment, subjectivity, and difference in Mumbai's slum frontier', *Antipode*, 45, 844–865.

Dovey, K. (2011) 'Uprooting critical urbanism', *City*, 15 (3–4), 347–354.

Drummond, L. and Labbé, D. (2013) '"We're a long way from Levittown, Dorothy": everyday suburbanism as a global way of life', in R. Keil (Ed) *Suburban Constellations: Governance, Land and Infrastructure in the 21st Century*, Berlin: Jovis Verlag.

Duany, A. and Plater-Zyberk, E. (1992) 'The second coming of the American small town', *Wilson Quarterly*, 16: 3–51.

Duneier, M. (1999) *Sidewalk*, New York: Farrar, Straus and Giroux.

Durkheim, E. (1893) [1964] *The Division of Labour in Society*, New York: The Free Press.

Dworkin, R. (1977) *Taking Rights Seriously*, Cambridge: Harvard University Press.

Dyer-Witheford, N. (2001) 'Empire, immaterial labor, the new combinations, and the global Worker', *Rethinking Marxism*, 13 (3–4): 70-80.

Dylan, T. (2006) *The Aesthetics of Decay: Nothingness, Nostalgia and the Absence of Reason*, New York: Peter Lang.

Edelman, L. (2004) *No Future: Queer Theory and the Death Drive*, Durham, NC: Duke University Press.

Edensor, T. (2015) 'Incipient ruination and the precarity of buildings: materiality, non-human and human agents, and maintenance and repair', in M. Bille and T. Sørensen (Eds) *Assembling Architecture: Archaeology, Affect and the Performance of Building Spaces*, London: Routledge.

Edensor, T. (2013) 'Vital urban materiality and its multiple absences: the building stone of Central Manchester', *Cultural Geographies*, 20 (4): 447–465.

Edensor, T. (2011a) 'Entangled agencies, material networks and repair in a building assemblage: the mutable stone of St Anns Church, Manchester', *Transactions of the Institute of British Geographers*, 36 (2): 238–252.

Edensor, T. (2011b) 'The rhythms of commuting', in T. Cresswell and P. Merriman (Eds) *Mobilities: Practices, Spaces, Subjects*, Aldershot: Ashgate.

Edensor, T. (2010) 'Walking in rhythms: place, regulation, style and the flow of experience', *Visual Studies*, 25(1): 69–79.

Edensor, T. (2000) 'Moving through the city', in D. Bell and A. Haddour (Eds) *City Visions*, London: Palgrave.

Edensor, T. and Jayne, M. (2012) (Eds) *Urban Theory Beyond 'The West': A World of Cities*, London: Routledge.

Ekers, M., Hamel, P., and Keil, R. (2012) 'Governing suburbia: modalities and mechanisms of suburban governance', *Regional Studies*, 46 (3): 405–422.

Ellis, M., Wright, R. and Parks, V. (2004) 'Work together, live apart? Geographies of racial and ethnic segregation at home and at work', *Annals of the Association of American Geographers,* 94 (2): 620–637.

Engles, F. (1844) [1887] *The Condition of the Working-Class in England*, London: Swann Sonnenschein.

Erel, U., Haritaworn, J., Rodriguez, E. G. and Klesse, C. (2008) 'On the depoliticisation of intersectionality talk: conceptualising multiple oppressions in critical sexuality studies', in

A. Kuntsman and M. Esperanza (Eds) *Out of Place: Interrogating Silences in Queerness/Raciality*, York: Raw Nerve Books.

Fainstein, S. (2010) *The Just City*, Ithaca, NY: Cornell University Press.

Faist, T. (2013) 'The mobility turn: A new paradigm for the social sciences?', *Ethnic and Racial Studies*, 36: 1637–46.

Fanon, F. (1961) *The Wretched of the Earth*, London: Grove Press.

Färber, A. (2014) 'Low-budget Berlin: towards an understanding of low-budget urbanity as assemblage', *Cambridge Journal of Regions, Economy and Society*, 7 (1): 119–136.

Farías, I. (2014) 'Planes maestros como cosmogramas: la articulación de fuerzas oceánicas y formas urbanas en Chile', *Revista Pléyade*, 14: 119–142.

Farías, I. (2011) 'The politics of urban assemblages', *City*, 15 (3–4): 365–374.

Farías, I. (2009) 'Introduction: decentering the object of urban studies', in I. Farías and T. Bender (Eds) *Urban Assemblages: How Actor-Network Theory Changes Urban Studies*, London: Routledge.

Farías, I. and Bender, T. (Eds) (2009) *Urban Assemblages: How Actor-Network Theory Changed Urban Studies*, London: Routledge.

Farley, P. and Symmons Roberts, M. (2011) *Edgelands: Journeys into England's True Wilderness*. London: Random House.

Fassin, D. (2010) *Enforcing Order: An Ethnography of Urban Policing*, London: Polity Press.

Featherstone, M. (1991) *Consumer Culture and Postmodernism*, London: Sage.

Fenster, T. (2005) 'The right to the gendered city: different formations of belonging in everyday life', *Journal of Gender Studies*, 14: 217–231.

Ferguson, J. (2006) *Global Shadows: Africa in the Neoliberal Order*, Durham (NC): Duke University Press.

Fernandes, E. (2002) 'The influence of de Soto's *The Mystery of Capital*: land lines', *Lincoln Institute of Land Policy*, 14: 5–8.

Fernandes, E. (2011) *Regularization of Informal Settlements in Latin America: Policy Focus Report*, Cambridge, MA.: Lincoln Institute of Land Policy.

Fincham, B., McGuiness, M. and Murray, L. (2010) *Mobile Methodologies*, Palgrave Macmillan.

Fischer, G. (2001) 'Communities of interest: learning through the interaction of multiple knowledge systems', in S. Bjornestad, R. Moe, A. Morch and A. Opdahl (Eds) *Proceedings of the 24th IRIS Conference, Ulvik*, Bergen: Department of Information Science.

Fleischer, F. (2010) *Suburban Beijing: Housing and Consumption in Contemporary China*, Minneapolis: University of Minnesota Press.

Florida, R. (2002) *The Rise of the Creative Class*, New York: Perseus.

Florida, R., Matheson, Z., Adler, P. and Brydges, T. (2014) *The Divided City: And the Shape of the New Metropolis*, Toronto: Martin Prosperity Institute.

Floud, R., Fogel, R. W., Harris, B., and Hong, S. C. (2011) *The Changing Body: Health, Nutrition, and Human Development in the Western World Since 1700*, Cambridge: Cambridge University Press.

Flyvberg, B. and Stewart, A. (2012) *Olympic Proportions: Cost and Cost Overrun at the Olympics 1960–2012*, Saïd Business School Working Papers, Oxford: University of Oxford.

Forsyth, A. (2012) 'Defining suburbs'. *Journal of Planning Literature*, 27 (3): 270–81.

Foucault, M. (2008) *The Birth of Biopolitics: Lectures at the Collège de France, 1978–79*, Basingstoke: Palgrave MacMillan.

Foucault, M. (1979) *The History of Sexuality, Vol. 1*, London: Allen Lane.

Foucault, M. (1979) *Discipline and Punish*, New York: Vintage.

Fox, L. (2007) *In the Desert of Desire*, Las Vegas: University of Nevada Press.

Franck, K. and Stevens, Q. (Eds) (2006) *Loose Space: Possibility and Diversity in Urban Life*, London: Routledge.

Frankenberg, R. (1993) *White Women, Race Matters: The Social Construction of Whiteness*, Minneapolis: University of Minnesota Press.

Franquesa, J. (2011) '"We've lost our bearings": place, tourism, and the limits of the "mobility turn"', *Antipode*, 43: 1012–33.

Fraser, N. (1997) *Justice Interruptus: Critical Reflections on the "Postsocialist" Condition*, New York: Routledge.

Freire, P. (1970) *Pedagogy of the Oppressed*, New York: Continuum.

Friedmann, J. (1986) 'The world city hypothesis', *Development and Change* 17, 69–88.

Friendly, A. (2013) 'The right to the city: theory and practice in Brazil', *Planning Theory and Practice*, 14: 158–179.

Frykman, J. and Löfgren, O. (Eds). (1996) 'Introduction', *Forces of Habit: Exploring Everyday Culture*, Lund: Lund University Press.

Fukuyama, F. (1989) 'The end of history?', *The National Interest,* (Summer).

Gallagher, M. and Prior, J. (2014) 'Sonic geographies: exploring phonographic methods', *Progress in Human Geography*, 38: 267–284.

Gallez, C. and Kaufmann, V. (2009) *Aux Racines de la Mobilité en Sciences Sociales, De l'histoire des Transports à L'histoire de la Mobilité?*, Rennes: Presses Universitaires de Rennes.

Gandy, M. (2005) 'Cyborg urbanization: complexity and monstrosity in the contemporary city', *International Journal of Urban and Regional Research*, 29 (1): 26–49.

Gandy, M. (2004) 'Rethinking urban metabolism: water, space and the modern city', *City*, 8 (3): 363–379.

Gandy, M. (2003) *Concrete and Clay: Reworking Nature in New York City*, Cambridge, MA: MIT Press.

Garcia Canclini, N. (2001) *Consumers and Citizens: Globalization and Multicultural Conflicts*, Minneapolis: University of Minnesota Press.

Geertz, C. (1988) *The Interpretation of Cultures: Selected Essays*, New York: Basic Books.

Geertz, C. (1972) 'Notes on the Balinese cockfight', *Daedelus*, 101: 1–37.

Gentleman, A. (2012) 'The woman who lives in a shed: how London landlords are cashing in', *Guardian*, 9th May.

Gertler, M. S. and Wolfe, D. A. (Eds) (2002) *Innovation and Social Learning: Institutional Adaptation in an Era of Technological Change*, Basingstoke, UK: Macmillan/Palgrave.

Gherardi, S. and Nicolini, D. (2000) 'To transfer is to transform: the circulation of safety knowledge', *Organization*, (7) 2: 329–348.

Ghertner, D. (2014) 'India's urban revolution: geographies of displacement beyond gentrification', *Environment and Planning A*, 46: 1554–1571.

Gibbs, D. (2002) *Local Economic Development and the Environment*, London: Routledge.

Gibson-Graham, J. K. (2006) *A Postcapitalist Politics*, Minneapolis: University of Minnesota Press.

Gidwani, V. and Baviskar, A. (2011) 'Urban commons', *Economic and Political Weekly,* 46 (50): 42–43.

Gilbert, A. (2007) 'The return of the slum: does language matter?', *International Journal of Urban and Regional Research*, 31 (4): 697–713.

Gilloch, G. (1996) *Myth and Metropolis: Walter Benjamin and the City*, Cambridge: Polity Press.

Gilroy, P. (2010) *Darker Than Blue: On the Moral Economies of Black Atlantic Culture*, Harvard: Harvard University Press.

Gilroy, P. (2012) '"My Britain is fuck all": zombie multiculturalism and the race politics of citizenship', *Identities*, 19: 380–397.

Glaeser, E. L. (1999) *Learning in Cities*, NBER Working Paper 6271, Cambridge, MA.

Gleeson, B. (2014) 'What role for social science in the "urban age"', in N. Brenner (Ed) *Implosions/Explosions: Towards a Study of Planetary Urbanization*, Berlin: Jovis Verlag.

Goffman, E. (1982) 'Where the action is: three essays' in E. Goffman (Ed.) *Interaction Ritual*, New York: Routledge.

Goffman, E. (1980) *Behaviour in Public Places*, Westport, CT: Greenwood.

Goffman, E. (1972) *Encounters: Two Studies in the Sociology of Interaction*, London: Martino Fine Books.

Goffman, E. (1959) *The Presentation of the Self in Everyday Life*, London: Anchor Books.

Goldberger, P. (1996) 'The rise of the private city' in M. J. Vitullo (Ed) *Breaking Away: The Future of Cities*, New York: The Twentieth Century Fund.

Goldfrank, B. and Schrank, A. (2009) 'Municipal neoliberalism and municipal socialism: urban political economy in Latin America', *International Journal of Urban and Regional Research*, 33 (3): 443–462.

Goodwin, M. (1997) 'The city as a commodity: the contested space of urban development', in G. Kearns and C. Philo (Eds) *Selling Places: The City as Cultural Capital, Past and Present*, Oxford: Pergamon Press.

Gottdiener, M. and Lagopoulos, A. (Eds) (1986) *The City and the Sign: An Introduction to Urban Semiotics*, Guilford, NY: Columbia University Press.

Gould, D. (2009) *Moving Politics: Emotion and Act Up's Fight Against AIDS*, Chicago: Chicago University Press.

Grabher, G. (2004) 'Temporary architectures of learning: knowledge governance in project ecologies', *Organization Studies*, 25 (9), 1491–1514.

Grabher, G. (1993) 'The weakness of strong ties: the lock-in of regional development in the Ruhr area', in G. Grabher (Ed.) *The Embedded Firm: on the Socio-economics of Industrial Networks*. London: Routledge.

Graham, S. and Marvin, S. (2001) *Splintering Urbanism: Networked Infrastructures, Technological Mobilities, and the Urban Condition*, London, New York: Routledge.

Graham, S. and Thrift, N. (2007) 'Out of order: understanding repair and maintenance', *Theory, Culture & Society*, 24(3), 1–25.

Greco, J. and Sosa, E. (Eds) (1999) *The Blackwell Guide to Epistemology*, Oxford: Blackwell Publishers.

Green, J. M. (1999) *Deep Democracy*, Lanham, MD and Oxford, UK: Rowman & Littlefield Publishers.

Gregg, M. and Seigworth, G. (2011) *The Affect Theory Reader*, Durham, NC: Duke University Press.

Gregory, D. (2004) *The Colonial Present: Afghanistan, Palestine, Iraq*, Oxford: Blackwell.

Grosz, L. (1992) 'Bodies-cities', in B. Colomina (Ed) *Sexuality and Space*, New York: Princeton Architectural Press.

Gruszczynska, R. (2009) '"I was as mad about it all, about the ban": emotional spaces of solidarity in Poznan March of Equality', *Emotion, Space and Society*, 2(1): 44–51.

Guattari, F. (1995) *Chaosmosis: An Ethico-Aesthetic Paradigm*, Sydney: Power Books.

Guggenheim, M. (2009) 'Mutable immobiles: building conversion as a problem of quasi-technologies', in I. Farías and T. Bender (Eds) *Urban Assemblages. How Actor-Network Theory Changes Urban Studies*, London: Routledge.

Gururani, S. (2013) 'On capital's edge: Gurgaon: India's millennial city', in R. Keil (Ed) *Suburban Constellations: Governance, Land and Infrastructure in the 21st Century*, Berlin: Jovis Verlag.

Hae, L. (2011) 'Legal geographies: the right to spaces for social dancing in New York City: a question of urban rights', *Urban Geography*, 32: 129–142.

Hagerstrand, T. (1977) 'The impact of social organization and environment upon the time-use of individuals and households', in A. Kuklinski (Ed.) *Social Issues and Regional Policy and Regional Planning*, Paris: Mouton.

Haldrup, M., Koefoed, L. and Simonsen, K. (2006) 'Practical orientalism-bodies, everyday life and the construction of otherness', *Geografiska Annaler: Series B, Human Geography*, 88 (2): 173–184.

Hall, P. (2004) 'European cities in a global world', in F. Eckhardt and D. Hassenpflug (Eds), *Urbanism and Globalization*, Frankfurt: Peter Lang.

Hall, T. (2010) 'Urban outreach and the polyrhythmic city', in T. Edensor (Ed) *Geographies of Rhythm: Nature, Place, Mobilities and Bodies*, Aldershot: Ashgate.

Hall, T. and Hubbard, P. (Eds) (1998) *The Entrepreneurial City: Geographies of Politics, Regime and Representation*, Chichester: John Wiley.

Hall, T. and Hubbard, P. (1996) 'The entrepreneurial city: new urban politics, new urban geographies?', *Progress in Human Geography*, 20 (2): 153–174.

Hall, S. (1973) 'Encoding/decoding', in Centre for Contemporary Cultural Studies (Ed) *Culture, Media, Language: Working Papers in Cultural Studies, 1972–79*, London: Hitchinson.

Halvorsen, S. (2015) 'Encountering Occupy London: boundary making and the territoriality of urban activism', *Environment and Planning D: Society and Space*, 33 (2): 314–330.

Hamel, P. and Keil, R. (Eds) (2015) *Suburban Governance: A Global View*, Toronto: University of Toronto Press.

Harding, A. (1994) 'Urban regimes and growth machines: towards a cross-national research agenda', *Urban Affairs Quarterly*, 29: 356–82.

Hardt, M. and Negri, A. (2009) *Commonwealth*, Harvard: Harvard University Press.

Harker, C. (2014) 'The only way is up? ordinary topologies of Ramallah', *International Journal of Urban and Regional Research*, 38: 318–335.

Harney, S. and Moten, F. (2013) *The Undercommons: Fugitive Planning and Black Study*, New York: Minor Compositions.

Harris, A. (2008) 'From London to Mumbai and back again: gentrification and public policy in comparative perspective', *Urban Studies*, 45 (3): 2407–2428.

Harris, A. and Moore, S. (2013) 'Planning histories and practices of circulating urban knowledge'. *International Journal of Urban and Regional Research*, 37: 1–27.

Harris, R. (2010) '"Meaningful types in a world of suburbs": suburbanization in global society', *Research in Urban Sociology*, 10: 15–47.

Hart, J. (2003) *Disabling Globalisation: Places of Power in Post-apartheid South Africa*. Berkeley: University of California Press.

Hart, K. (1973) 'Informal income opportunities and urban employment in Ghana', *Modern African Studies*, 11 (1): 61–89.

Hartman, C. (1974) *Yerba Buena: Land Grab and Community Resistance in San Francisco*, San Francisco: Glide Publications.

Harvey, D. (2013) 'New David Harvey interview on class struggle in urban spaces', *Critical-Theory*, August 7th.

Harvey, D. (2013) *Rebel Cities: the Current Outlook*, Millercom Lecture, University of Illinois at Urbana-Champaign, 13th May.

Harvey, D. (2012) *Rebel Cities: From the Right to the City to the Urban Revolution*, London: Verso.

Harvey, D. (2008) 'The right to the city', *New Left Review*, 53: 23–40.

Harvey, D. (2006) *Spaces of Global Capitalism*, New York: Verso Books.

Harvey, D. (2005) *A Brief History of Neoliberalism*, Oxford and New York: Oxford University Press.

Harvey, D. (2003) *The New Imperialism*, Oxford: Oxford University Press.

Harvey, D. (2000) *Spaces of Hope*, Berkeley: University of California Press.

Harvey, D. (1989a) 'From managerialism to entrepreneurialism: The transformation in urban governance in late capitalism', *Geografiska Annaler: Series B, Human Geography*, 71 (1): 3–17.

Harvey, D. (1989b) *The Condition of Postmodernity: An Enquiry into the Origins of Cultural Change*, Oxford: Basil Blackwell.

Harvey, D. (1989c) *The Urban Experience*, Oxford: Blackwell.

Harvey, D. (1985) *The Urbanization of Capital*, Baltimore: Johns Hopkins University Press.

Harvey, D. (1982) *The Limits to Capital*, Oxford: Blackwell.

Harvey, D. (1979) 'Monument and myth', *Annals of the Association of American Geographers*, 69: 362–381.

Harvey, D. (1974) 'What kind of Geography for what kind of public policy?', *Transactions of the Institute of British Geographers*, 63: 18–24.

Harvey, D. (1973) *Social Justice and the City*, Oxford: Blackwell.

Harvey, D. (1969) *Explanation in Geography*, London: Arnold.

Hasan, A. (2015) 'Value extraction from land and real estate in Karachi', in L. Lees, H. Shin and E. López-Morales (Eds) *Global Gentrifications: Uneven Development and Displacement*, Bristol: Policy Press.

Hasan, A. (2012) *The Gentrification of Karachi's Coastline*, 'Towards an Emerging Geography of Gentrification in the Global South' 23–24th March, London.

Hayek, F. (1945) 'The use of knowledge in society', *American Economic Review*, 35 (4): 519–30.

Hayden, D. (1997) 'Housing and American life', in L. McDowell (Ed.) *Space, Gender, Knowledge: Feminist Readings*, London: Arnold.

Hayward K. (2004) *City Limits: Crime, Consumer Culture and the Urban Experience*, London: GlassHouse.

Hebdige, D. (1979) *Subculture: The Meaning of Style*, London: Methuen.

Helms, G. (2008) *Towards Safe City Centres? Remaking the Spaces of an Old-Industrial City*, Aldershot: Ashgate.

Hemmings, C. (2011) *Why Stories Matter: The Political Grammar of Feminist Theory*, Durham, NC and London: Duke University Press.

Hentschel, C. (2014) 'Review essay', *International Journal of Urban and Regional Research*, 38 (1), 364–374.

Herbert, S. (2006) *Citizens, Cops, and Power: Recognizing the Limits of Community*, Chicago, IL: University of Chicago Press.

Heynen, N. C. (2003) 'The scalar production of injustice within the urban forest', *Antipode*, 35 (5): 980–998.

Heynen, N. C., Kaika, M. and Swyngedouw, E. (Eds) (2006) *In the Nature of Cities: Urban Political Ecology and the Politics of Urban Metabolism*, London: Routledge.

Highmore, B. (2002) 'Street life in London: towards a rhythmanalysis of London in the late nineteenth century', *New Formations*, 24: 171–193.

Hill, R. C. and Kim, J. W. (2000) 'Global cities and developmental states: New York, Tokyo and Seoul', *Urban Studies*, 37 (3): 2167–2195.

Hirsch, A. R. (1983) *Making the Second Ghetto: Race and Housing in Chicago, 1940–1960*, Cambridge and New York: Cambridge University Press.

Hitchings, R. (2011) 'Researching air-conditioning addiction and ways of puncturing practice: professional office workers and the decision to go outside', *Environment and Planning A*, 43, 12 (3): 28–38.

Hodkinson, S. (2012) 'The return of the housing question', *Ephemera: Theory and Politics in Organization*, 12 (4): 423–444.

Hodkinson, S. (2011) 'Housing regeneration and the private finance initiative in England: unstitching the neoliberal urban straitjacket', *Antipode*, 43: 358–383.

Hodkinson, S. and Chatterton, P. (2006) 'Autonomy in the city? Reflections on the social centres movement in the UK', *City*, 103: 305–315.

Höhne, S. (2011) 'Tokens, suckers und der great New York token war', *Zeitschrift für Medien-und Kulturforschung*, 1: 143–157.

Holifield, R. (2009) 'Actor-network theory as a critical approach to environmental justice: a case against synthesis with urban political ecology', *Antipode*, 41 (4): 637–658.

Hollands, R. (2008) 'Will the real smart city please stand up? Intelligent, progressive or entrepreneurial?' *City*, 12 (3): 303–320.

Holloway, J. (2010) *Crack Capitalism*, London: Pluto Press.

Holston, J. (2009) 'Insurgent citizenship in an era of global peripheries', *City and Society*, 21: 245–267.

Holston, J. (Ed) (1999) *Cities and Citizenship*, Durham, NC: Duke University Press.

Hopkins, J. (1990) 'West Edmonton mall: landscape and myth and elsewhereness', *The Canadian Geographer*, 34: 2–17.

Horkheimer, M. and Adorno, T. W. (1947) *Dialectic of Enlightenment*, Stanford University Press: Stanford, California.

Houdart, S. (2008) 'Copying, cutting and pasting social spheres: computer designers' participation in architectural projects', *Science Studies*, 21: 47–63.

Hubbard, P. (2012) *Cities and Sexualities*, London: Routledge.

Hubbard, P. (2006) *City*, London: Routledge.

Hubbard, P. (2004) 'Revenge and injustice in the neoliberal city: uncovering masculinist agendas', *Antipode*, 36 (4): 665–689.

Hubbard, P. (2003) 'Fear and loathing at the multiplex: everyday anxiety at the multiplex', *Capital and Class*, 27 (2): 51–75.

Huchzermeyer, M. (2014) 'Invoking Lefebvre's "right to the city" in South Africa today: a response to Walsh', *City*, 18: 41–49.

Huchzermeyer, M. (2011) *Cities with 'Slums': From Informal Settlement Eradication to a Right to the City in South Africa*, Claremont: UCT Press.

Huchzermeyer, M. (2007) 'Elimination of the poor in KwaZulu-Natal', *Pambazuka News: Weekly Forum for Social justice in South Africa*, Available from: www.pambazuka.org/. Accessed 27 May 2009.

Hughes, G. (1999) 'Urban revitalization: the use of festive time strategies', *Leisure Studies*, 18 (2): 119–34.

Hughes, T. P. (1983) *Networks of Power: Electrification in Western Society, 1880–1930*, Baltimore: John Hopkins University Press.

Huizinga, J. (1949) *Homo Ludens: A Study of the Play Element in Culture*, London: Routledge and Kegan Paul.

Hulchanski, D. (2010) *The Three Cities Within Toronto. Income Polarization Among Toronto's Neighbourhoods, 1970–2005*, Toronto: University of Toronto.

Hutchinson, S. (2000) 'Waiting for the bus', *Social Text*, 18: 107–120.

Hyra, D. (2008) *The New Urban Renewal: the Economic Transformation of Harlem and Bronzeville*, Chicago: University of Chicago Press.

Imrie, R. (1996) *Disability and the City*, London: Sage.

Imrie, R. and Raco, M. (1999) 'How new is the new local governance? Lessons from the United Kingdom', *Transactions of the Institute of British Geographers*, 24: 45–64.

Ingold, T. (2008) 'Bindings against boundaries: entanglements of life in an open world', *Environment and Planning A*, 40: 1796–1810.

Ingold, T. (2000) *The Perception of the Environment: Essays in Livelihood, Dwelling and Skill*, London: Routledge.

Irvine, J. (2014) 'Is sexuality research "dirty work"? Institutionalized stigma in the production of sexual knowledge', *Sexualities*, 17: 632–656.

Isin, E. (Ed.) (2000) *Democracy, Citizenship and the Global City*, New York: Routledge.

Isoke, Z. (2013) *Urban Black Women and the Politics of Resistance*, Macmillan: Basingstoke.

Iveson, K. (2013) 'Cities within the city: do-it-yourself urbanism and the right to the city', *International Journal of Urban and Regional Research*, 37: 941–956.

Iveson, K. (2007) *Publics and the City*, Oxford: Wiley-Blackwell.

Jackson, J. B. (1953) 'The westward moving house', *Landscape*, 2 (3): 8–21.

Jacobs, J. (1961) *The Death and Life of Great American Cities*, New York: Random House.

Jacobs, J. M. (2012a) 'Commentary: comparing comparative urbanisms', *Urban Geography*, 33 (3): 904–14.

Jacobs, J. M. (2012b) 'Urban geographies I: still thinking cities relationally', *Progress in Human Geography*, 36 (3): 412–422.

Jacobs, J. M., Cairns, S. and Strebel, I. (2010) 'Materialising vision: performing a high-rise view', in D. Richardson, S. Daniels and D. Delyser (Eds) *Envisioning Landscapes, Making Worlds: Geography and the Humanities*, London: Routledge.

Jacobs, J. M. (2006) 'A geography of big things', *Cultural Geographies*, 13: 1–27.

Jacobs, J. M., Cairns, S. and Strebel, I. (2007) '"A tall storey ... but, a fact just the same": the red road high-rise as a black box', *Urban Studies*, 44: 609–629.

Jameson, F. (1991) *Postmodernism of the Cultural Logic of Late Capitalism*, Durham, NY: Duke University Press.

Janoschka, M., Sequera, J. and Salinas, L. (forthcoming) 'Gentrification in Spain and Latin America: a critical dialogue', *International Journal of Urban and Regional Research*.

Jauhiainen, J. S. (2013) 'Suburbs', in P. Clark (Ed) *The Oxford Handbook of Cities in World History*, Oxford: Oxford University Press.

Jayne, M. (2016) 'Sharon Zukin', in A. Latham and R. Koch (Eds) *Key Thinkers on Cities*, London: Sage.

Jayne, M. (2013) 'Ordinary urbanism - neither trap nor tableaux', *Environment and Planning A*, 45, 2305–2313.

Jayne, M. (2005) *Cities and Consumption*, London: Routledge.

Jayne, M., Hubbard, P. and Bell, D. (2013) 'Twin cities: territorial and relational geographies of worldly Manchester', *Urban Studies*, 50 (2): 239–254.

Jayne, M., Hubbard, P. and Bell, D. (2011) 'Worlding a city: twinning and urban theory', *City: Analysis of Urban Trends, Culture, Theory, Policy and Action*, 15 (1), 25–41.

Jazeel, T. (2012) 'Spatializing difference beyond cosmopolitanism: rethinking planetary futures', *Theory, Culture and Society*, 28 (2): 75–97.

Jensen, O. B. (2007) *Biking in the Land of the Car: Clashes of Mobility Cultures in the USA*, Paper presented at the Trafikdage, in Aalborg University.

Jensen, O. B. (2009) 'Flows of meaning, cultures of movements: urban mobility as meaningful everyday life practice', *Mobilities*, 4: 139–58.

Jessop, B. (2011) 'Rethinking the diversity and varieties of capitalism: On variegated capitalism in the world market', G. Wood and Lane, E. (Ed) *Capitalist Diversity and Diversity Within Capitalism*, London: Routledge.

Jessop, B. (1998) 'The narrative of enterprise and the enterprise of narrative: place marketing and the entrepreneurial city', in T. Hall and P. Hubbard (Eds), *The Entrepreneurial City*, Chichester: Wiley.

Jessop, B., Peck, J. A. and Tickell, A. (1999) 'Retooling the machine: economic crisis, state restructuring, and urban politics', in A. E. G. Jonas and D. Wilson (Eds) *The Urban Growth Machine: Critical Perspectives Two Decades Later*, Albany NY, State University Press of New York.

Johnson, S. (2001) *The Connected Life of Ants, Brains, Cities and Software*, London: Penguin.

Johnson, H. and Wilson, G. (2009) *Learning for Development*, London: Zed Books.

Jonas, A. E. G. (2013) 'City-regionalism as a "contingent geopolitics of capitalism"', *Geopolitics*, 18: 284–298.

Jonas, A. E. G., Gibbs, D. and While, A. (2011) 'The new urban politics as a politics of carbon control', *Urban Studies*, 48: 2537–2544.

Jonas, A. E. G., Goetz, A. R. and Battarcharjee, S. (2014) 'City-regionalism and the politics of collective provision: regional transportation infrastructure in Denver, USA', *Urban Studies*, 51: 2444–2465.

Jonas, A. E. G. and Wilson, D. (Eds) (1999) *The Urban Growth Machine: Critical Perspectives Two Decades Later*, Albany: State University Press of New York.

Jones, G. A. and Ward, P. M. (1998) 'Privatizing the commons: reforming the Ejido and urban development in Mexico', *International Journal of Urban and Regional Research*, 22: 76–93.

Jones, P. and Evans, J. (2006) *Urban Regeneration in the UK*, London: Sage.

Joseph, M. (2002) *Against the Romance of Community*, Minneapolis: University of Minnesota Press.

Judd, D. R. and Swanstrom, T. (2014) *City Politics: The Political Economy of Urban America*, Boston: Pearson.

Judd, D. and Simpson, D. (Eds) (2011) *The City, Revisited: Urban Theory from Chicago, Los Angeles, New York*, Minneapolis: University of Minnesota Press.

Kantor, P. and Savitch, H. V. (2005) 'How to study comparative urban development politics: a research note', *International Journal of Urban and Regional Research*, 29 (2): 135–151.

Karagianni, M. (2013) *Homelessness, Housing Informality and Policies for the Homeless in Thessaloniki, Greece during the Financial Crisis*, Unpublished dissertation, University of Manchester.

Kärrholm, M. (2012) *Retailising Space: Architecture, Retail and the Territorialisation of Public Space*, Aldershot: Ashgate.

Kärrholm, M. (2009) 'To the rhythm of shopping: on synchronisation in urban landscapes of consumption', *Social and Cultural Geography*, 10 (4): 421–440.

Kasarda, J. and Crenshaw, E. (1991) 'Third World urbanization: dimensions, theories, and determinants', *Annual Review of Sociology*, 17: 467–501.

Katz, J. (1999) *How Emotions Work*, Chicago: University of Chicago Press.

Kaufmann, V. (2002) *Re-thinking Mobility: Contemporary Sociology*, Aldershot: Ashgate.

Keil, R. (Ed) (2013) *Suburban Constellations: Governance, Land and Infrastructure in the 21st Century*, Berlin: Jovis Verlag.

Keil, R. (2011) '*Cities in Waiting*', Suburbanicity blog post. 11th December; available at http://suburbs.apps01.yorku.ca/2011/12/12/cities-in-waiting/; accessed 31 October 2014.

Keil, R. (2007) 'Empire and the global city: perspectives of urbanism after 9/11', *Studies in Political Economy*, 79: 167–92.

Keil, R. (1998) *Los Angeles: Globalization, Urbanization and Social Struggles*, Chichester: Wiley.

Keil, R. and Ronneberger, K. (1994) 'Going up the country: internationalization and urbanization on Frankfurt's northern fringe', *Environment and Planning D: Society and Space*, 12 (2): 137–166.

Keil, R. and Young, D. (2014) 'In-between mobility in Toronto's new (sub)urban neighbourhoods', in P. Watt and P Smets (Eds) *Mobilities and Neighbourhood Belonging in Cities and Suburbs*, Houndmills, Basingstoke: Palgrave Macmillan.

Kennedy, R. (2004) *Subwayland: Adventures in the World Beneath New York*, New York: St Martin's Press.

Kern, L. and Mullings, B. (2013) 'Feminism, urban knowledge and the killing of politics', in L. Peake and M. Rieker (Eds) *Rethinking Feminist Interventions into the Urban*, London: Routledge.

King, A. (1986) *The Bungalow*, London: Routledge.

King, A. D. (2004) *Spaces of Global Cultures: Architecture, Urbanism, Identity*, London: Routledge.

King, A. D. (1990) *Global Cities: Post-Imperialism and the Internationalization of London*, London: Routledge.

King, K. and McGrath, S. (2004) *Knowledge for Development? Comparing British, Japanese, Swedish and World Bank Aid*, London: Zed Books.

Knox, P. (2010) *Metroburbia, USA*, New Brunswick: Rutgers University Press.

Kofman, E. and Lebas, E. (1996) 'Lost in transposition: time, space and the city', in E. Kofman and E. Lebas (Eds) *Writings on Cities*, Oxford: Blackwell Publishing.

Kracauer, S. (1926) *The Mass Ornament*, Cambridge, MA: Harvard University Press.

Kraftl, P. (2010) 'Geographies of architecture: the multiple lives of buildings', *Geography Compass*, 4: 402–415.

Kraftl, P. and Adey, P. (2008) 'Architecture/affect/inhabitation: geographies of being-in buildings', *Annals of the Association of American Geographers*, 98: 213–231.

Krueger, R. and Buckingham, S. (2012) 'Towards a 'consensual'urban politics? Creative planning, urban sustainability and regional development', *International Journal of Urban and Regional Research*, 36 (3): 486–503.

Krueger, R., and Gibbs, D. (Eds) (2007) *The Sustainable Development Paradox: Urban Political Economy in the United States and Europe*, New York: Guilford Press.

Krugman, P. R. (1996) 'Making sense of the competitiveness debate', *Oxford Review of Economic Policy*, 12: 17–25.

Kundnani A. (2014) *The Muslims are Coming! Islamophopia, Extremism, and the Domestic War on Terror*, London: Verso.

Kuymulu, M. B. (2013) 'The vortex of rights: "right to the city", at a crossroads', *International Journal of Urban and Regional Research*, 37: 923–940.

Labbé, D. and Boudreau, J. A. (2011) 'Understanding the causes of urban fragmentation in Hanoi: the case of new urban areas', *International Development Planning Review*, 33: 273–91.

Lachmund, J. (2013) *Greening Berlin: The Co-Production of Science, Politics, and Urban Nature*, Cambridge, MA: MIT Press.

Laclau, E. (2006) 'Ideology and post-Marxism', *Journal of Political Ideologies*, 11: 103–114.

Lamb, M. (2014) 'Misuse of the monument: the art of Parkour and the discursive limits of a disciplinary architecture', *Journal of Urban Cultural Studies*, 1(1): 107–126.

Landry, C. (2006) *The Art of City Making*, London: Earthscan.

Latham, A. (2015) 'The history of a habit: jogging as a palliative to sedentariness in 1960s America', *Cultural Geographies*, 22(1): 103–126.

Latham, A. (2011) 'Topologies and the multiplicities of space-time', *Dialogues in Human Geography*, 1 (3): 312–315.

Latham, A. (2003a) 'Research, performance, and doing human geography: some reflections on the diary-photo diary-interview method', *Environment and Planning A*, 35: 1993–2017.

Latham, A. (2003b) 'Urbanity, lifestyle and making sense of the new urban cultural economy: notes from Auckland, New Zealand', *Urban Studies*, 40(9): 1699–1724.

Latham, A. and McCormack, D. P. (2004) 'Moving cities: rethinking the materialities of urban geographies', *Progress in Human Geography*, 28 (6): 701–724.

Latour, B. (2005) *Reassembling the Social*, Oxford: Oxford University Press.

Latour, B. (1993) *We Have Never Been Modern*, Hemel Hempstead: Harvester, Wheatsheaf.

Lauria, M. (Ed.) (1997) *Reconstructing Urban Regime Theory: Regulating Urban Politics in a Global Economy*, Thousand Oaks CA: Sage Publications.

Laurier, E. (2010) 'Being there/seeing there', in B. Fincham, M. McGuinness and L. Murray (Eds) *Mobile Methodologies*, Aldershot: Ashgate.

Laurier, E. and Philo, C. (2006) 'Cold shoulders and napkins handed: gestures of responsibility', *Transactions of the Institute of British Geographers*, 31: 193–207.

Laurier, E., Lorimer, H., Brown, B., Jones, O. Juhlin, O., Noble, A., Perry, M., Pica, D. Sormani, P. and Strebel, I. (2008) 'Driving and "passengering": notes on the ordinary organization of car travel', *Mobilities*, 3: 1–23.

Lees, L. (forthcoming) 'Doing comparative urbanism in gentrification studies: fashion or progress?', in H. Silver (Ed.) *Comparative Urban Studies*, New York: Routledge.

Lees, L. (2014a) 'Gentrification in the Global South?', in S. Parnell and S. Oldfield (Eds) *The Routledge Handbook on Cities of the Global South*, New York: Routledge.

Lees, L. (2014b) 'The urban injustices of New Labour's "new urban renewal": the case of the Aylesbury Estate in London', *Antipode*, 46: 921–947.

Lees, L. (2012) 'The geography of gentrification: thinking through comparative urbanism', *Progress in Human Geography*, 36 (2): 155–171.

Lees, L. (2001) 'Towards a critical geography of Architecture: the case of an ersatz colosseum', *Ecumene: A Journal of Cultural Geographies*, 8: 51–86.

Lees, L. (2003) 'Super-gentrification: the case of Brooklyn Heights, New York City', *Urban Studies*, 40 (12): 2487–2509.

Lees, L. and Baxter, R. (2011) 'A 'building event' of fear: thinking through the geography of architecture, *Social and Cultural Geography*, 12: 107–122.

Lees, L. and Demeritt, D. (1998) 'Envisioning the livable city: the interplay of "Sin City" and "Sim City" in Vancouver's planning discourse', *Urban Geography*, 19: 332–359.

Lees, L. and Ferrerai, M. (forthcoming) 'Resisting gentrification on its final frontiers: the case of the Heygate Estate in London (1974–2013)', *Cities*.

Lees, L., Shin, H. and López-Morales, E. (Eds) (2015) *Global Gentrifications: Uneven Development and Displacement*, Bristol: Policy Press.

Lees, L., Shin, H. and López-Morales, E. (forthcoming) *Planetary Gentrification*, Cambridge: Polity Press.

Lees, L., Slater, T. and Wyly, E. (2010) *The Gentrification Reader*, London: Routledge.

Lees, L., Slater, T. and Wyly, E. (2008) *Gentrification*, New York: Routledge.

Lee, S. and Webster, C. (2006) 'Enclosure of the urban commons', *GeoJournal*, 66 (1–2): 27–42.

Lefebvre, H. (2004) *Rhythmanalysis: Space, Time and Everyday Life*, London: Continuum.

Lefebvre, H. (2003) *The Urban Revolution*, Minneapolis: University of Minnesota Press.

Lefebvre, H. (1996) *Writings on Cities* (translated by E. Kofman and E. Lebas), Oxford: Blackwell.

Lefebvre, H. (1991) *The Production of Space*, Oxford: Basil Blakewell.

Lefebvre, H. (1984) *Critique of Everyday Life*, London: Verso.

Lefebvre, H. (1968) *La Droit a La Ville*, Paris: Anthropos.

Le Galès, P. (1998) 'Regulation and governance in European cities', *International Journal of Urban and Regional Research*, 22 (3), 482–506.

Leitner, H. (2012) 'Spaces of encounters: immigration, race, class, and the politics of belonging in small-town America', *Annals of the Association of American Geographers*, 102: 828–846.

Lemanski, C. (forthcoming) 'Hybrid gentrification in South Africa: theorising across southern and northern cities', *Urban Studies*.

Lemke, T. (2001) 'The birth of bio-politics: Michel Foucault's lecture at the Collège de France on neo-liberal governmentality, *Economy and Society*, 30: 190–207.

Lerup, L. (2001) *After the City*, Cambridge, MA: MIT Press.

Lévy, J. (2003) 'Capital spatial', in L, Lévy and M. Lussault (Eds) *Dictionnaire de la Géographie et de l'espace des Sociétés*, Paris: Belin.

Lewis, O. (1967) *La Vida: A Puerto Rican Family in the Culture of Poverty: San Juan and New York*, London: Secker and Warburg.

Ley, D. (1993) 'Co-operative housing as a moral landscape: re-examining the post- modern city', in J. Duncan and D. Ley (Eds) *Place/Culture/Representation*, London and New York: Routledge.

Ley, D. and Teo, S. Y. (2014) 'Gentrification in Hong Kong? Epistemology vs ontology', *International Journal of Urban and Regional Research*, 38 (4): 1286–1303.

Lieberman, D. (2013) *The Story of the Human Body: Evolution, Health and Disease*, London: Allen Lane.

Lim, J. (2007) 'Queer critique and the politics of affect', in K, Browne, J, Lim, and G, Brown (Eds) *Geographies of Sexualities: Theory, Practices and Politics*, Aldershot: Ashgate.

Lindner, R. (1996), *The Reportage of Urban Culture: Robert Park and the Chicago School*, Cambridge: Cambridge University Press,

Linebaugh, P. (2014) *Stop, Thief! The Commons, Enclosures, and Resistance*, Oakland: PM Press.

Linebaugh, P. (2008) *The Magna Carta Manifesto: Liberties and Commons for All*. Berkeley: University of California Press.

Logan, J. R. and Molotch, H. (1987) *Urban Fortunes: The Political Economy of Place*, Berkeley and Los Angeles: University of California Press.

Lombard, M. (2014) 'Constructing ordinary places: place-making in urban informal settlements in Mexico', *Progress in Planning*, 94: 1- 53.

Longhurst, R. (2012) 'Becoming smaller: autobiographical spaces of weight loss', *Antipode*, 44 (3), 655–62.

Longhurst, R. (2001) *Bodies: Exploring Fluid Boundaries*, London: Routledge.

Longhurst, R. (1993) 'Pregnant corporeality and a postmodern mall in Hamilton, New Zealand', *Postmodern Cities Conference Proceedings*, 14–16th April 1993, University of Sydney.

López-Morales, E. (2011) 'Gentrification by ground rent dispossession: the shadows cast by large scale urban renewal in Santiago de Chile', *International Journal of Urban and Regional Research*, 35 (2), 1–28.

López-Morales, E. (2010) 'Real estate market, state-entrepreneurialism, and urban policy in the "gentrification by ground rent dispossession" of Santiago de Chile', *Journal of Latin American Geography*, 9 (1): 145–173.

Lorimer, J. (2015) *Wildlife and the Anthropocene: Conservation After Nature*, Minneapolis: University of Minnesota Press.

Lloyd, P. (1979) *Slums of Hope? Shanty Towns of the Third World*, Manchester: Manchester University Press.

Lynch, K. (1960) *The Image of the City*, Cambridge, MA: MIT Press.

Mabey, R. (1973) *The Unofficial Countryside*. London: Vintage.

Mabin, A. (2014) 'Grounding southern city theory in time and place', in S. Parnell and S. Oldfield (Eds) *The Routledge Handbook on Cities of the Global South*, Abingdon: Routledge.

Mabin, A. (2013) 'Suburbanisms in Africa?' in R. Keil (Ed) *Suburban Constellations: Governance, Land and Infrastructure in the 21st Century*, Berlin: Jovis Verlag.

Machlup, F. (1962) *Production and Distribution of Knowledge*, Princeton: Princeton University Press.

MacLeod, G. (2011) 'Urban politics reconsidered: growth machine to post-democratic city?', *Urban Studies*, 48: 2629–2660.

MacLeod, G. (2002) 'From urban entrepreneurialism to a "revanchist city"? On the spatial injustices of Glasgow's renaissance', *Antipode*, 34, 3: 602–624.

MacLeod, G. and Jones, M. (2007) 'Territorial, scalar, connected, networked: in what sense a 'regional' world', *Regional Studies*, 41, 1177–1791.

MacLeod, G. and Goodwin, M. (1999) 'Space, scale and state strategy: rethinking urban and regional governance', *Progress in Human Geography*, 23: 503–527.

Maloutas, T. (2012) 'Contextual diversity in gentrification research', *Critical Sociology*, 38 (1): 33–48.

Manalansan IV, M. (2015) 'Queer worldings: the messy art of being global in Manila and New York' 17 (3): 566–579.

Mangin, W. (1967) 'Latin American squatter settlements: A problem and a solution', *Latin American Studies*, 2: 65–98.

Marcuse, H. (1932) [2005] 'Neue quellen zur grundlegung des historischen materialismus' [New sources on the foundation of historical materialism], R. Wolin and Abromeit, J. (Eds), *Heideggerian Marxism*, Lincoln, NE, and London: University of Nebraska Press.

Marcuse, P. (2009) 'From critical urban theory to the right to the city', *City*, 13: 185–197.

Marcuse, P. (1985) 'Gentrification, abandonment and displacement: connections, causes and policy responses in New York City', *Journal of Urban and Contemporary Law*, 28: 195–240.

Marcuse, P., Imbroscio, D., Parker, S., Davies, J. S. and Magnusson, W. (2014) 'Critical urban theory versus critical urban studies: a review debate', *International Journal of Urban and Regional Research*, 38 (5): 1904–1907.

Marcuse, P. and Van Kempen, R. (Eds) (2008) *Globalizing Cities*, London: John Wiley and Sons.

Marston, S. (2000) 'The social construction of scale', *Progress in Human Geography*, 24: 219–42.

Marston, S. A., Jones, J. P. and Woodward, K. (2005) 'Human geography without scale', *Transactions of the Institute of British Geographers*, 30 (4): 416–432.

Martin, D. G. (2013) 'Up against the law: legal structuring of political opportunities in neighborhood opposition to group home siting in Massachusetts', *Urban Geography*, 34(4): 523–540.

Martin, D. G. (2003a) 'Enacting neighborhood', *Urban Geography*, 24 (5): 361–385.

Martin, D. G. (2003b) '"Place-framing" as place-making: constituting a neighborhood for organizing and activism', *Annals of the Association of American Geographers*, 93 (3): 730–750.

Marx, K. (1973) *Grundrisse* (Translated by M. Nicolaus), New York: Vintage.

Masser, I. and Williams, R. H. (1986) *Learning from Other Countries: The Cross-national. Dimension in Urban Policy-making*, Norwich: Geobooks/Elsevier.

Massey, D. (2011) 'A counterhegemonic relationality of place', E. McCann and K. Ward (Eds) *Mobile Urbanism: Cities and Policymaking in the Global Age*, Minneapolis: University of Minnesota Press.

Massey, D. (2005) *For Space*, London: Sage.

Massey, D. (1991) 'A global sense of place', *Marxism Today*, 35: 24–29.

Massey, D. (1984) *Spatial Divisions of Labour*, London: Macmillan.

Massey, D., Allen, J. and Pile, S. (1999) *City Worlds*, London: Routledge.

Massumi, B. (2002) *Parables for the Virtual: Movement, Affect, Sensation*, London: Duke University Press.

Matejskova, T. and Leitner, H. (2011) 'Urban encounters with difference: the contact hypothesis and immigrant integration projects in eastern Berlin', *Social and Cultural Geography*, 12: 717–741.

Mauss, M. (1923) 'Essai sur le don forme et raison de l'échange dans les sociétés archaïques', *L'Année Sociologique*, 30: 186.

Mayer, M. (2009) 'The "right to the city" in the context of shifting mottos of urban social movements', *City*, 13: 362–374.

Mbembe, A. and Nuttall, S. (2004) 'Writing the world from an African metropolis', *Public Culture*, 16: 347–372.

McCann, E. (2013) 'Policy boosterism, policy mobilities, and the extrospective city', *Urban Geography*, 34: 5-29.

McCann, E. (2011) 'Urban policy mobilities and global circuits of knowledge: toward a research agenda', *Annals of the Association of American Geographers*, 101: 107–30.

McCann, E. (2007) 'Expertise, truth, and urban policy mobilities: global circuits of knowledge in the development of Vancouver, Canada's "four pillar" drug strategy', *Environment and Planning A*, 40 (4): 805–904.

McCann, E. and Ward, K. (2013) 'A multi-disciplinary approach to policy transfer research: geographies, assemblages, mobilities and mutations', *Policy Studies*, 34 (1): 2–18.

McCann, E. and Ward, K. (2012) 'Assembling urbanism: following policies and studying through 'the sites and situations of policy making', *Environment and Planning A*, 44 (1): 42–51.

McCann, E. and K. Ward (Eds) (2011) *Mobile Urbanism: Cities and Policymaking in the Global Age*, Minneapolis: University of Minnesota Press.

McCann, E. and Ward, K. (2010) 'Relationality/territoriality: toward a conceptualization of cities in the world', *Geoforum*, 41 (3): 175–184.

McCormack, D. (2013) *Refrains for Moving Bodies: Experience and Experiment in Affective Spaces,* Durham, NC, and London: Duke University Press.

McCormack, D. (2012a) 'Geography and abstraction towards an affirmative critique', *Progress in Human Geography,* 36 (6): 715–734.

McCormack, D. (2012b) 'Governing economic futures through the war on inflation', *Environment and Planning A,* 44: 1536–1553.

McDonough, T. (Ed.) (2002) *Guy Debord and the Situationist International: Texts and Documents,* Cambridge, MA: MIT Press.

McFarlane, C. (2012) 'Rethinking informality: politics, crisis, and the city', *Planning Theory and Practice,* 13 (1): 89–108.

McFarlane, C. (2011a) 'Assemblage and critical urbanism', *City,* 15 (2): 204–224.

McFarlane, C. (2011b) 'The city as assemblage: dwelling and urban space', *Environment and Planning D: Society and Space,* 29 (4): 649–671.

McFarlane, C. (2011c) *Learning the City: Knowledge and Translocal Assemblage,* Oxford: Wiley-Blackwell.

McFarlane, C. and Robinson, J. (2012) 'Experiments in comparative urbanism', *Urban Geography,* 33 (6): 765–773.

McGee, T. (2015) 'Deconstructing the decentralized urban spaces of the mega-urban regions in the Global South', in P. Hamel and R. Keil (Eds) *Suburban Governance: A Global View.* Toronto: University of Toronto Press.

McGintie, A. (2014) *Discussion with Planner,* City of Glasgow, 6 June.

McGovern, B. J. (2013) *A Life Lived on the Corner,* PhD thesis, Sociology, Victoria University of Wellington, New Zealand.

McGuirk, J. (2014) *Radical Cities: Across Latin America in Search of a New Architecture,* London: Verso.

McGuirk, P. and Dowling, R. (2011) 'Governing social reproduction in masterplanned estates: urban politics and everyday life in Sydney', *Urban Studies,* 48: 2611–28.

McKinnon, D. (2008) 'Evolution, path-dependence and economic geography', *Geography Compass,* 2 (5): 1449–1463.

McNeill, D. (2009) *The Global Architect: Firms, Fame and Urban Form,* London: Routledge.

McNeill, D. (2007) 'Office buildings and the signature architect: Piano and Foster in Sydney', *Environment and Planning A,* 39: 487–501.

McNeill, W. H. (1976) *Plagues and Peoples,* Garden City, NY: Anchor Press.

Merrifield, A. (2014a) *The New Urban Question,* London: Pluto Press.

Merrifield, A. (2014b) 'The entrepreneurs new clothes', *Geografiska Annaler: Series B, Human Geography,* 96 (4): 389–391.

Merrifield, A. (2013) *The Politics of the Encounter: Urban Theory and Protest Under Planetary Urbanization,* Athens, GA: University of Georgia Press.

Merrifield, A. (2012) *Whither Urban Studies,* cities@manchester (http://citiesmcr.wordpress.com/2012/12/10/whither-urban-studies/), 10 December; viewed, 5 May 2014.

Merrifield, A. (2011) 'The right to the city and beyond: notes on a Lefebvrian re-conceptualization', *City,* 15: 473–481.

Merrifield, A. (2002) *Dialectical Urbanism: Social Struggles in the Capitalist City,* London: Monthly Review Press.

Merrifield, A. (2002) *Metromarxism: A Marxist Tale of the City,* New York: Routledge.

Merriman, P. (2010) 'Architecture/dance: choreographing and inhabiting spaces with Anna and Lawrence Halprin', *Cultural Geographies,* 17: 427–449.

Merriman, P. (2007) *Driving Spaces: A Cultural-Historical Geography of England's M1 Motorway,* Oxford: Wiley-Blackwell.

Merriman, P. (2005) 'Materiality, subjectification, and government: the geographies of Britain's motorway code', *Environment and Planning D: Society and Space*, 23: 235–50.

Meth, P. (2013a) 'Viewpoint: millennium development goals and urban informal settlements: unintended consequences', *International Development Planning Review*, 35 (1): v–xiii.

Meth, P. (2013b) 'Parenting in informal settlements: an analysis of place, social relations and emotions', *International Journal of Urban and Regional Research*, 37 (2): 537–55.

Meth, P. (2009) 'Marginalised emotions: men, emotions, politics and place', *Geoforum*, 40 (5): 853–863.

Mickelson, W. (1970) *Man and His Urban Environment: A Sociological Approach*, Reading, MA: Addison-Wesley.

Miles, S. (2010) *Spaces for Consumption: Pleasure and Placelessness in the Post-Industrial City*, London: Sage.

Millar, K. (2014) 'The precarious present: wageless labour and disrupted life in Rio de Janeiro, Brazil', *Cultural Anthropology*, 29, 1: 32–53.

Miller, J. (2014) '"Malls without stores": the affectual spaces of a Buenos Aires shopping mall', *Transactions of the Institute of British Geographers*, 39 (1): 14–25.

Miraftab, F. (2009) 'Insurgent planning: situating radical planning in the global south', *Planning Theory*, 8 (1): 32–50.

Mitchell, D. (2003) *The Right to the City: Social Justice and the Fight for Public Space*, New York: Guilford Press.

Mitchell, D. and Heynen, N. (2009) 'Geography of survival and the right to the city: speculations on surveillance, legal innovation and the criminalization of intervention', *Urban Geography*, 30: 611–632.

Mitchell D. and Villanueva J. (2010) 'The right to the city', in R. Hutchison (Ed) *Encyclopedia of Urban Studies*, Thousand Oaks, CA: Sage.

Miyazaki, H. (2006). 'Economy of dreams: hope in global capitalism and its critiques', *Cultural Anthropology*, 4: 71–88.

Mol, A. (2002) *The Body Multiple: Ontology in Medical Practice*, Durham, NC: Duke University Press.

Mollenkopf, J. (1975) 'The postwar politics of urban development', *Politics and Society*, 5: 247–95.

Molotch, H. (2014) *Against Security: How We Go Wrong at Airports, Subways, and Other Sites of Ambiguous Danger*, Princeton: Princeton University Press.

Monbiot, G. (2014) *Feral: Rewilding the Land, the Sea, and Human Life*, Chicago: University of Chicago Press.

Monk, J. (1994) 'Place matters: comparative international perspectives on feminist geography', *The Professional Geographer*, 46 (3): 277–288.

Moore, M. and Prain, L. (2009) *Yarn Bombing: The Art of Crochet and Knit Graffiti*, Vancouver: Arsenal Pulp Press.

Moraga, C. and Anzaldua, G. (Eds) (1981) *This Bridge Called My Back: Writings by Radical Women of Colour*, Watertown, MA: Persephone.

Moreno, L. (2014) 'The urban process under financialised capitalism', *City*, 18: 244–68.

Moser, C. (2009) *Ordinary Families, Extraordinary Lives: Assets and Poverty Reduction in Guayaquil, 1978–2004*, Washington, D.C.: The Brookings Institution.

Moser, C. (1996) *Confronting Crisis: A Comparative Study of Household Responses to Poverty and Vulnerability in Four Poor Urban Communities*, Washington, D.C.: The World Bank.

Moser, C. (1994) 'The informal sector debate, Part 1: 1970–1983', in C. Rakowski (Ed) *Contrapunto: The Informal Sector Debate in Latin America*, Albany, NY: State University of New York Press.

Moser, C., and Peake, L. (Eds) (1987) *Women, Human Settlements and Housing*, London: Tavistock Publications.

Mouffe, C. (2005) *On the Political*, London: Routledge.

Mumford, L. (1934) *Technics and Civilization*, New York: Harcourt, Brace and Company.

Munoz, J. E. (2009) *Cruising Utopia: The Then and There of Queer Futurity*, Durham, NC. and London: Duke University Press.

Muschamp, H. (1995) 'Architecture: remodelling New York for the bourgeoisie', *The New York Times*, 24 September.

Myers, G. (2011) 'Why African cities matter', *African Geographical Review*, 31 (1): 101–106.

Myrdahl, T. (2013) 'Ordinary (small) cities and LGBTQ lives', *ACME-An International E-Journal for Critical Geographies*, 12 (2): 279–304.

Nash, C. and Gorman-Murray, A. (2014) 'LGBT neighbourhoods and 'new mobilities': towards understanding transformations in sexual and gendered urban landscapes', *International Journal of Urban and Regional Research*, 38 (3): 756–772.

Nasr, J. and Volait, M. (2003) (Eds) *Urbanism: Imported or Exported?*, Oxford: Wiley-Blackwell.

Nast, H. (2006) 'Loving … whatever: alienation, neoliberalism and pet-love in the twenty-first century', *ACME: An International E-Journal for Critical Geographies*, 5: 300–27.

Nast, H. and Pile, S. (1998) 'Introduction: making places/bodies', in H. Nast, and S. Pile (Eds) *Places Through the Body*, London: Routledge.

Ndjio, B. (2009) ''Shanghai beauties' and African desires: migration, trade and Chinese prostitution in Cameroon', European Journal of Development Research, *European Journal of Development Research*, 21 (3): 606–621.

Neiyyar, D. (2013) *Arrests After Raids on 'Slums' in Southall*, BBC News, 8 February.

Newman, A. (2013) 'Gatekeepers of the urban commons? Vigilant citizenship and neoliberal space in multiethnic Paris', *Antipode*, 45: 947–964.

Newman, O. (1972) *Defensible Space: People and Design in the Violent City*, London: Architectural Press.

Ngai, S. (2005) *Ugly Feelings*, Cambridge, MA: Harvard University Press.

Ng, S. and Popkin, B. (2012) 'Time use and physical activity: a shift away from movement across the globe', *Obesity Reviews*, 13, 659–680.

Nicol, F., Humphreys, M. and Roaf, S. (2012) *Adaptive Thermal Comfort: Principles and Practice*, London: Routledge.

Nijman, J. (2007) 'Place-particularity and "deep analogies": A comparative essay on Miami's rise as a world city', *Urban Geography*, 28: 92–107.

NION (2010) 'Brand Hamburg initiative not in our name! Jamming the gentrification machine: a manifesto', *City*, 14 (3): 323–325.

Noterman, E., and Pusey, A. (2012) 'Inside, outside, and on the edge of the academy: experiments in radical pedagogies', in R. Haworth (Ed) *Anarchist Pedagogies: Collective Actions, Theories, and Critical Reflections on Education*, Oakland: PM Press.

Novy, J. and Colomb, C. (2013) 'Struggling for the right to the (creative) city in Berlin and Hamburg: new urban social movements, new 'spaces of hope"?', *International Journal of Urban and Regional Research*, 37 (5): 1816–1838.

O'Rourke, M. (2011) 'Series editor's preface - Europe's faltering project or infinite task (some other headings for queer theory)', in L. Downing and R. Gillett (Eds) *Queer in Europe: Contemporary Case Studies*, Aldershot: Ashgate.

Olds, K. (2001) *Globalization and Urban Change: Capital, Culture and Pacific Rim Mega-Projects*, Oxford: Oxford University Press.

Olsen, B. (2010) *In Defence of Things: Archaeology and the Ontology of Objects*, Plymouth, MA: Rowman Altamira.

Pain, R. (1991) 'Space, sexual violence and social control: integrating geographical and feminist analyses of women's fear of crime', *Progress in Human Geography*, 15 (4): 415–431.

Painter, J. and Jeffrey, A. (2009) *Political Geography* (second edition), London: Sage. Pantheon.

Parker, S. (2015) *Urban Theory and the Urban Experience: Encountering the city*, 2nd Edition, London: Routledge.

Park, R E. (1915) 'The city: suggestions for the investigation of human behaviour in the city environment', *American Journal of Sociology*, 20 (5): 577–612.

Park, R. E., Burgess, E. W. and McKenzie, R. D. (1925) *The City*, Chicago: University of Chicago Press.

Parkinson, J. (2012) *Democracy and Public Space: The Physical Sites of Democratic Performance*, Oxford: Oxford University Press.

Parnell, S. and Robinson, J. (2012) '(Re)theorizing cities from the global South: looking beyond neo-liberalism', *Urban Geography*, 33 (4), 593–617.

Parnell, S., Pieterse, E. and Watson, V. (2009) 'Planning for cities in the global South: an African research agenda for sustainable human settlements', *Progress in Planning*, 72: 233–240.

Patel, S., Burra, S and D'Cruz, C. (2001) 'Slum/shack dwellers international (SDI): foundations to treetops'. *Environment and Urbanization*, 13 (2): 45–59.

Paterson, M. (2011) More-than visual approaches to architecture: vision, touch, technique', *Social and Cultural Geography*, 12: 263–281.

Peake, L. and Rieker, M. (Eds) (2013) *Rethinking Feminist Interventions into the Urban*, London: Routledge.

Peck, J. (2014a) 'Entrepreneurial urbanism: between uncommon sense and dull compulsion', *Geografiska Annaler: Series B, Human Geography*, 96 (4): 396–401.

Peck, J. (2014b) 'Pushing austerity: state failure, municipal bankruptcy and the crises of fiscal federalism in the USA', *Cambridge Journal of Regions, Economy and Society*, 7: 17–44.

Peck, J. (2011) 'Neoliberal suburbanism: frontier space', *Urban Geography*, 32 (6): 884–919.

Peck, J. (2010) *Constructions of Neoliberal Reason*, Oxford and New York: Oxford University Press.

Peck, J. (2006) 'Liberating the city: between New York and New Orleans', *Urban Geography*, 27: 222–232.

Peck, J. (2005) 'Struggling with the creative class', *International Journal of Urban and Regional Research*, 29 (4): 740–770.

Peck, J. (2001) 'Neoliberalizing states: thin policies/hard outcomes', *Progress in Human Geography*, 25: 445–455.

Peck, J. (1995) 'Moving and shaking: business elites, state localism and urban privatism', *Progress in Human Geography*, 19: 16–46.

Peck, J. and Theodore, N. (2012) 'Follow the policy: a distended case approach', *Environment and Planning A*, 44 (3): 21–30.

Peck, J. and Theodore, N. (2010) 'Mobilizing policy: models, methods and mutations', *Geoforum*, 41 (2): 169–174.

Peck, J. and Theodore, N. (2007) 'Variegated capitalism', *Progress in Human Geography*, 31: 731–772.

Peck, J. and Tickell, A. (2002) 'Neoliberalizing space', *Antipode*, 34: 380–404.

Peck, J. and Tickell, A. (1995) 'Business goes local: dissecting the 'business agenda' in Manchester', *International Journal of Urban and Regional Research*, 19: 55–78.

Peck, J., Siemiatycki, E. and Wyly, E. (2014) 'Vancouver's suburban involution', *City: Analysis of Urban Trends, Culture, Theory, Policy, Action*, 18 (4-5): 386–415.

Peck, J., Theodore, N. and Brenner, N. (2009) 'Neoliberal urbanism: models, moments, mutations', *SAIS Review*, XXIX: 49–66.

Pendlebury, J., Short, M. and While, A. (2009) 'Urban world heritage sites and the problem of authenticity', *Cities*, 26: 349–58.

Phelps, N. A. and Wood, A. M. (2011) 'The new post-suburban politics?', *Urban Studies*, 48 (12): 2591–2610.

Phelps, N. A. and Wu, F. (Eds) (2011) *Perspectives on Suburbanization: A Post-Suburban World?*, Houndmills, Basingstoke: Palgrave Macmillan.

Philo, C. and Wilbert, C. (Eds) (2000) *Animal Spaces, Beastly Places: New Geographies of Human-Animal Relations*, London: Routledge.

Pickering, A. (2001) 'Practice and posthumanism', in T. Schatzki, K Knorr-Cetina, and E von Savigny (Eds) *The Practice Turn in Contemporary Theory*, London: Routledge.

Pickvance, C. and Preteceille, E. (Eds) (1991) *State Restructuring and Local Power: a Comparative Perspective*, London: Pinter.

Pile, S. (2011) 'Emotions and affect in recent human geography', *Transactions of the Institute of British Geographers*, 35: 5–20.

Pile, S. (1996) *The Body and the City: Psychoanalysis, Space and Subjectivity*, London: Routledge.

Pile, S. (2001) 'The un(known) city…or, an urban geography of what lies buried below the surface', in I. Borden, J. Kerr, J. Rendell and A. Pivaro (Eds) *The Unknown City: Contesting Architecture and Social Space*, Cambridge, MA: MIT Press.

Pile, S. and Thrift, N. (Eds) (1995) *Mapping the Subject: Geographies of Cultural Transformation*, London: Routledge.

Pimlott, M. (2007) *Without and Within: Essays on Territory and the Interior*, Rotterdam: Episode.

Pink, S. (2009) *Doing Sensory Ethnography*, London: Sage.

Pink, S. (2007) 'Sensing Cittàslow: slow living and the constitution of the sensory city', *Senses and Society*, 2 (1): 59–78.

Piven, F. F. and Friedland, R. (1984) 'Public choice and private power: a theory of the urban fiscal crisis', in A. Kirby, P. Knox and S. Pinch (Eds) *Public Service Provision and Urban Development*, New York: Croom Helm/St Martin's.

Plann, S. (2012) *Discussion with Planner*, City of Chicago, 8 June.

Plummer, K. (2010) 'Generational sexualities, subterranean traditions, and the hauntings of the sexual world: some preliminary remarks', *Symbolic Interaction*, 33 (2): 163–190.

Power, E. R. (2009) 'Border-processes and homemaking: encounters with possums in suburban Australian homes', *Cultural Geographies*, 16: 29–54.

Pradel, M. (2014) *Crisis and (re-)informalisation processes: the case of Barcelona*, XVIII ISA World Congress of Sociology, Yokohama.

Pratt, G. (2004) *Working Feminism*, Philadelphia: Temple University Press.

Pratt, M. L. (1991) 'Arts of the contact zone', *Profession*, 33–40.

Pred, A. (1996) 'Interfusions, consumption, identity and the practices of power relations in everyday life', *Environment and Planning A*, 28: 11–24.

Prentice, R. (2001) 'Experiential cultural tourism: museums and the marketing of the new romanticism of evoked authenticity', *Museum Management and Curatorship*, 19 (1): 5–26.

Prorok, C.V. (2000) 'Boundaries are made for crossing: the feminized spatiality of Puerto Rican Espiritismo in New York City', *Gender, Place and Culture*, 7: 57–79.

Puar, J. (2004) 'Queer times, queer assemblages', *Social Text*, 84–85: 121–139.

Puig de la Bellacasa, M. (2011) 'Matters of care in technoscience: assembling neglected things', *Social Studies of Science*, 41 (1): 85–106.

Purcell, M. (2002) 'Excavating Lefebvre: The right to the city and its urban politics of the inhabitant', *GeoJournal*, 58: 99–108.

Pusey, A. (2010) 'Social centres and the new cooperativism of the common', *Affinities: A Journal of Radical Theory, Culture, and Action*, 4: 1.

Pusey, A., Russell, B. and Sealey-Huggins, L. (2012) 'Movements and moments for climate justice: from Copenhagen to Cancun via Cochabamba', *ACME*, 11 (3): 15–32.

Quinby, R. (2011) *Time and the Suburbs: The Politics of Built Environments and the Future of Dissent*, Winnipeg: Arbeiter Ring Publishing.

Raco, M. (1999) 'Researching the new urban governance: an examination of closure, access and complexities of institutional research', *Area*, 31: 271–279.

Rakowski, C. (1994) 'The informal sector debate, Part 2: 1984–1993', in C. Rakowski (Ed) *Contrapunto: The Informal Sector Debate in Latin America*, Albany: State University of New York Press.

Rancière, J. (2006) *Hatred of Democracy*, London: Verso.

Rancière, J. (1999) Disagreement: Politics and Philosophy (translated by Rose, J.), Minneapolis: University of Minnesota Press.

Rantisi, N. M. and Leslie, D. (2010) 'Materiality and creative production: the case of the Mile End neighborhood in Montréal', *Environment and Planning A*, 42: 2824–2841.

Rao, V. (2006) 'Slum as theory: the South/ Asian city and globalisation', *International Journal of Urban and Regional Research* 30 (1): 225–232.

Renkin, H. (2015) 'Perverse frictions: pride, dignity and the Budapest LGBT march', *Ethnos*, 80 (3): 409–432.

Rerat, P. and Lees, L. (2011) 'Spatial capital, gentrification and mobility: lessons from Swiss Core Cities', *Transactions of the Institute of British Geographers*, 36 (1): 126–142.

Reynolds, R. (2008) *On Guerrilla Gardening: A Handbook for Gardening Without Boundaries*, London: Bloomsbury.

Rich, A. (1986) 'Notes towards a politics of location', in A. Rich (Ed) *Blood, Bread and Poetry: Selected Prose 1979–1985*, New York: W. W. Norton and Co.

Roberts, I. (2010) *The Energy Glut: The Politics of Fatness in an Overheating World*, London and New York: Zed Books.

Robinson, J. (2016) 'Thinking cities through elsewhere: tactics for a more global urban studies', *Progress in Human Geography*, 40 (1): 3–29.

Robinson, J. (2013) ''Arriving at' urban policies/the urban: traces of elsewhere in making city futures', in O. Söderström (Ed.) *Critical Mobilities*, London: Routledge.

Robinson, J. (2011) 'Cities in a world of cities: the comparative gesture', *International Journal of Urban and Regional Research*, 35 (3): 1–23.

Robinson, J. (2006) *Ordinary Cities: Between Modernity and Development*, London: Routledge.

Rodgers, D., Beall, J., and Kanbur, R. (2012) 'Re-thinking the Latin American city', in D. Rodgers, D. Beall, and R. Kanbur (Eds) *Latin American Urban Development into the 21st Century: Towards a Renewed Perspective on the City*, Basingstoke: Palgrave Macmillan.

Rodgers, S., Barnett, C. and Cochrane, A. (2014) 'Where is urban politics?', *International Journal of Urban and Regional Research*, 38: 1551–1560.

Rorty, R. (1996) 'What's wrong with "rights"', *Harper's Magazine*, 202: 15–18.

Rose, G. (1993) *Feminism and Geography: The Limits of Geographical Knowledge*, Minneapolis: University of Minnesota Press.

Rose, G., Degen, M. and Basdas, B. (2010) 'More on "big things": building events and feelings', *Transactions of the Institute of British Geographers*, 35: 334–349.

Rose, G., Degen, M. and Melhuish, C. (2014) 'Networks, interfaces, and computer-generated images: learning from digital visualisations of urban redevelopment projects', *Environment and Planning D: Society and Space*, 32: 386–403.

Rose, N. and Miller, P. (1992) 'Political power beyond the state: problematics of government', *British Journal of Sociology*, 43: 173–205.

Rossi, U. (2015) 'The variegated economics and the potential politics of the smart city', *Territory, Politics, Governance*, 4 (3): 332–357.

Rossi, U. and Vanolo, A. (2012) *Urban Political Geographies. A Global Perspective*, London and Los Angeles: Sage.

Routledge, P. (2011) 'Translocal climate justice solidarities', in J. S. Dryzek, R. B Norgaard and Schlosberg, D. (Eds) *The Oxford Handbook of Climate Change and Society*, Oxford: Oxford University Press.

Roy, A. (2015) 'Governing the postcolonial suburbs', in P. Hamel and R. Keil (Eds) *Suburban Governance: A Global View*, Toronto: University of Toronto Press.

Roy, A. (2013) 'Book Review Forum: learning the city', *Urban Geography*, 34 (1): 131–134.

Roy, A. (2011a) 'Conclusion: Postcolonial urbanism: speed, hysteria, mass dreams', in A. Roy and A. Ong (Eds) *Worlding Cities*, Oxford: Wiley-Blackwell.

Roy, A. (2011b) 'Slumdog cities: rethinking subaltern urbanism', *International Journal of Urban and Regional Research*, 35 (2): 223–238.

Roy, A. (2009) 'The 21st century metropolis: new geographies of theory', *Regional Studies*, 43: 819–830.

Roy, A. (2005) 'Urban informality: toward an epistemology of planning', *Journal of the American Planning Association*, 71 (2): 147–158.

Roy, A. (2003) 'Paradigms of propertied citizenship: transnational techniques of analysis', *Urban Affairs Review*, 38: 463–490.

Roy, A. and Ong, A. (2011) (Eds) *Worlding Cities: Asian Experiments and the Art of Being Global*, Oxford: Wiley-Blackwell.

Russell, B. (2014) 'Beyond activism/academia: militant research and the radical climate and climate justice movement(s)', *Area*, 3 (4):15–31.

Sack, R. D. (1992) *Place, Modernity and the Consumer's World*, London: John Hopkins University Press.

Said, E. (1978) *Orientalism*, New York: Pantheon Books.

Samara, T., He, S. and Chen, G. (Eds) (2013) *Locating Right to the City in the Global South*, New York: Routledge.

Sandercock, L. (1998) 'Death of modernist planning: radical praxis for a postmodern age', in M. Douglas and J. Friedmann (Eds) *Cities for Citizens: Planning and the Rise of Civil Society in a Global Age*, New York: John Wiley and Sons.

Sassen, S. (2011) *Cities in a World Economy*, Beverly Hills: Sage.

Sassen, S. (2006) *Territory, Authority, Rights: From Medieval to Global Assemblages*, Princeton: Princeton University Press.

Sassen, S. (Ed) (2002) *Global Networks, Linked Cities*, London: Routledge.

Sassen, S. (1994) *Cities in a World Economy*, Thousand Oaks, CA: Pine Forge Press.

Satterthwaite, D. (Ed) (1999) *The Earthscan Reader in Sustainable Cities*, London: Earthscan.

Saunders, D. (2010) *Arrival City: The Final Migration and Our Next World*, Toronto: Knopf.

Saunders, P. (1986) *Social Theory and the Urban Question*, London: Hutchison.

Savage, M. (1995) 'Walter Benjamin's urban thought', *Environment and Planning D*, 13: 201–216.

Savitch, H. and Kantor, P. (2004) *Cities in the International Marketplace: The Political Economy of Urban Development in North America and Western Europe,* Princeton: Princeton University Press.

Schafran, A. (2013) 'Discourse and dystopia, American style: The rise of "slumburbia" in a time of crisis', *City*, 17, 130–148.

Schechner, R. (2013) *Performance Studies: An Introduction* (3rd edition), Abingdon: Routledge.

Schechner, R. (1993) *The Future of Ritual: Writings on Culture and Performance*, London: Routledge.

Schuermans, N. (2013) 'Ambivalent geographies of encounter inside and around the fortified homes of middle class whites in Cape Town', *Journal of Housing and the Built Environment*, 28: 679–688.

Schumpeter, J. (1934) *Theory of Economic Development*, London: Harvard University Press.

Scott, A. J. (2006) 'A perspective of economic geography', *Journal of Economic Geography*, 4 (5), 479–499.

Scott, A. and Storper, M. (2015) 'The nature of cities: the scope and limits of urban theory', *International Journal of Urban and Regional Research*, 39 (1): 1–15.

Scott, A. J., Agnew, J., Soja, E. W. and Storper, M. (2001) 'Global city-regions', in Scott, A. J. (Ed) *Global City-Regions: Trends, Theory, Policy*, Oxford: Oxford University Press.

Scott, A. J., Allen, J. and Soja, E. W. (Eds) (1996) *The City Los Angeles and Urban Theory at the End of the Twentieth Century*, Berkeley and Los Angeles: University of California Press.

Sennett, R. (2008) *The Craftsman*, London: Allen Lane.

Sennett, R. (1994) *Flesh and Stone: The Body and the City in Western Civilization*, New York: W. W. Norton.

Sennett, R. (1978) *The Fall of Public Man: On the Social Psychology of Capitalism*, London: Penguin.

Sennett, R. (1971) *The Uses of Disorder: Personal Identity and City Life*, Harmondsworth: Penguin.

Serres, M. (1982) *The Parasite*, Minneapolis: University of Minnesota Press.

Seto, K. C., Güneralp, B. and Hutyra, L.R. (2012) 'Global forecasts of urban expansion to 2030 and direct impacts on biodiversity and carbon pools', *PNAS Early Edition*, www.pnas.org/cgi/doi/10.1073/pnas.1211658109.

Shapiro, M. J. (2010) *The Time of the City: Politics, Philosophy and Genre*, London: Routledge.

Shatkin, G. (1998) '"Fourth world" cities in the global economy: the case of Phnom Penh', *International Journal of Urban and Regional Research*, 22 (3): 378–393.

Shefter, M. (1985) *Political Crisis/Fiscal Crisis: The Collapse and Revival of New York City*, New York: Basic Books.

Sheller, M. and Urry, J. (2006) 'The new mobilities paradigm', *Environment and Planning A*, 38: 207–26.

Shields, R. (1991) *Places on the Margin: Alternative Geographies of Modernity*, London: Routledge.

Shin, H. B. (2013) 'The right to the city and critical reflections on China's property rights activism', *Antipode*, 45 (5): 1167–1189.

Shin. H. B. (2009) 'Property-based redevelopment and gentrification: the case of Seoul, South Korea', *Geoforum*, 40 (5): 906–917.

Shiva, V. (1997) *Biopiracy: The Plunder of Nature and Knowledge*, San Francisco: South End Press.

Shove, E. (2003) *Comfort, Cleanliness and Convenience*, Oxford: Berg.

Sieverts, T. (2003) *Cities Without Cities: An Interpretation of the Zwischenstadt*, London and New York: Spoon Press.

Silva, J. M. and Viera, P. J. (2014) 'Geographies of sexualities in Brazil: between national invisibility and subordinate inclusion in postcolonial networks of knowledge production', *Geography Compass*, 8 (10): 767–777.

Silva, J. M. and Viera, P. J. (2010) 'Fucking geography: an interview with David Bell', *Revista Latino-Americana de Geografia e Genero*, 1 (2): 326–332.

Simmel, G. (1950) 'The metropolis and mental life', in K. Wolff (Ed) *The Sociology of Georg Simmel*, London: Collier-Macmillan.

Simmel, G. (1908) *Soziologie: Untersuchungen über die formen der Vergesellschaftung*, Leipzig: Duncker and Humblot.

Simmel, G. (1903) 'The metropolis and mental life', in P. Kasinitz (Ed) *Metropolis: Center and Symbol for Our Times*, New York: New York University Press.

Simone, A. (2011) *City Life from Jakarta to Dakar: Movements at the Crossroads*, London: Routledge.

Simone, A. (2004) *For the City Yet to Come: Changing African Life in Four Cities*, London: Duke University Press.

Simone, A. (2001) 'Straddling the divides: remaking associational life in the informal African city', *International Journal of Urban and Regional Research*, 25 (2): 102–117.

Sintomer, Y., Herzberg, C. and Rocke, A. (2008) 'Participatory budgeting in Europe: potentials and challenges', *International Journal of Urban and Regional Research*, 32 (1): 164–178.

Slater, D. (1997) *Consumer Culture and Modernity*, Cambridge: Polity Press.

Slater, T. (forthcoming) 'Planetary rent gaps', *Antipode*.

Sloterdijk, P. (2004) *Sphären III: Schäume*, Frankfurt: Suhrkamp.

Smith, C. (2012) *Discussion with Planner*, City of Chicago, 14 June.

Smith, H. (Ed) (1999) *Economic Geography: Past, Present and Future*, London: Routledge.

Smith, N. (2011) *The City and the Global*, Millercom Lecture, 9 November.

Smith, N. (2002) 'New globalism, new urbanism: gentrification as global urban strategy', *Antipode*, 34, 3: 427–450.

Smith, N. (1996) *The New Urban Frontier: Gentrification and the Revanchist City*, London and New York: Routledge.

Smith, N. (1992a) 'Contours of a spatialized politics: homeless vehicles and the production of geographical scale', *Social Text*, 55–81.

Smith, N. (1992b) 'Geography, difference and the politics of scale', in J. Doherty, E. Graham, and M. Malek (Eds) *Postmodernism and the Social Sciences*, London: Macmillan.

Smith, N. (1984) *Uneven Development: Nature. Capital, and the Production of Space*, New York: Blackwell.

Smith, N. (1982) 'Gentrification and uneven development', *Economic Geography*, 58 (2): 139–155.

Smith, N. (1979) 'Toward a theory of gentrification: a back to the city movement by capital, not people', *Journal of the American Planning Association*, 45 (4): 538–548.

Smith, R. and Hetherington, K. (Eds) (2013) *Urban Rhythms: Mobilities, Space and Interaction in the Contemporary City*, Oxford: Wiley-Blackwell.

Söderström, O. (2014) *Cities in Relations: Trajectories of Urban Change in Hanoi and Ouagadougou*, Oxford: Wiley-Blackwell.

Söderström, O. (2013) 'What traveling urban types do: postcolonial modernization in two globalizing cities', in O. Söderström (Ed.) *Critical Mobilities*, London.

Söderström, O. and Geertman, S. (2013) 'Loose threads: the translocal making of public space policy in Hanoi', *Singapore Journal of Tropical Geography*, 34: 244–60.

Soja, E. (2010) *Seeking Spatial Justice*, Minnesota University Press: Minneapolis.

Soja, E. (1996) *Thirdspace: Journeys to Los Angeles and Other Real-and-Imagined Places*, Cambridge, Mass: Blackwell.

Soja, E. (1989) *Postmodern Geographies: The Reassertion of Space in Critical Social Theory*, London: Verso.

Souza, M. L. D. (2010) 'Which right to which city? In defence of political-strategic clarity', *Interface*, 2: 315–333.

Spariosu, M. (1989) *Dionysus Reborn: Play and the Aesthetic Dimension in Modern Philosophical and Scientific Discourse*, Ithaca, NY: Cornell University Press.

Spicker, P. (1987) 'Poverty and depressed estates: a critique of utopia on trial', *Housing Studies*, 2: 283–292.

Spinney, J. (2011) 'A chance to catch a breath: using mobile video ethnography in cycling research', *Mobilities*, 6: 161–82.

Spinney, J. (2010) 'Performing resistance? Re-reading practices of urban cycling on London's South Bank', *Environment and Planning A*, 42: 2914–37.

Spinoza, B. (1992) *Ethics: with The Treatise on the Emendation of the Intellect and Selected Letters*, London: Hackett Publishing.

Spivak, G. (1993) *Outside in the Teaching Machine*, London: Routledge.

Staeheli, L. and Dowler, L. (2002) 'Social transformation, citizenship, and the right to the city' in, *GeoJournal*, 58: 73–75.

Star, S. L. and Ruhleder, K. (1996) 'Steps toward an ecology of infrastructure: design and access for large information spaces', *Information Systems Research*, 7 (1): 111–134.

Stengers, I. (2010) *Cosmopolitics*, Minneapolis: University of Minnesota Press.

Stevens, Q. (2010) 'The German "city beach" as a new waterfront development model', in G. Desfor, J. Laidley, Q. Stevens and D. Schubert (Eds) *Transforming Urban Waterfronts: Fixity and Flow*, New York: Routledge.

Stevens, Q. (2007) *The Ludic City: Exploring the Potential of Public Spaces*, London: Routledge.

Stewart, K. (2007) *Ordinary Affects*, Durham, NC, and London: Duke University Press.

Stickells, L. (2011) 'The right to the city: rethinking architecture's social significance', *Architectural Theory Review*, 16: 213–227.

Stone, C. N. and Sanders, H. T. (Eds) (1987) *The Politics of Urban Development*, Lawrence: University of Kansas Press.

Storper, M. and Scott, A. J. (2009) 'Rethinking human capital, creativity and urban growth', *Journal of Economic Geography*, 9: 147–167.

Storper, M. and Venables, A. (2004) *Buzz: Face-to-Face Contact and the Urban Economy*, paper presented at 'The Resurgent City' Conference, LSE, London, April.

Strange, I. (1997) 'Directing the show? Business leaders, local partnership and economic regeneration in Sheffield', *Environment and Planning C: Government and Policy*, 15: 1–17.

Streeck, W. and Schäfer, A. (Eds) (2013) *Politics in the Age of Austerity*, Cambridge: Polity Press.

Sutcliffe, A. (1981) *Towards the Planned City, Germany, Britain, the United States and France, 1780–1914,* Oxford: Blackwell.

Sutton-Smith, B. (1997) *The Ambiguity of Play*, Cambridge, MA: Harvard University Press.

Swanton, D. (2010) 'Sorting bodies: race. affect, and everyday multiculture in a mill town in Northern England', *Environment and Planning A*, 42: 2332–2350.

Swyngedouw, E. (2015) *Urban Theory for the Twenty-first Century*, paper presented at cities@ manchester Summer School, 28 June–2 July

Swyngedouw, E. (2009) 'The antinomies of the postpolitical city: in search of a democratic politics of environmental production', *International Journal of Urban and Regional Research*, 33: 601–620.

Swyngedouw, E. (2007) 'Impossible sustainability and the post-political condition', in Krueger, R. and Gibbs, D. (Eds) *The Sustainable Development Paradox: Urban Political Economy in the United States and Europe*, New York: Guilford.

Swyngedouw, E. (2006a) 'Circulations and metabolisms: (hybrid) natures and (cyborg) cities', *Science as Culture*, 15 (2): 105–121.

Swyngedouw, E. (2006b) 'Metabolic urbanization: the making of cyborg cities', in N. Heynen, M. Kaika and E. Swyngedouw (Eds) *In the Nature of Cities: Urban Political Ecology and the Politics of Urban Metabolism*, New York: Routledge.

Swyngedouw, E. (2004) 'Globalisation or "glocalisation"? Networks, territories and rescaling', *Cambridge Review of International Affairs*, 17 (1): 25–48.

Swyndegouw, E. (1989) 'The heart of the place: the resurrection of locality in an age of hyperspace', *Geografiska Annaler*, 71: 31–42.

Talen, E. (1999) 'Sense of community and neighbourhood form: an assessment of the social doctrine of new urbanism', *Urban Studies*, 36: 1361–1379.

Tampio, N. (2009) 'Assemblages and the multitude: Deleuze, Hardt, Negri, and the postmodern left', *European Journal of Political Theory*, 8 (3): 383–400.

Taylor, P. (2013) *Extraordinary Cities: Millennia of Moral Syndromes, World-Systems and City/State Relations*, London: Edward Elgar.

Taylor, P. (2004) *World City Network: A Global Urban Analysis*, London: Routledge.

Taylor, P. J., Ni, P., Derruder, B., Hoyler, M., Huang, J., and Wilcox, J. (2012) (Eds) *Global Urban Analysis: A Survey of Cities in Globalization*, London: Routledge.

Temenos, C. and McCann, E. (2012) 'The local politics of policy mobility: learning, persuasion, and the production of a municipal sustainability fix', *Environment and Planning A*, 2 (4): 21–34.

Thien, D. (2005) 'After or beyond feeling? A consideration of affect and emotion in geography', *Area*, 37: 450–454.

Thomas, M. (2002) Out of control: emergent cultural landscapes and political change in urban Vietnam', *Urban Studies*, 39: 1611–1624.

Thornley, A. and Newman, P. (2011) *Planning World Cities: Globalization and Urban Politics*, New York: Palgrave Macmillan.

Thrift, N. (2014) 'The "sentient" city and what it may portend', *Big Data and Society*, 1(1), 20–41.

Thrift, N. (2008) *Non-Representational Theory: Space, Politics, Affect*, London and New York: Routledge.

Thrift, N. (2004a) 'Driving in the city', *Theory, Culture and Society*, 21: 41–59.

Thrift, N. (2004b) 'Intensities of feeling: towards a spatial politics of affect', *Geografiska Annaler: Series B, Human Geography*, 86: 57–78.

Thrift, N. (1993) 'An urban impasse?', *Theory, Culture and Society*, 10 (2): 229–238.

Till, J. (2000) 'Thick time', in I. Borden (Ed) *Intersections*, London: Routledge.

Tilly, C. (1984) *Big Structures, Large Processes, Huge Comparisons*, New York: Russell Sage Foundation.

Tironi, M. (2014) '(De) politicising and ecologising bicycles: the history of the Parisian Vélib' system and its controversies', *Journal of Cultural Economy*, 8 (2): 166–183.

Tironi, M. (2009) 'Gelleable spaces, eventful geographies: the case of Santiago's experimental music scene', in I. Farías and T. Bender (Eds) *Urban Assemblages. How Actor-Network Theory Changes Urban Studies*, London and New York: Routledge.

Tonkis, F. (2005) *Space, the City and Social Theory: Social Relations and Urban Forms*, Cambridge: Polity.

Tönnies, F. (1887) *Community and Society*, New York: Harper and Row.

Tönnies, F. (1955) *Community and Association (Gemeinschaft and Gesellschaft)*, London: Routledge and Kegan Paul.

Tornaghi, C. (2014) 'Critical geography of urban agriculture', *Progress in Human Geography*, 21 (2): 1–17.

Tornaghi, C. (2012) 'Public space, urban agriculture and the grassroots creation of new commons', *Sustainable Food Planning: Evolving Theory and Practice*, 2 (4): 21–34.

Tucker, A. (2009) *Queer Visibilities: Space, Identity and Interaction in Cape Town*, Chichester: Wiley-Blackwell.

Turner, J. (1972) 'Housing as a verb', in J. Turner and R. Fichter (Eds) *Freedom to Build: Dweller Control of the Housing Process*, New York: Collier-Macmillan.

Turner, V. (1982) *From Ritual to Theatre: the Human Seriousness of Play*, New York: Performing Arts Journal Publications.

Tushnet, M. (1984) 'An essay on rights', *Texas Law Review*, 62: 1363–1412.

Uitermark, J., Nicholls, W. and Loopmans, M. (2012) 'Cities and social movements: theorizing beyond the right to the city', *Environment and Planning A*, 44 (2): 2546.

UNDP (2003) *Partnership for Local Capacity Development: Building on the Experiences of City-to-City Cooperation*, New York: UNDP.

UN-Habitat (2003). *The Challenge of the Slums: Global Report on Human Settlements*, London: Earthscan.

United States Census (2014) *History Through the Decades: Fast Facts*, www.census.gov/ history/www/through_The_decades/fast_facts/. Accessed 16 November 2014.

Ureta, S. (2014) 'The shelter that wasn't there: on the politics of co-ordinating multiple urban assemblages in Santiago, Chile', *Urban Studies*, 51 (2): 231–246.

Urry, J. (2007) *Mobilities*, Cambridge: Polity Press.

Urry, J. (2006) 'Inhabiting the car', *The Sociological Review*, 54: 17–31.

Urry, J. (2000) *Sociology Beyond Societies: Mobilities for the 21st Century*, London: Routledge.

Urry, J. (1998) 'Contemporary ransformations of time and space' in P. Scott (Ed) *The Globalization of Higher Education*, Buckingham: SRHE/Open University Press.

Vaiou, D. (2013) 'Transnational city lives: changing patterns of care and neighbouring', in L. Peake and M. Rieker (Eds) *Rethinking Feminist Interventions into the Urban*, London: Routledge.

Valentine, G. and Waite, L. (2012) 'Negotiating difference through everyday encounters: the case of sexual orientation and religion and belief', *Antipode*, 44: 474–492.

Valentine, G. (2008) 'Living with difference: reflections on geographies of encounter', *Progress in Human Geography*, 32: 323–337.

Valentine, G. (1989) 'The geography of women's fear', *Area*, 21: 385–390.

Valentine, G. (1997) 'Making space: separation and difference', in J-P. Jones III, H. J. Nast H. J. and S. M. Roberts (Eds) *Thresholds in Feminist geography: Difference, Methodology, Representation*, Lanham, MD: Rowman and Littlefield.

Van den Berg, A. E., Hartig, T. and Staats, H. (2007) 'Preference for nature in urbanized societies: stress, restoration, and the pursuit of sustainability', *Journal of Social Issues*, 63 (1), 79–96.

Varley, A. (2013) 'Postcolonialising informality?', *Environment and Planning D*, 31 (1): 4–22.

Vasudevan, A., McFarlane, C. and Jeffrey, A. (2008) 'Spaces of enclosure', *Geoforum*, 39 (5):1641–1646.

Veblen, T. (1899) *The Theory of the Leisure Class*, London: Constable.

Velásquez Atehortúa, J. (2014) 'Barrio women's invited and invented spaces against urban elitisation in Chacao, Venezuela', *Antipode*, 46 (3): 835–856.

Vertovec, S. (2015) *Diversities Old and New: Migration and Socio-spatial Patterns in New York, Singapore and Johannesburg*, Palgrave Macmillan: Basingstoke.

Vertovec, S. (1999) 'Conceiving and researching transnationalism', *Ethnic and Racial Studies*, 22 (2): 447–62.

Villanueva, J. (2013) *The Territorialization of the 'Republican Law': Judicial Presence in Seine-Saint-Denis, France*, unpublished Ph.D. thesis, Syracuse University.

Visser, G. (2003a) 'Gay men, tourism and urban space: reflections on Africa's "gay capital"', *Tourism Geographies*, 5 (2): 168–189.

Visser, G. (2003b) 'Gay men, leisure space and South African cities: the case of Cape Town.' *Geoforum*, 34 (1): 123–137.

Vogel, R. K., Savitch, H. V., Xu, J, Yeh, A.G.O, Wu, W, Sancton, A, Kantor, P. and Newman, P. (2010) 'Governing global city regions in China and the West', *Progress in Planning*, 73, 1–75.

Vox, E. (2015) *Discussion with Glasgow Council Officer*, 10 October.

Wachsmuth, D. (2012) 'Three ecologies: urban metabolism and the society-nature opposition', *The Sociological Quarterly*, 53 (4): 506–523.

Wacquant, L. (2008) *Urban Outcasts: A Comparative Sociology of Advanced Marginality*, Cambridge: Polity Press.

Wacquant, L., Slater, T. and Pereria, V. (2014) 'Territorial stigmatization in action', *Environment and Planning A*, 46: 1270–1280.

Walby, S. (1997) *Gender Transformation*, London: Routledge.

Waley, P. (2012) 'Japanese cities in Chinese perspective: towards a contextual, regional approach to comparative urbanism', *Urban Geography*, 33: 816–828.

Walker, R. (1981) 'A theory of suburbanization: capitalism and the construction of urban space in the United States', in M. Dear and Scott, A (Eds) *Urbanization and Urban Planning Under Advanced Capitalist Societies*, New York: Methuen.

Walks, A. (2013) 'Suburbanism as a way of life, slight return', *Urban Studies*, 50 (8): 1471–1388.

Walz, S. and Deterding, S. (Eds) (2014) *The Gameful World: Approaches, Issues, Applications*, Cambridge MA: MIT Press.

Walzer, M. (1984) 'Liberalism and the art of separation', *Political Theory*, 12: 315–30.

Ward, C. (1989) *Welcome Thinner City: Urban Survival in 1990s*, London: Bedford Square Press.

Ward, K. (Ed) (2014) *Researching the City*, London: Sage.

Ward, K. (2012) 'Policy transfer in space: entrepreneurial urbanism and the making up of 'urban' politics', in A.E.G. Jonas and A. Wood (Eds) *Territory, the State and Urban Politics*, Aldershot: Ashgate.

Ward, K. (2010) 'Towards a relational comparative approach to the study of cities', *Progress in Human Geography* 34 (1): 471–487.

Ward, K. (2007) 'Geography and public policy: activist, participatory and policy geographies', *Progress in Human Geography* 31: 695–705.

Ward, K. (2006) 'Policies in motion, urban management and state restructuring: the translocal expansion of Business Improvement Districts', *International Journal of Urban and Regional Research*, 30: 54–70.

Ward, K. (2005) 'Geography and public policy: a recent history of "policy relevance"', *Progress in Human Geography*, 29 (3): 310–319.

Ward, K. G. (2000) 'A critique in search of a corpus: re-visiting governance and re-interpreting urban politics'. *Transactions of the Institute of British Geographers*, 25 (2): 169–185.

Ward, P. (Ed.) (1982) *Self-help Housing: A Critique*, London: Mansell.

Wardrop, A. and Withers, D. M. (Eds.) (2014) *The Para-academic Handbook: A Toolkit for Making-Learning-Creating-Acting*, London: HammerOn Press.

Watson, S. (2009) 'The magic of the marketplace: sociality in a neglected public space'. *Urban Studies*, 46 (2): 1577–1591.

Watson, S. (2006) *City Publics: The (Dis)enchantments of Urban Encounters*, London: Routledge.

Watson, V. (2009) 'Seeing from the south: refocusing urban planning on the globe's central urban issues', *Urban Studies*, 46 (3): 2259–2275.

Watt, P. (2009) 'Housing stock transfers, regeneration and state-led gentrification in London', *Urban Policy and Research*, 27 (3): 229–242.

Weber, M. (1921) 'Die nichtlegitime Herrschaft (Typologie der Städte)', in *Wirtschaft und Gesellschaft*. Tübingen: J. C. B. Mohr (Paul Siebeck).

Weeks, A. (2013) *Discussion with Chicago Planner*, 16 June.

Weinstein, L. and Ren, X. (2009) 'The changing right to the city: urban renewal and housing rights in globalizing Shanghai and Mumbai', *City and Community*, 8 (4): 407–432.

Weizman, E. (2007) *Hollow Land: Israel's Architecture of Occupation*, London: Verso.

Wessel, T. (2009) 'Does diversity in urban space enhance intergroup contact and tolerance?', *Geografiska Annaler: Series B*, 91: 5–17.

Whatmore, S. (2002) *Hybrid Geographies: Natures, Cultures, Spaces*, London: Sage.

Whitehead, A. N. (1920) *The Concept of Nature*, Cambridge: Cambridge University Press.

Williams, R. (1973) *Country and the City*, London: Hogarth.

Williams, R. J. (2004) *The Anxious City: British Urbanism in the Late Twentieth Century*, London: Routledge.

Williamson, D. (2005) 'West Side neighbours live in fear: group home for girls looks at Highland St.', *Worcester Telegram and Gazette*, April 24.

Willis, A., Canavan, S. and Prior, S. (2015) 'Searching for safe space: the absent presence of childhood sexual abuse in human geography', *Gender, Place and Culture: A Journal of Feminist Geography*, Volume 22, Issue 10: 1481–1492.

Wilson, D. (2014) *New Discursive Times and Redevelopment in Chicago*, unpublished manuscript, available from author.

Wilson, D. (2007) *Cities and Race: America's New Black Ghetto*, London: Routledge.

Wilson, D. (2004) 'Toward a contingent urban neoliberalism', *Urban Geography*, 25 (8): 771–783.

Wilson, H. F. (2013) 'Collective life: parents, playground encounters and the multicultural city', *Social and Cultural Geography*, 6 (14): 625–648.

Wilson, H. F. (2011) 'Passing propinquities in the multicultural city: the everyday encounters of bus passengering', *Environment and Planning A*, 43: 634–649.

Winn, J. (2014) 'The University as a hackerspace', *Friction: An Interdisciplinary Conference on Technology and Resistance*, 8th–9th May, Nottingham.

Wirth, L. (1938) 'Urbanism as a way of life', *American Journal of Sociology*, 44: 1–24.

Wise, A. (2010) 'Sensuous multiculturalism: emotional landscapes of inter-ethnic living in Australian suburbia', *Journal of Ethnic and Migration Studies*, 36 (2): 917–937.

Wise, J. M. (2005) 'Assemblage', in C. J. Stivale (Ed) *Gilles Deleuze: Key Concepts*, Montreal, Kingston, IT: McGill-Queen's University Press.

de Witte, M. (2016) 'Encountering religion through Accra's urban soundscape', in J. Darling and H. F. Wilson (Eds) *Encountering the City: Urban Encounters from Accra to New York*, Aldershot: Ashgate.

Wolch, J. (2002) 'Anima urbis', *Progress in Human Geography*, 26 (2): 721–742.

Wood, P. and Landry, C. (2008) *The Intercultural City: Planning for Diversity Advantage*, London: Earthscan.

Woodyer, T. (2012) 'Ludic geographies: not merely child's play', *Geography Compass*, 6 (6): 313–326.

Woolf, P. (1988) 'Symbol of the second empire: cultural politics and the Paris opera house', in S. Cosgrove and S. Daniels (Eds) *The Iconography of Landscape: Essays on the Symbolic Representation, Design and Use of Past Environments*, Cambridge: Cambridge University Press.

Wu, F. (2013) 'Chinese suburban constellations: the growth machine, urbanization and middle-class dreams', in R. Keil (Ed) *Suburban Constellations: Governance, Land and Infrastructure in the 21st Century*, Berlin: Jovis Verlag.

Wu, F. and Phelps, N.A. (2011) 'From suburbia to post-suburbia in China? Aspects of the transformation of the Beijing and Shanghai global city regions', *Build Environment*, 34 (4): 464–481.

Wu, F. and Shen, J. (2015) 'Suburban development and governance in China', in P. Hamel and R. Keil (Eds) *Suburban Governance: A Global View*, Toronto: University of Toronto Press.

Yaneva, A. (2009) *The Making of a Building*, Oxford: Peter Lang.

Yeo, J. H. and Neo. H. (2010) 'Monkey business: human-animal conflicts in urban Singapore', *Social and Cultural Geography*, 11: 681–699.

Yeoh, B. S. and Huang, S. (2010) 'Transnational domestic workers and the negotiation of mobility and work practices in Singapore's home-spaces', *Mobilities*, 5: 219–36.

Yiftachel, O. (2009a) 'Theorizing "grey space": the coming of urban apartheid?', *Planning Theory*, 8: 88–100.

Yiftachel, O. (2009b) 'Critical theory and "gray space": mobilization of the colonized', *City*, 13 (2–3): 240–256.

Yiftachel, O. (2000) 'Social control, urban planning and ethno-class relations: Mizrahim in Israel's development towns', *International Journal of Urban and Regional Research*, 24 (2): 417–434.

Young, D. and Keil, R. (2014) 'Locating the urban in-between: tracking the urban politics of infrastructure in Toronto', *International Journal of Urban and Regional Research*, 38 (5): 1589–1608.

Young, I. (1990) *Justice and the Politics of Difference*, Princeton: Princeton University Press.

Zhang, Y. and Fang, K. (2004) 'Is history repeating itself? From urban renewal in the United States to inner-city redevelopment in China', *Journal of Planning Education and Research*, 23: 286–298.

Žižek, S. (1989) *The Sublime Object of Ideology*, London: Verso.

Zukin, S. (2011) 'Reconstructing the authenticity of place', *Theory and Society*, 40: 161–165.

Zukin, S. (2010) *Naked City: The Death and Life of Authentic Urban Places*. Oxford: Oxford University Press.

Zukin, S. (2009) 'Changing landscapes of power: opulence and the urge for authenticity', *International Journal of Urban and Regional Research*, 33 (2): 543–553.

Zukin, S. (1998) 'Urban lifestyles: diversity and standardisation in spaces of consumption', *Urban Studies*, 35 (5/6): 825–40.

Zukin, S. (1988) *Loft Living: Culture and Capital in Urban Change*. London: Radius.

INDEX

Note: Page numbers in **bold** type refer to figures

Made in the USA
Monee, IL
22 August 2025

23978767R00208